ADVANCES IN
NUCLEAR DYNAMICS

Proceedings of the 10th Winter Workshop on Nuclear Dynamics

ADVANCES IN
NUCLEAR DYNAMICS

Snowbird, Utah, USA   16 – 22 January 1994

Editors

J. Harris
Lawrence Berkeley Laboratory

A. Mignerey
University of Maryland

W. Bauer
Michigan State University

World Scientific
Singapore • New Jersey • London • Hong Kong

Published by

World Scientific Publishing Co. Pte. Ltd.

P O Box 128, Farrer Road, Singapore 9128

USA office: Suite 1B, 1060 Main Street, River Edge, NJ 07661

UK office: 73 Lynton Mead, Totteridge, London N20 8DH

ADVANCES IN NUCLEAR DYNAMICS
Proceedings of the 10th Winter Workshop on Nuclear Dynamics

ISBN 981-02-1802-8

Printed in Singapore.

FOREWORD

The dynamics of nuclear systems span a broad range of research interests. The most exciting of these is perhaps the anticipated use of ultra-relativistic collisions of nuclei to search for a predicted phase transition from normal nuclear matter to a new state of matter, the quark-gluon plasma. Theoretical research has focused on identifying characteristics and possible signatures of the quark-gluon plasma, and on the dynamical evolution of the collision from normal nuclei to plasma formation and its subsequent de-excitation. With accelerators under construction to provide beams with the potential of forming a quark-gluon plasma, experimentalists are presently designing, testing and constructing detectors to search for and study the plasma.

Current studies of relativistic nucleus-nucleus collisions have focused on very hot and dense baryonic matter through the measurement of produced particles and anti-particles, and scattered nucleons. Sophisticated theoretical approaches have been developed and are being refined to better understand the dynamical evolution of these collisions. At lower energies, near and below particle production thresholds, studies of the decay of hot nuclear systems have blossomed with the observation of multifragmentation of heavy systems and detailed studies of the temporal and spatial extent of the system emitting fragments. These studies are of particular interest in the continuing search for the nuclear liquid-gas phase transition. In addition, new reaction studies are being developed through the use of unstable (radioactive) nuclear beams.

These proceedings of the Tenth Winter Workshop on Nuclear Dynamics attempt to provide an overview of the present status and future directions of the field over the entire range of energies of interest.

John Harris
Berkeley, CA

Alice Mignerey
College Park, MD

Wolfgang Bauer
East Lansing, MI

PREVIOUS WORKSHOPS

The following table contains a list of the dates and locations of the previous Winter Workshops on Nuclear Dynamics as well as the members of the organizing committees. The chairpersons of the conferences are underlined.

1. Granlibakken, California, 17-21 March 1980
   W.D. Myers, J. Randrup, G.D. Westfall

2. Granlibakken, California, 22-26 April 1982
   W.D. Myers, J.J. Griffin, J.R. Huizenga, J.R. Nix, F. Plasil, V.E. Viola

3. Copper Mountain, Colorado, 5-9 March 1984
   W.D. Myers, C.K. Gelbke, J.J. Griffin, J.R. Huizenga, J.R. Nix, F. Plasil, V.E. Viola

4. Copper Mountain, Colorado, 24-28 February 1986
   J.J. Griffin, J.R. Huizenga, J.R. Nix, F. Plasil, J. Randrup, V.E. Viola

5. Sun Valley, Idaho, 22-26 February 1988
   J.R. Huizenga, J.I. Kapusta, J.R. Nix, J. Randrup, V.E. Viola, G.D. Westfall

6. Jackson Hole, Wyoming, 17-24 February 1990
   B.B. Back, J.R. Huizenga, J.I. Kapusta, J.R. Nix, J. Randrup, V.E. Viola, G.D. Westfall

7. Key West, Florida, 26 January - 2 February 1991
   B.B. Back, W. Bauer, J.R. Huizenga, J.I. Kapusta, J.R. Nix, J. Randrup

8. Jackson Hole, Wyoming, 18-25 January 1992
   B.B. Back, W. Bauer, J.R. Huizenga, J.I. Kapusta, J.R. Nix, J. Randrup

9. Key West, Florida, 30 January - 6 February 1993
   B.B. Back, W. Bauer, J. Harris, J.I. Kapusta, A. Mignerey, J.R. Nix, G.D. Westfall

10. Snowbird, Utah, 16-22 January 1994
    B.B. Back, W. Bauer, J. Harris, A. Mignerey, J.R. Nix, G.D. Westfall

CONTENTS

ELASTOPLASTICITY IN DISSIPATIVE HEAVY-ION COLLISIONS

M. RHEIN[a,1], R. BARTH[a], E. DITZEL[a], H. FELDMEIER[b], E. KANKELEIT[a],
V. LIPS[a], C. MÜNTZ[a], W. NÖRENBERG[a,b], H. OESCHLER[a],
A. PIECHACZEK[a], W. POLAI[a], I. SCHALL[a]

[a] Institut für Kernphysik, Technische Hochschule, D-6100 Darmstadt, Germany
[b] Gesellschaft für Schwerionenforschung, D-6100 Darmstadt, Germany

ABSTRACT

From the measured yield of δ-electrons with energies between 3 MeV and 8 MeV, emitted in the heavy-ion reaction Pb + Pb at 12 MeV/u incident energy, a very fast deceleration of the nuclei in the approach phase is inferred. This stopping is much faster than predicted by the microscopic one-body dissipation model. Since the deceleration time is smaller than or comparable to the thermal equilibration time, the dissipation process is non-Markovian and therefore memory effects have to be included. This is performed on the basis of dissipative diabatic dynamics, which ascribes elastoplastic properties to nuclear matter. The new diabatic one-body dissipation model describes the fast deceleration and the long nuclear contact time, both deduced from the experimental data.

In the past decade it has been demonstrated that δ-electron spectroscopy serves as a method to determine the time evolution of dissipative heavy-ion collisions. The dependence of the lepton spectra on the nuclear trajectories was discussed in Ref. [1,2] and is best illustrated when using the scaling model [2]. In this model the transition amplitude a_{if} of a bound electron to be excited into the continuum, is given by the Fourier transform

$$a_{if} \propto \frac{1}{E_{if}} \int_{-\infty}^{\infty} dt \, \frac{\dot{R}(t)}{R(t)} \exp\left(i\frac{E_{if}}{\hbar}t\right), \qquad (1)$$

where E_{if} is the difference between the initial and final energies of the electron. The quantity $R(t)$ denotes twice the time-dependent root-mean-square radius of the charge distribution of the two nuclei and is equivalent to the center-to-center distance for large R-values [3]. The ratio $\dot{R}(t)/R(t)$ contains the information about the trajectory. The spectral shapes calculated within the scaling model are in good agreement with the more elaborate coupled-channels calculations [4] which are used in the following analysis.

The shape of the δ-electron spectra is mainly determined by two features of the trajectories. One is the interplay of conservative and frictional forces during the separation of the nuclei, which causes a time delay compared to Coulomb trajectories and leads to a steeper decrease of the spectra for electron energies below 2 MeV. This has

---

[1] Present Adress: Physics Division, Argonne National Laboratory, Argonne, IL 60439

been shown in several experiments [5-8]. The other feature is the fast deceleration of the nuclei at the beginning of the collision, which leads to a rapid change of $\dot{R}/R$ and thus to the emission of high-energy δ-electrons [9]. So far these electrons have only been measured up to 3.5 MeV. To obtain sufficient yield in the high-energy part of the δ-electron spectra large relative velocities of the nuclei are needed when the deceleration due to nuclear forces sets in. Therefore, we performed an experiment at 12.0 MeV/u bombarding energy for the collision system Pb+Pb and measured the electron energy spectra up to 8 MeV.

The experiment was carried out with the TORI spectrometer [10], installed at the UNILAC accelerator of GSI Darmstadt. About 10^9 ions of ^{208}Pb per second were impinging on ^{12}C-backed ^{208}Pb targets of $\approx 350\mu$g/cm^2 thickness. The TORI spectrometer is a magnetic transport system designed to measure simultaneously electrons and positrons in coincidence with the scattered heavy ions. To extend the energy range of electrons up to 8 MeV, a new detector [11] is placed in the solenoidal field of the TORI-spectrometer. The energy of the electrons is measured with a 5 cm long plastic cylinder (NE102) of 7.5 cm diameter. The rather poor energy resolution of $\Delta E/E \approx 10\%$ for energies above 2.0 MeV is tolerated in order to have the advantage of a good time resolution ($\Delta t \approx 1$ ns). To separate electrons from γ-rays and neutrons, a silicon surface barrier ΔE-detector of 500μm thickness is placed in front of the plastic block. Due to its very low detection efficiency for gamma rays and neutrons this counter is used as a trigger for electrons. Furthermore, we profit from the high electron transport efficiency to the detector in the solenoidal field. This yields a solid-angle ratio of 122 for electrons compared to γ-rays and neutrons. The additional suppression obtained by the ΔE-counter results in a detection-efficiency ratio of $\varepsilon_{e-}/\varepsilon_\gamma \approx 4500$. A cylindrical aluminum baffle mounted in front of the detector provides a good suppression of the very high yield of low-energy electrons which spirale close to the magnetic axis. With an additional time-of-flight analysis a complete suppression of neutron events is obtained. The response function is determined by using high-energetic electrons delivered by an electron accelerator [11] and by the β-source ^{106}Ru.

The other counters are similar as in previous experiments [8]. Si(Li) counters detect electrons and positrons in the energy range of 0.8 MeV$\leq$ E$_{e-}$ $\leq$ 2.5 MeV. A pair of position-sensitive heavy-ion counters is used to study kinematical coincidences by measuring the angles of the outgoing fragments and to distinguish sequential fission events from scattered heavy ions. These parallel-plate avalanche detectors work with delay-line technique. The delay lines are read out on both sides in order to recognize double-hit events. They originate from cases where one reaction partner undergoes fission. The total kinetic-energy loss (TKEL) of the collisions is calculated from the scattering angles of the heavy ions with a resolution of about 50% due to angular resolution and neutron evaporation.

For high TKEL-values a significant amount of the measured electrons originate from nuclear processes. These contributions have to be subtracted because only the atomic part of the spectra contains the information about the nuclear trajectories. To determine the contribution of leptons from nuclear origin, the γ-ray spectra are measured with a 10 cm × 15 cm BaF_2 crystal. The γ-detector is mounted perpendicular to the beam axis leading to maximum Doppler broadening without a significant mean Doppler shift. As for the plastic detector, the good time resolution of the BaF_2 crystal is necessary to separate γ-rays from neutron-induced background events. For the conversion of the measured γ-spectra into lepton spectra theoretical conversion coefficients have been used [12,13]. The validity of the multipolarity decomposition of the γ-spectra is controlled by the comparison of measured and converted positron spectra. This procedure is very accurate for high TKEL-values where only a small fraction of positrons ($\lesssim 10$ %) originate from atomic processes. This contribution is subtracted.

To compare the measured δ-electron spectra with theoretical predictions, the impact parameter distributions have to be known. This is achieved by a Monte-Carlo simulation [8] using the deflection functions calculated with the trajectory models discussed below. Applying the same cuts in the simulated and measured TKEL versus scattering-angle correlations we obtain the impact-parameter distributions for the selected event class.

In Fig. 1 the δ-electron spectra are shown which have been measured in coincidence with Pb+Pb collisions. The correction for the conversion-electron background amounts to less than 10%. As discussed above, the data exhibit the influence of the nuclear trajectories on the spectral shape of the emitted δ-electrons. With increasing TKEL the spectra show a steeper descent in the region up to 2.0 MeV and an increasing yield of high-energy δ-electrons compared to calculations based on Rutherford trajectories (dotted lines). The lines in Fig. 1 are the results obtained with coupled-channels calculations [14] where only R(t) enters for the trajectories. All theoretical spectra are multiplied by a factor 1.5 in order to match the absolute height in the spectrum for elastic collisions.

For the description of the nuclear trajectories we start out from the microscopic semiclassical one-body dissipation model as explained in detail in Ref. [15]. Without adjustable parameters this model describes a large body of experimental data on nucleus-nucleus collisions [16]. Its friction coefficients are calculated microscopically, based on the ideas of the window and wall dissipation [17]. In particular, for the system Pb+Pb at an incident energy of 8.6 MeV/u [8], the δ-electron spectra up to 2 MeV are well accounted for within the one-body dissipation model. The observed long contact times are caused by the formation of a pronounced neck in the outgoing phase of the reaction.

In Fig. 1 results from calculations with trajectories obtained with the one-body dissipation model are shown by the dashed lines. Statistical fluctuations of the nuclear trajectories [8,18] can be neglected in our calculations as rather broad TKEL windows are used. The underestimation of the measured yield at high electron energies within this model evidences that there is not sufficient deceleration in the approach phase of the reaction. One could try to increase the stopping power by multiplying the microscopically calculated friction tensor by a factor. This procedure, however, does not work because it results in a shortening of the nuclear contact times, and hence leads to unacceptable discrepancies for the low-energy part of the δ-electron spectra. Of course, there are other friction models [18,19] which by a proper fit of the friction coefficients give faster deceleration and may even be consistent with the large contact times. However, already in the one-body dissipation model the deceleration is so fast (although still too slow compared to the experimental result of this paper) that the Markov assumption implied in this and all other friction models is questionable [20]. Indeed, the radial deceleration time $\tau_{decel} \approx 2 \cdot 10^{-22}$s (cf. lower part of Fig. 2) while the thermal equilibration time

$$\tau_{therm} \approx 2 \cdot 10^{-22} \text{s} \cdot \text{MeV}/\varepsilon^*,$$

where ε^* denotes the excitation energy per nucleon [21]. Thus, $\tau_{decel} \lesssim \tau_{therm}$ during the whole deceleration phase ($0 < \varepsilon^* \lesssim 1$ MeV) with the consequence that the dissipation process is essentially non-Markovian. By measuring high-energy δ-electrons one is sensitive to time scales of the order of 10^{-22}s and hence memory effects can be investigated. Such memory effects have been extensively studied within the diabatic approach to dissipative collective nuclear motion [20,22]. This theory of

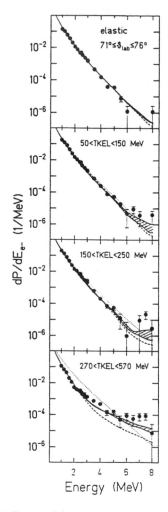

Fig. 1: Spectra of electron emission probability per collision of Pb+Pb at 12.0 MeV/u incident energy. The curves are results of coupled-channels calculations based on the nuclear trajectories predicted by various reaction models (dotted: Rutherford trajectories, dashed: one-body dissipation model, hatched area in between the solid lines: elastoplastic model with α=6 - 12 and β=0.15).

dissipative diabatic dynamics (DDD) is closely related to the microscopic one-body dissipation model [20] which, therefore, is easily extended to include the memory effects as will be explained in the following.

In the one-body dissipation model the friction force for the macroscopic degrees of freedom q_j is given by $F_i^{markov} = -\sum_j \gamma_{ij}(\vec{q})\dot{q}_j$, where $\gamma_{ij}(\vec{q})$ denotes the friction tensor. Guided by the importance of diabatic single-particle motion in dissipative heavy-ion collisions we have, for the shape degrees of freedom, replaced the Markovian friction force of the one-body dissipation model by a retarded friction force $\vec{F}$ as introduced in Ref. [20]. Its components are determined by the differential equations

$$\frac{dF_i}{dt} = -\frac{1}{\tau_{intr}(t)}F_i(t) - \sum_j C_{ij}(\vec{q})\,\dot{q}_j \;, \qquad (2)$$

where $\tau_{intr}(t)$ and C_{ij} denote the intrinsic equilibration time and the stiffness tensor, respectively. In the limit of large τ_{intr}, such that the first term on the r.h.s. can be neglected, the force does neither depend explicitly on time nor on $\dot{q}$, and hence is conservative. In this case, the collective motion is elastic without any dissipation. In the opposite limit of small τ_{intr}, such that $F(t)$, C_{ij} and $\dot{q}_j$ do not change considerably during time intervals of the order τ_{intr}, dF/dt can be neglected in eq. (2), and hence the force becomes a pure friction force given by $-\sum_j C_{ij}\tau_{intr}\,\dot{q}_j$. For intermediate values of τ_{intr}, the system behaves like a damped oscillator with a frequency-dependent friction coefficient. The elastic response on fast deformations and the dissipative response on slow deformations is typical for elastoplastic materials like glass, glycerine and "Silly Putty" (a plastic toy).

There is a close relationship between dissipative diabatic dynamics and the one-body dissipation model. Both approaches are based on the same microscopic picture which determines the collective motion from the distortion of the nucleonic Fermi distribution by the time-dependent shapes of the nuclear system. This coupling between intrinsic and shape degrees of freedom yields the stiffness tensor C_{ij} in the diabatic approach and it provides the friction tensor γ_{ij} in the one-body dissipation model via the additional assumption that the distortions in the nucleonic momentum distribution relax instantaneously. Indeed, for the quadrupole motion of a cube of matter the stiffness parameter C is related to the friction coefficient γ in the one-body dissipation model (wall formula) by [20]

$$C = \frac{24}{5} \cdot \frac{2}{9} \cdot \frac{\gamma}{\tau_{s.p.}} \;, \qquad (3)$$

where $\tau_{s.p.}$ denotes the time of flight of a nucleon with Fermi velocity v_F through the cube. We assume the general validity of this relation in the form

$$C_{ij}(q) = \alpha\,\gamma_{ij}(q)\,/\,\tau_{s.p.} \qquad (4)$$

with $\tau_{s.p.} = 2R_o/v_F$ and $R_o = 1.2 \times (A_1 + A_2)^{1/3}$ fm the radius of the compound nucleus. Here, a parameter α is introduced to correct for shortcomings of the simple cube model with respect to realistic shapes, single-particle potentials and effective nucleon masses m_{eff}. According to the microscopic expression [20,23], the stiffness tensor C_{ij} is proportional to $m_{eff}^{-5/2}$ and to the level density which experimentally is twice its Fermi-gas value. Since $\gamma_{ij}/\tau_{s.p.}$ is deduced from the Fermi-gas model with $m_{eff} = m_{nucleon}$, we expect $\alpha \approx 2 m_{eff}^{-5/2}$ which yields $\alpha \approx 5$ for a realistic value of $m_{eff} = 0.7 m_{nucleon}$. Furthermore, compression is not allowed in the collective model, and hence we expect even larger values (larger by roughly a factor of 2 corresponding to $\alpha \approx 10$) for the stiffness coefficient to be effective in the approach phase.

For the intrinsic equilibration rate τ_{intr}^{-1} we consider three major contributions,

$$\tau_{intr}^{-1} = \tau_2^{-1} + \tau_x^{-1} + \tau_a^{-1}. \tag{5}$$

The first term results from two-body collisions and is estimated by Bertsch [21] as $\tau_2^{-1} = \varepsilon^*/t^*$ with ε^* denoting the excitation energy per particle and $t^* = 2 \times 10^{-22}$ MeV s. The rate τ_x^{-1} is associated with the decay of diabatic states due to quasicrossings. This rate is given in Ref. [24] as $2\pi|H'|^2/\hbar\delta_x$ with H' the mean coupling matrix element between the diabatic states and δ_x the mean distance in energy between quasicrossings along a diabatic level. For realistic values $\delta_x \approx 2$ MeV and $H' \approx 0.5$ MeV we obtain $\tau_x \approx 0.8 \times 10^{-21}$ s. The last term accounts for the decay of diabatic states due to couplings proportional to the acceleration $\ddot{q}$. From the golden rule and the microscopic expression [25] for the coupling hamiltonian we write $\tau_a^{-1} = \beta \cdot \ddot{q}^2 (10^{-21}s)^3$ fm^{-2} with β estimated to 0.1 using a mean absolute value of $0.2 R_o |\dot{q}|$ for the single-particle matrix element of the velocity potential.

In the new diabatic one-body dissipation model the parameters α and β, estimated above, influence the stopping power and the nuclear interaction times almost independently. Due to the retardation of the frictional force, a large repulsive potential (proportional to α) is built up in the approach phase which leads to a fast deceleration. Due to this fast deceleration, the decay rate of the diabatic potential is mainly determined by τ_a^{-1}, and hence sensitive to the parameter β. A fast enough decay is decisive for a long contact time, because it prevents the nuclei from bouncing back elastically. The hatched area in between the solid lines in Fig. 1 are the theoretical predictions obtained with the elastoplastic model using $\beta = 0.15$ and α-values between 6 and 12. Values of α larger than 12 do not change the result significantly. At high TKEL-values the intrinsic equilibration rate is mainly determined by two-body collisions, leading to a weak sensitivity on the parameter β. Details of the calculations, their sensitivity, and also a comparison with the data obtained at 8.6 MeV/u can be found in Ref. [26]. There it will be shown that the diabatic extension also describes the experimental correlation between scattering angles and TKEL-values better than the original one-body dissipation model.

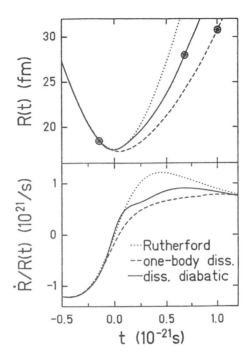

Fig. 2: Nuclear trajectories in terms of twice the root-mean-square radius, R(t), and $\dot{R}/R(t)$ as a function of time for the highest TKEL-window. The spoke-wheel symbols denote the instants of touching and separating shapes [15].

Figure 2 illustrates the nuclear trajectories (with R being twice the root-mean-square radius) for the highest TKEL-window. They have been calculated using the mean impact parameter deduced from the impact parameter distribution shown in the insert. In contradistinction to the smooth $\dot{R}/R(t)$ curves of the Rutherford and the one-body dissipation trajectories the elastoplastic trajectory shows a faster deceleration connected with additional weak oscillations, leading to the increased yield of high-energy δ-electrons seen in Fig.1.

In summary, studying the Pb + Pb collisions at 12.0 MeV/u a very fast radial deceleration is deduced from the yield of the high-energetic part of the δ-electron spectra. It is much stronger than predicted by the one-body dissipation model [15]. The extension of this model by including diabatic effects according to Ref. [20] allows

8

to account for the two aspects of the experimental data, the fast stopping and the long contact times. This strongly supports the existence of an elastoplastic behaviour of nuclear matter in dissipative heavy-ion collisions.

1. References

[1] G. Soff et al. Phys. Rev. Lett. 43, 1981 (1979).
[2] E. Kankeleit, Nukleonika 25, 253 (1980).
[3] E. Kankeleit, in International Advanced Courses on "Physics of Strong Fields", ed. by W. Greiner (Plenum, New York, 1987).
[4] U. Müller et al., Phys. Rev. C30, 1199 (1984).
[5] R. Krieg et al., Phys. Rev. C34, 562 (1986).
[6] M. Krämer et al., Phys. Lett. B201, 215 (1988).
[7] J. Stroth et al., Ric. Sci. Educ. Permanente Suppl. 63, 659 (1988).
[8] M. Krämer et al., Phys. Rev. C40, 1662 (1989).
[9] Th. de Reus et al., Z. Phys. A321, 589 (1985).
[10] E. Kankeleit et al., Nucl. Instrum. Meth. A234, 81 (1985).
[11] M. Rhein, Ch. Müntz, GSI Annual Report 1989, 293 (1989).
[12] R.S. Hager, E.C. Seltzer, V.F. Trusov in Atomic and Nuclear Data Reprints, Internal Conversion Coefficients, ed. by K. Way (Academic, New York, 1973).
[13] P. Schlüter, G. Soff, W. Greiner, Phys. Rep. 75, 327 (1981).
[14] Th. de Reus et al., Phys. Rev. C40, 752 (1989).
[15] H. Feldmeier, Rep. Progr. Phys. 50, 915 (1987).
[16] W.U. Schröder and J.R. Huizenga, in Treatise on Heavy-Ion Science, Vol. 2, ed. by D.A. Bromley (Plenum, New York, 1984), and references therein; J. Randrup, Nucl. Phys. A327, 490 (1979).
[17] J. Blocki et al., Ann. Phys. (N.Y.) 113, 330 (1978).
[18] P. Fröbrich and J. Stroth, Phys. Rev. Lett. 64, 629 (1990).
[19] R. Schmidt, V.D. Toneev and G. Wolschin, Nucl. Phys. A311, 247 (1978).
[20] W. Nörenberg, Phys. Lett. B104, 107 (1981); W. Nörenberg, in Heavy Ion Reaction Theory, ed. by W.Q. Shen, J.Y. Li and L.X. Ge (World Scientific, Singapore, 1989), and in New Vistas in Nuclear Dynamics, ed. by P.J. Brussaard and J.H. Koch (Plenum, New York, 1986).
[21] G.F. Bertsch, Z. Phys. A289, 103 (1987).
[22] P. Rozmej and W. Nörenberg, Phys. Lett. B117, 278 (1986).
[23] A. Lukasiak and W. Nörenberg, Z. Phys. A326, 79 (1987).
[24] W. Cassing and W. Nörenberg, Nucl. Phys. A433, 467 (1985).
[25] A. Lukasiak, W. Cassing, W. Nörenberg, Nucl. Phys. A426, 181 (1984).
[26] M. Rhein et al., Phys. Rev. C49, 250 (1994).

NUCLEAR DISASSEMBLY IN SYMMETRIC HEAVY-ION COLLISIONS AT INTERMEDIATE ENERGIES

W. J. LLOPE and the MSU 4π GROUP[§]

National Superconducting Cyclotron Laboratory
Michigan State University
East Lansing, MI 48824-1321, USA

ABSTRACT

We have experimentally studied small impact parameter heavy-ion collisions in the (nearly) symmetric entrance channels ^{12}C+^{12}C, ^{20}Ne+^{27}Al, ^{40}Ar+^{45}Sc, ^{84}Kr+^{93}Nb, and ^{129}Xe+^{139}La, each at many intermediate beam energies. Several analysis strategies based on the "shapes" of the experimental events are used to investigate the relative efficiencies of various experimental methods for the selection of the most central collisions, and to search for possible beam energy-dependent transitions from sequential binary disassembly to multifragmentation in the central events. Comparisons to dynamical and hybrid model code calculations will be discussed. The average shapes of subsets of the central events, in particular the intermediate mass fragments (IMFs, for which $3 \leq Z \lesssim 20$), are presented. Critical behavior, attributed to a transition from sequential binary disassembly to multifragmentation, is observed in all of these analyses. The transitional beam energies for the central ^{40}Ar+^{45}Sc, ^{84}Kr+^{93}Nb, and ^{129}Xe+^{139}La reactions are near $\sim$55, $\sim$40, and $\sim$40 MeV/nucleon, respectively.

1. Introduction

It is possible to form excited nuclear systems in the laboratory by colliding atomic nuclei. The impact parameter, as well as the predominant reaction mechanisms at each impact parameter, can be inferred from the experimentally measured characteristics of the particle emission. Given an efficient experimental selection of the most central collisions, beam energies from $\sim$10 to $\sim$150 MeV/nucleon can result in the formation of single nuclear systems with excitation energies from several to tens of MeV/nucleon. In such a range of excitation energies, previous experiments have indicated possible transitions between sequential binary (SB) disassembly mechanisms and multifragmentation (MF) (see Ref. 1 for a recent review). Detailed theoretical calculations[2-5] have predicted that the equivalent of a proper liquid-gas phase transition in finite nuclear systems occurs at excitation energies on the order of 10

[§]National Superconducting Cyclotron Laboratory, Michigan State University; Department of Chemistry, State University of New York - Stony Brook; Department of Physics, U. of Michigan - Dearborn; Cyclotron Institute, Texas A&M University; Department of Physics and Astronomy, U. of Iowa.

MeV/nucleon for systems of mass ~ 100. The possibility that transitions in disassembly mode from SB to MF are an artifact of such a liquid-gas phase transition is, however, only one of many. Systematic experimental studies of the system mass and excitation energy dependence of the predominant disassembly mechanisms and the applicability of the various theoretical descriptions are therefore necessary. In this contribution, such studies based on event shape analyses of a comprehensive set of experimental data are described.

The experimental data was collected using the MSU 4π Array[6] at the National Superconducting Cyclotron Laboratory. Reactions in the entrance channels $^{12}C+^{12}C$, $^{20}Ne+^{27}Al$, $^{40}Ar+^{45}Sc$, $^{84}Kr+^{93}Nb$, and $^{129}Xe+^{139}La$ were measured with a minimum bias trigger in 5-10 MeV/nucleon steps in beam energy, up the maximum energy available from the K1200 Cyclotron for each projectile: 155, 140, 115, 75, and 60 MeV/nucleon, respectively. A detailed description of the data collection can be found in Ref. 7.

The analyses of these data will proceed via the study of the average event shapes, which summarize aspects of the three-dimensional average patterns of the particle emission as viewed from the center of momentum (CM) frame. The cartesian components of the CM frame particle momenta, $p^{(k)}$, are used to fill a tensor,[9] $F_{ij} = \sum_k^N [p_i^{(k)} p_j^{(k)} / 2m_k]$, in each event. The normalization of the eigenvalues of this tensor, t_i, via $q_i = t_i^2 / \sum_{i=1}^3 t_i^2$, allows the calculation of the sphericity[8] using $S = \frac{3}{2}(1 - q_3)$, where q_3 is the largest normalized eigenvalue. All shape observables extracted from F_{ij}, e.g. S, depend strongly[12] on the number of particles, N, included in the sum in F_{ij}. For a given value of N, particle emission patterns that are isotropic in a momentum space coordinate system that spatially coincides with the CM frame have the largest possible sphericities, while otherwise deformed emission patterns have smaller sphericities.

Given methods to remove the dependence of S on N, it is possible to extract information concerning the impact parameter and the characteristics of the predominant reaction mechanisms from shape analyses. The sensitivity of the sphericity to the impact parameter is provided by the increasing probability for the emission of particles from spectator-like sources as the impact parameter is increased. Particles emitted from such sources have relatively large momenta and forward/backward focussed emission angles when viewed from the CM frame, and hence strongly suppress sphericity. Also, in the most central collisions, SB disassembly of the system at rest in the CM frame results in emission patterns that are more elongated in momentum space than those expected for MF reactions.[8,11] The finite values of the multiplicities, N, and inefficiencies in the central event selection conspire to suppress $\langle S \rangle$, and to decrease the distinctions in average shape between SB and MF events.

2. The Selection of Central Events

The selection of the most central experimental collisions constrains the mass of the excited nuclear system via the projectile and target masses and results in a monotonic relationship between the beam energy and the excitation energy in this system. Software cuts on global observables, i.e. centrality variables, that are assumed to be correlated with the impact parameter are used to select samples of the most central events. One must ensure, however, that the specific cut used to select these events is relatively inefficient at selecting larger impact parameter events with significant topological fluctuations, and does not autocorrelate with subsequent stages of the analysis.

The relative efficiencies of various cuts for small impact parameter events were evaluated under the assumption that particle emission from projectile and target-like sources severely elongates that shape of the event in the CM frame.[10] For each entrance channel and beam energy, five different samples of events were produced for which the value of a given centrality variable was in the maximal $\sim 10\%$ of it's minimum bias spectrum. For the actual thresholds placed on the different centrality variables for this study (see Ref. 13), impact parameters $\langle b \rangle / [R_P + R_T] \sim 0.26$–$0.32$ are expected in each sample from (approximate) geometrical arguments. In each sample separately, the average sphericity $\langle S \rangle$ versus the total charged particle multiplicity was extracted. At a specific total charged particle multiplicity in these samples of selected events, the most(least) efficient small impact parameter cut is assumed to result in the largest(smallest) $\langle S \rangle$.

To more clearly depict the relative efficiencies of different centrality variables studied with these data, these average sphericities are "reduced" in a way that preserves the insensitivity of the present comparisons to the finite multiplicity distortions. At each total charged particle multiplicity, the values of $\langle S \rangle$ are divided by the semi-inclusive average sphericities, $\langle S_{inc} \rangle$, which are obtained for each entrance channel, beam energy, and multiplicity without a centrality cut. These ratios are then averaged over the multiplicity in each sample of selected events, using weights for each multiplicity obtained from the number of counts in each bin.

These "multiplicity-averaged reduced (central) event-averaged sphericities" are shown versus the beam energy for all five entrance channels in Figure 1. The centrality variables compared are the total charged particle multiplicity, N_{chgd} (solid), the total proton multiplicity, N_p (long dashed), the total charge of hydrogen and helium isotopes, Z_{LCP} (dot dashed), the total charge in a software gate centered at mid-rapidity, Z_{MR} (short dashed), and the total transverse kinetic energy, KE_T (dotted). The variables Z_{MR} and KE_T are defined as described in Ref. 14.

In the two lightest entrance channels, ^{12}C+^{12}C and ^{20}Ne+^{27}Al, centrality cuts on the variables KE_T and N_p lead to the largest average reduced sphericities, and hence are thus the most efficient for the selection of small impact parameters in these reactions. The results for the ^{40}Ar+^{45}Sc reactions are similar to those from the ^{12}C+^{12}C and ^{20}Ne+^{27}Al reactions for the same range of beam energies. However, for ^{40}Ar+^{45}Sc reactions at beam energies below about 45 MeV/nucleon, small impact parameter cuts based on the multiplicities of light particles, i.e. on Z_{LCP} and N_p, are apparently the most efficient. These light particle multiplicity variables are also

12

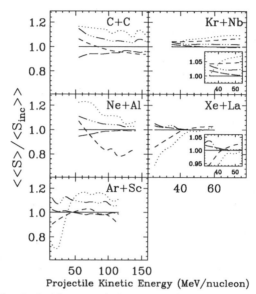

Fig. 1: The average reduced sphericity versus the beam energy for separate samples of events passing ~10% cuts on the different centrality variables listed in the text.

the most efficient for beam energies below ~40 MeV/nucleon in the ^{129}Xe+^{139}La entrance channel. In the ^{84}Kr+^{93}Nb and ^{129}Xe+^{139}La reactions above about ~45 MeV/nucleon, the most efficient centrality variables are KE_T and Z_{MR}, although the efficiencies of all of the centrality variables compared in Figure 1 become similar for increasing entrance channel mass.

The wiggles in these lines are related to that noted in Ref. 13. Autocorrelations between the average sphericity and the different centrality variables were checked via the study of $\Delta S = \sqrt{\langle S^2 \rangle - \langle S \rangle^2}$ in the different samples of selected events. If, at a specific multiplicity, an increasingly strict cut results in an increasing suppression of the width ΔS, it is assumed that a significant autocorrelation biases the observed values of $\langle S \rangle$ in the events selected by that cut. Significant suppressions of these widths relative to that obtained with lower thresholds on each centrality variable were observed for cuts generally stricter than ~1–3%, and were negligible for the ~10% cuts[13] used herein.

The beam energy-dependent transitions in the most efficient means of selecting the central events seen in Figure 1 are assumed to be related to similarly beam energy-dependent transitions in the predominant reaction mechanisms at these impact parameters. Copious light particle emission is expected from excited systems decaying by a sequential binary mechanism, due to the importance of the Coulomb and angular momentum barriers during such decays. Thus, the most central collisions for an entrance channel and beam energy for which SB disassembly dominates would be those for which the largest light particle multiplicities were observed. The extent to which the transitions apparent in Figure 1 are an artifact of transitions from SB to MF disassembly in the small impact parameter collisions is further investigated in the following Sections. Events were generated using SB and MF model codes and an accurate software replica of the apparatus, and direct comparisons to the experimental central events were made.

In order to limit the selection of larger impact parameter collisions with significant topological fluctuations, a two-dimensional centrality cut will be used for all that follows. For each entrance channel and beam energy, this cut selects events in which the two most efficient centrality variables (from Fig. 1) exceeded the ~10% thresholds[13] used above. This generally allowed ~4–8% of the minimum bias events ($\langle b \rangle / [R_P + R_T] \sim 0.2$–0.28 geometrically). The comparisons to software events described in the next Section will also provide a means of evaluating whether $\langle b \rangle / [R_P + R_T]$ is really in the range 0.2–0.28 in the selected events.

3. Comparisons to Model Code Events

In this Section, events generated by a number of different model codes which embody either SB or MF disassembly are filtered through an detailed software replica of the MSU 4π Array and compared directly to the data. The event generation was performed in both dynamic (FREESCO[8]) and hybrid approaches, for which BUU[16] and QMD[17] calculations were used to describe the initial stages of the reactions. The "after-burners" used in the hybrid event generation were the Berlin[18] and Copenhagen[2,19] MF codes, as well as the SB codes GEMINI[20] and SEQUENTIAL.[21]

All of these codes were run with the default parameters with the exception of the charge, mass, and excitation energy in the composite system, which was extracted from the BUU calculations in the same manner as described in Ref. 22. A soft equation of state was assumed, and the calculations were terminated when the radial density profile of the composite system most closely resembled that of a ground state nucleus.[16] The Berlin and GEMINI codes also require a cut-off angular momentum, which was taken as the maximum angular momentum that can be supported by the predicted composite system formed for each entrance channel and beam energy.

According to the BUU and QMD calculations for the impact parameters predicted geometrically for the present two-dimensional cuts, some fraction of the nucleons in the entrance channel are not found in an excited residue at rest in the CM frame. In the present event generation, these particles are thermally emitted from projectile and target-like sources, using reasonable assumptions for the velocities and temperatures of these spectator-like sources for each reaction.

For a given excited nucleus and at a specific final state multiplicity, the values of $\langle S \rangle$ for the events generated using the SEQUENTIAL(Berlin) code should agree with those from the GEMINI(Copenhagen) codes, and this was found to be true to about the ~10% level. Thus, for clarity in the Figures below, only the results using the MF code

14

Berlin[18] and SB code SEQUENTIAL[21] will be plotted. The average sphericities from the dynamic MF model FREESCO, and the hybrid MF models QMD+Copenhagen, and BUU+Copenhagen also agree to within ~10% at specific multiplicities, implying a relative insensitivity of the present shape comparisons to the method chosen for the specification of the input parameters to the decay codes. By assumption, the average shapes of the selected experimental events cannot be below the predictions of the SB model calculations, or above the predictions of the MF models.

The average sphericities of the central experimental(generated) events are plotted versus the measured(filtered) total charged particle multiplicity in Figure 2 for the ^{40}Ar+^{45}Sc and ^{129}Xe+^{139}La reactions. The crossed points are the results from the central experimental events, while the solid(dashed) lines are the results from the filtered BUU+MF(BUU+SB) events.

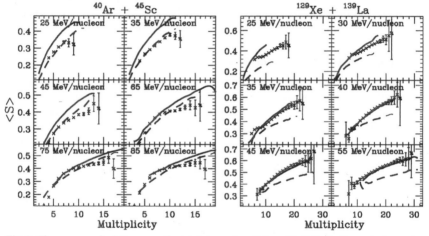

Fig. 2: The average sphericity versus the total charged particle multiplicity for the central experimental events (crossed points), the filtered BUU+SB events (dotted lines), and the filtered BUU+MF events (solid lines), for ^{40}Ar+^{45}Sc and ^{129}Xe+^{139}La reactions at the noted beam energies.

The average sphericities of the central ^{40}Ar+^{45}Sc reactions at beam energies near and below 45 MeV/nucleon are in agreement with those predicted by the BUU+SB codes. The data jump from the BUU+SB to the BUU+MF predictions for multiplicities on the order of eight between the beam energies of 45 and 65 MeV/nucleon. At larger multiplicities in this entrance channel, a dramatic suppression to below the SB model predictions is visible. This implies a failure in one or more of the assumptions used in the event generation described above, and is further investigated in the next Section.

In the central ^{129}Xe+^{139}La reactions, the average sphericities from the data are between the predictions of the filtered SB and MF models for all available beam energies. The average sphericities increase, relative to the SB predictions, with the beam energy up to ∼40 MeV/nucleon, above which they are in agreement with the MF model predictions.

4. Subset Shapes

The previous Section noted that the average sphericities of the largest multiplicity central events in the ^{40}Ar+^{45}Sc (and lighter) reactions were significantly below the predictions of the SB model calculations. In this Section, the sphericity of particular subsets of an event, i.e. the IMFs, will be distinguished from the sphericity of all of the particles in this event.

For all of the available entrance channels and beam energies, the largest average multiplicities of IMFs are found in the events with the largest total charged particle multiplicities. This is depicted in Figure 3, where each point style corresponds to a

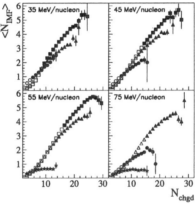

Fig. 3: The average number of IMFs versus the total charged particle multiplicity for ^{20}Ne+^{27}Al (stars), ^{40}Ar+^{45}Sc (circles), ^{84}Kr+^{93}Nb (triangles), and ^{129}Xe+^{139}La (squares) reactions at four representative beam energies. The N_{chgd} values above a threshold set for each system and beam energy allowing the maximal ∼10% of the events are depicted with the solid points.

particular entrance channel. The solid points for each system and beam energy are those above a ∼10% cut[13] on the inclusive spectrum of N_{chgd}. These solid points give a rough indication of the IMF multiplicities in samples of events of similar centrality for the different entrance channels and beam energies shown. A roughly universal dependence of $\langle N_{IMF} \rangle$ on N_{chgd} is noted for the more peripheral collisions (open points).

As depicted in Figure 4, the average IMF sphericities, $\langle S_{IMF} \rangle$, are generally well below the those predicted by the filtered SB model calculations for all IMF multiplicities. As the emission of IMFs is the most important at the largest charged particle multiplicities (Fig. 3), these suppressed IMF sphericities effect the suppression noted in Fig. 2. A similar suppression of the IMF sphericities to below the SB model predictions was also observed for the central ^{12}C+^{12}C and ^{20}Ne+^{27}Al reactions.

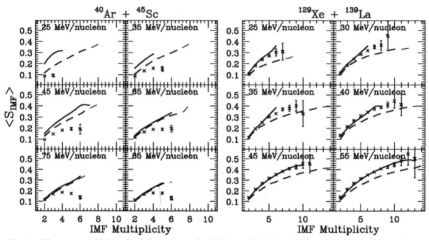

Fig. 4: The average IMF sphericity versus the IMF multiplicity for the central experimental events (crossed points), the filtered SB events (dotted lines), and the filtered MF events (solid lines), for ^{40}Ar+^{45}Sc and ^{129}Xe+^{139}La reactions at the noted beam energies.

For the central ^{129}Xe+^{139}La reactions, the average IMF sphericities are always between the SB and MF model predictions. Good agreement between the experimental IMF shapes and the MF model predictions are observed for beam energies above $\sim$40 MeV/nucleon.

The IMF emission patterns in the ^{40}Ar+^{45}Sc entrance channel are far more deformed than that expected from the SB (or MF) model calculations. It is possible to imagine several possible causes for this effect. The first concerns impact parameter fluctuations. In the lightest entrance channels, there is simply not much information available upon which to base a centrality cut. An (integer) centrality variable's binwidth may be a significant fraction of the maximum value in the spectrum populated by the minimum bias events. As noted in Ref. 14, this leads to relatively larger fluctuations in the impact parameters deduced for each event. If this were the only possibility, a breakdown in the geometrical assumption that the maximal 4–8% of the minimum bias events corresponds to $\langle b \rangle / [R_P + R_T] \sim 0.2$–0.28 in the light entrance channels would be evident. The second possibility assumes that fluctuations in the initial stages of the reactions increase in importance as the entrance channel mass is decreased. In those collisions for which the equilibration of the excited system at mid-rapidity is particularly incomplete, some knowledge of the initial trajectories of the projectile and target nuclei could be retained by the particles in the final state. Prolate shapes oriented along the beam direction would be expected for such events, in similarity to that expected given significant contaminations to the samples of selected central events from more peripheral collisions. The third possibility assumes the formation of non-compact geometries other than bubbles, e.g. toroids,[24] at freeze-out. These result in coplanar IMF emission patterns with the IMF flow angles $\theta_2 = \cos^{-1}(\mathbf{t}_2 \cdot \hat{\mathbf{z}})$ and $\theta_3 = \cos^{-1}(\mathbf{t}_3 \cdot \hat{\mathbf{z}})$ near 90°, where $\hat{\mathbf{z}}$ is the incident beam axis and $\mathbf{t}_2(\mathbf{t}_3)$ is the IMF eigenvector corresponding to the second largest(largest) eigenvalue.

Other aspects of the average shapes of the IMFs in the central events were studied to investigate these possibilities. The IMF shapes in the selected ^{40}Ar+^{45}Sc events (and in the lighter entrance channels) are manifestly prolate. For all values of N_{IMF}, the central event averaged probability that $t_3 - t_2 > t_2 - t_1$, corresponding to prolate IMF emission patterns, increases significantly for decreasing entrance channel mass. The CM frame ratios $\langle KE \rangle / 1.5 \langle KE_T \rangle$, calculated for each particle species in the selected events, also increase in this way. These observations point to an increase in the probability that the IMF emission patterns are elongated along the beam direction as the entrance channel mass is decreased, despite the event selection described in Section 2. To search specifically for toroidal IMF freeze-out configurations, the average IMF coplanarity, $C_{IMF} = \frac{\sqrt{3}}{2}(q_2 - q_1)$, and flatness $F_{IMF} = \frac{\sqrt{3}}{2}(t_2 - t_1)/(\sum_{i=1}^{3} t_i)$, were studied versus the IMF flow angle, θ_3, and the sum $\theta_3 + \theta_2$ for specific values of N_{IMF}. The average charge of the IMFs, the dispersion of the IMF charges, and the average IMF kinetic energy was plotted versus the values of the IMF flow angles and the shape observables C_{IMF} and F_{IMF}. No strong signal for toroidal geometries was seen using these observables, although these observables are, like S, strongly affected by the finite multiplicity distortions. Further analyses along these lines are presently underway.

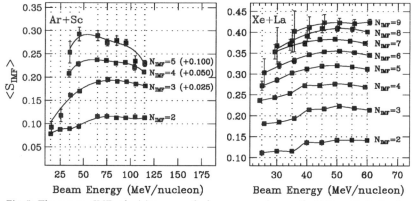

Fig. 5: The average IMF sphericity versus the beam energy for specific IMF multiplicities in central ^{40}Ar+^{45}Sc and ^{129}Xe+^{139}La reactions.

The beam energy dependence of the average IMF sphericities are shown in Figure 5 for specific IMF multiplicities in the central ^{40}Ar+^{45}Sc and ^{129}Xe+^{139}La reactions. Some points in Figure 5 have been vertically offset by the amounts shown for clarity, while the solid lines are included only to guide the eye.

In the central ^{40}Ar+^{45}Sc reactions, a significant increase in the average IMF sphericities is observed at beam energies near 55 MeV/nucleon for all IMF multiplicities. A similar increase of the IMF sphericities is also observed in the central ^{129}Xe+^{139}La reactions at beam energies near 40 MeV/nucleon. In the ^{84}Kr+^{93}Nb entrance channel, the average IMF sphericity for each value of N_{IMF} increases by ~10–25% when going from 35 to 45 MeV/nucleon in beam energy, and is constant (to <5%) from 45 to 75 MeV/nucleon. Similar, although less pronounced, maxima in the average sphericities of all of the particles in the selected events are also observed near the beam energies that lead to maximal IMF sphericities.

5. Conclusions

The beam energy dependence of various projections of the shapes of the small impact parameter events in the available ^{40}Ar+^{45}Sc and ^{129}Xe+^{139}La reactions indicated critical phenomena at beam energies near 55±10, and 40±10 MeV/nucleon, respectively. The definition of a critical beam energy in the central ^{84}Kr+^{93}Nb reactions is made difficult by the lack of beam energies below 35 MeV/nucleon for this entrance channel. However, there are indications that MF disassembly dominates in the central ^{84}Kr+^{93}Nb reactions for beam energies near and above 45 MeV/nucleon. No critical behavior was observed in the central ^{12}C+^{12}C(^{20}Ne+^{27}Al) reactions for beam energies from 55 to 155(140) MeV/nucleon.

The beam energies at which the various analyses described above indicated critical phenomena in the central events in each entrance channel are depicted in Figure 6. Also included in this Figure is the critical beam energy observed by Cebra et al. for central ^{40}Ar+^{51}V reactions,[21] and the critical beam energies observed in a charge correlations analysis of the present data.[23] In this analysis, a clear transition in the relative sizes of the three largest fragments in central events is observed at beam energies of 47±10, 35±10, and 32±5 MeV/nucleon in the ^{40}Ar+^{45}Sc, ^{84}Kr+^{93}Nb, and ^{129}Xe+^{139}La entrance channels, respectively.

The various analyses indicate similar critical beam energies for each entrance channel. The transitional beam energies observed for the central ^{40}Ar+^{45}Sc reactions are significantly larger than those observed for the central ^{84}Kr+^{93}Nb and ^{129}Xe+^{139}La reactions.

The most obvious possibilities for this trend concern the open questions raised in Sections 3 and 4. These would involve an increasing importance of fluctuations in impact parameter, or in the degree of equilibration of the excited systems formed at one impact parameter, for decreasing entrance channel mass. Two other possibilities are noted, which should not, however, be considered until the two above have been fully explored. The third assumes an increase in the importance of quantum mechanical finite-size effects for the excited systems formed in increasingly lighter entrance channels. The fourth involves the fact that increasingly heavier entrance channels result in excited systems that are increasingly more proton-rich.

The investigation of these open questions is possible only after the remeasurement of these reactions with an improved experimental acceptance. More efficient impact

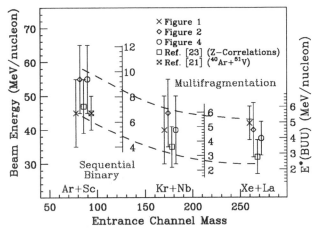

Fig. 6: A summary of the results of the present shape analyses indicating transitions in various characteristics of the predominant reaction mechanisms for the systems formed in the central collisions in the different entrance channels. Predictions of the BUU calculations for the approximate excitation energies reached in the selected events are given for each entrance channel on the right side axes. The dashed lines are included only to guide the eye.

parameter selection and studies of the characteristics of pre-equilibrium particle emission would then be possible. A more sensitive search for non-compact geometries would be feasible. Larger total event and subset multiplicities would be available for event shape analyses. Recently, important upgrades to the MSU 4π Array were finished, which dramatically improve the forward acceptance and substantially lower the overall particle kinetic energy thresholds. The entrance channels $^{40}Ar+^{45}Sc$, $^{129}Xe+^{139}La$, and $^{197}Au+^{197}Au$ were recently (re)measured with the improved apparatus, each for a systematic set of beam energies, while systematic experiments in the entrance channels $^{28}Si+^{27}Al$, $^{58}Ni+^{58}Ni$, and $^{84}Kr+^{93}Nb$ are scheduled.

6. Acknowledgements

We gratefully acknowledge helpful discussions with P. Danielewicz, L. Phair, W. Bauer, and J. Krieger. This work was supported by the U.S. National Science Foundation under Grants No. PHY 89-13815 and No. PHY 92-14992.

20

7. References

[1] L.G. Moretto and G.J. Wozniak, *Ann. Rev. Nucl. Part. Sci.* (May, 1993).

[2] J.P. Bondorf *et al.*, *Nucl. Phys.* **A448**, 753 (1986), and references therein.

[3] D.H.E. Gross, *Prog. Part. Nucl. Phys.* **30**, 155 (1993), and references therein.

[4] A.S. Botvina *et al.*, *Nucl. Phys.* **A475**, 663 (1987).

[5] W. Bauer, *Phys. Rev. C* **38**, 1297 (1988).

[6] G.D. Westfall *et al.*, *Nucl. Inst. and Methods* **A238**, 347 (1985).

[7] G.D. Westfall *et al.*, "Advances in Nuclear Dynamics, Proc. of the 8th Winter Workshop on Nuclear Dynamics", eds.: W. Bauer and B. Back, Jackson Hole, Wyoming (1992);
G.D. Westfall *et al.*, *Phys. Rev. Lett.* **71**, 1986 (1993);
E. Bauge *et al.*, *Phys. Rev. Lett.* **70**, 3705 (1993);
R.A. Lacey *et al.*, *Phys. Rev. Lett.* **70**, 1224 (1993).

[8] G. Fái and J. Randrup, *Nucl. Phys.* **A404**, 551 (1983).

[9] M. Gyulassy, K.A. Frankel, and H. Stöcker, *Phys. Lett.* **110B**, 185 (1982).

[10] H. Stöcker *et al.*, *Nucl. Phys.* **A387**, 205c (1982).

[11] J. A. López and J. Randrup, *Nucl. Phys.* **A491**, 477 (1989).

[12] P. Danielewicz and M. Gyulassy, *Phys. Lett.* **129B**, 283 (1983);
J.P. Bondorf *et al.*, *Phys. Lett. B* **240**, 28 (1990).

[13] It is not always possible to select a specific fraction of the impact parameter inclusive events, as often the case for centrality variables that are integer quantities. The threshold on a centrality variable chosen was the lowest possible bin for which <10% of the minimum bias events are in and above this bin. Generally, 7–10% of the events are selected by these cuts.

[14] L. Phair *et al.*, *Nucl. Phys.* **A548**, 489 (1992).

[15] G. Fái *et al.*, *Nucl. Phys.* **A381**, 557 (1982).

[16] W. Bauer and C.M. Mader, private communication;
C.M. Mader, Ph.D. Thesis, Michigan State University, unpublished (1993).

[17] G. Peilert, private communication;
J. Achelin *et al.*, *Phys. Rev. C* **37**, 2451 (1988).

[18] A.R. DeAngelis and D.H.E. Gross, private communication;
X.-Z. Zhang *et al.*, *Nucl. Phys.* **A461**, 668 (1987), and references therein.

[19] H.W. Barz *et al.*, *Phys. Lett. B* **267**, 317 (1991).

[20] R.J. Charity *et al.*, *Nucl. Phys.* **A483**, 371 (1988).

[21] D.A. Cebra *et al.*, *Phys. Rev. Lett.* **64**, 2246 (1990).

[22] D.R. Bowman *et al.*, *Phys. Rev. C* **46**, 1834 (1992).

[23] N. Stone, W.J. Llope, and G.D. Westfall, Preprint MSUCL-916, for submission.

[24] L.G. Moretto *et al.*, *Phys. Rev. Lett.* **69**, 1884 (1992);
W. Bauer, G.F. Bertsch, and H. Schulz, *ibid*, 1888 (1992).

Evolution of Multifragmentation in ^{84}Kr+^{197}Au Collisions at 35-400 AMeV

C. Schwarz, D. R. Bowman[a] , J. Dinius, C. K. Gelbke, T. Glasmacher, D. O. Handzy,
W. C. Hsi, M. J. Huang, W. S. Huang, M.A. Lisa[b], W.G. Lynch, G.F. Peaslee[c],
L. Phair[b], M.B. Tsang, C. Williams
National Superconducting Cyclotron Laboratory
and Department of Physics and Astronomy,
Michigan State University, East Lansing, MI 48824-1321, USA

M.-C. Lemaire, S. R. Souza
Laboratoire National SATURNE, CEN Saclay, 91191 Gif-sur-Yvette, France

G. Van Buren, R. J. Charity, and L. G. Sobotka
Department of Chemistry, Washington University, St. Louis, MO 63130, USA

G. J. Kunde, U. Lynen, J. Pochodzalla, H. Sann, W. Trautmann
Gesellschaft für Schwerionenforschung, D-64220 Darmstadt, Germany

D. Fox, R. T. de Souza
Department of Chemistry and IUCF, Indiana University,
Bloomington, IN 47405, USA

G. Peilert
Lawrence Livermore National Laboratory, Livermore, CA 94550, USA

and

N. Carlin
Instituto de Física, Universidade de São Paulo, CEP 01498, São Paulo

ABSTRACT

The relationship between observed intermediate mass fragment (IMF) and total charged particle multiplicities has been measured for ^{84}Kr+^{197}Au collisions at energies between E/A= 35 and 400 MeV. Fragment multiplicities are greatest for central or near-central collisions. For these collisions, fragment production increases up to $E/A \approx 100$ MeV, and then decreases at higher energies.

Highly excited nuclear systems have been observed [1-7] to decay by multifragment emission. There is accumulating evidence that multifragment decays are favored for systems that expand to subnormal densities [5-10], and it has been suggested that multifragment decays might provide key information about a liquid-gas phase transition in nuclear matter [11-15]. Rather general phase space and barrier penetrability arguments [16-20] lead to the expectation that the multifragment emission probability will exhibit a strong initial rise as a function of temperature. At very high temperatures, on the other hand, the entropy of the system becomes so high that fragment production is suppressed. Hence, fragment production should exhibit a maximum at some intermediate temperature, which may depend on the total charge of the fragmenting system.

Measurements of the impact-parameter dependence of fragment multiplicities in projectile fragmentation reactions of Au nuclei at E/A=600 MeV have revealed the qualitative features of this "rise and fall" of fragment production [3,21]. Focusing on central collisions, a rise in the intermediate mass fragment [IMF] multiplicity with incident energy has been observed for central Ar+Au collisions over the energy range E/A=35-110 MeV [6]. A decline in IMF multiplicity with incident energy has been observed for central Au + Au collisions over a higher energy range E/A=100-400 MeV [22]. To identify the incident energy with the peak fragment multiplicity, however, requires heretofore nonexistent measurements of both the rise and the decline of multifragmentation with a single system. To address this issue we have performed measurements of ^{84}Kr+^{197}Au collisions over the incident energy range E/A=35-400 MeV with a low-threshold 4π detector [23].

Measurements with ^{84}Kr ions at beam energies of E/A=35, 55, and 70 MeV were performed with beams from the K1200 cyclotron of the National Superconducting Cyclotron Laboratory of Michigan State University. Typical beam intensities were $1 \cdot 10^8 - 2 \cdot 10^8$ particles per second (intensities at E/A=70 MeV were lower by a factor of two). Measurements at E/A=100, 200, and 400 MeV were performed at the Laboratoire National SATURNE at Saclay, with typical beam intensities of $10^6 - 10^7$ particles per spill. The gold target thicknesses were: 1.3 mg/cm^2 at E/A=35 and 55 MeV, 4 mg/cm^2 at E/A=70 MeV and 5 mg/cm^2 at E/A = 100, 200 and 400 MeV. The emitted charged particles were detected with the combined MSU Miniball/Washington University Miniwall 4π phoswich detector array. This detector system consisted of 276 low-threshold plastic- scintillator-CsI(Tl) phoswich detectors, covering polar angles of $\theta_{lab} = 5.4° - 160°$, corresponding to a total geometric efficiency of approximately 90% of 4π. For the experiment at lower incident energies, (E/A¡100 MeV), 268 plastic- scintillator-CsI(Tl) phoswich detectors were used. An ion chamber substituted one Miniball detector in each ring for $\theta_{lab} > 25°$ [24]. Data taken with these ion chambers were not included in the present analysis; this omission results in an estimated 3-4% reduction in the fragment multiplicities for $E/A < 100$ MeV. Wall detectors located at forward angles, $\theta_{lab} = 5.4° - 25°$, used plastic scintillator foils of 80 μm thickness and CsI(Tl) crystals of 3 cm thickness. The thresholds for particle identification in these detectors were $E_{th}/A \approx 4$ MeV (6 MeV) for Z=3

(Z=10) particles, respectively. For the higher incident energies ($E/A > 100$ MeV), the energy thresholds in the wall were set somewhat higher at about 7 MeV (7.5 MeV) for Z=3 (10) particles. Ball detectors at larger angles, $\theta_{lab} = 25° - 160°$, used 40 μm scintillator foils and 2 cm thick CsI(Tl) crystals; the corresponding thresholds were $E_{th}/A \approx 2$ MeV (4 MeV) for Z=3 (Z=10) particles, respectively. To avoid contamination from low energy electrons, hardware discriminator thresholds of 5 MeV were imposed on the Z=1 particles for the Miniball and 10 MeV for the Miniwall. For incident energies with $E/A \geq 100 MeV$, the Z=1 thresholds for the Miniwall were higher, typically 20 MeV. Unit charge resolution up to $Z \approx 10$ was routinely achieved for particles that traversed the fast plastic scintillator. Lithium ions that punched through the CsI(Tl) crystals were not counted as IMF's because they were not distinguished from light particles. Double hits consisting of a light particle and an IMF were identified as single IMF, double hits consisting of two light particles were identified as a single light particle, and double hits consisting of two IMF's were identified as a single IMF. Typically, multiple hits reduced the charge particle multiplicity in central collisions by an estimated 15-25% and the IMF multiplicity by 1.5-2.5% , depending on incident energy.

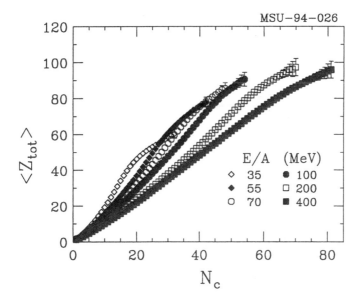

Fig. 1: The correlation between the measured mean total charge, $< Z_{tot} >$, and the measured charged particle multiplicity, N_C, is shown for the six incident energies. Due to coincidence summing effects, the systematic uncertainty in $< Z_{tot} >$ can be of order 10%.

Similar to other measurements [5,6,25], the measured charged particle multiplicity distributions exhibit a rather structureless plateau and a near-exponential fall-off at the highest multiplicities. The multiplicity where one observes the exponential fall-off increases from $N_C \approx 30$ to 65 as the beam energy is increased from $E/A = 35$ to 400 MeV. As in previous work [26], we constructed a "reduced" impact parameter scale from the charged particle multiplicity by means of the geometric formula [26,27]:

$$\hat{b} = \frac{b}{b_{max}} = \left[\int_{N_C(b)}^{\infty} dN_C \, P(N_C) \right]^{1/2}. \tag{1}$$

Here, $P(N_C)$ is the probability distribution for detecting the N_C charged particles and b_{max} is the impact parameter where $N_C \approx 4$. The reduced impact parameter assumes values of $\hat{b}=1$ for the most peripheral collisions and $\hat{b}=0$ for the most central collisions.

To illustrate the detection capabilities of the experimental setup, the mean total charge , $< Ztot >$, is shown in Fig. 1 as a function of the incident energy and the detected charged particle multiplicity, N_C. At each energy, the measured mean total charge is a monotonic function of the charged particle multiplicity; the maximum detected charge is observed for central collisions and increases from about 60 at E/A=35 MeV to more than 80 at E/A=100 MeV out of a total of 115, and it remains roughly constant thereafter. Losses in efficiency are most significant for beam velocity particles emitted to $\theta_{lab} < 5.4°$ and for heavy target-like residues which do not penetrate the scintillator foils of the phoswich detectors and are, hence, not identified.

Figure 2 shows the observed mean IMF multiplicity, $< N_{IMF} >$, as a function of detected charged particle multiplicity, N_C. For measurements at E/A=35-100 MeV, the data display a rather similar dependence of $< N_{IMF} >$ upon N_C. At the higher two energies, much higher charged particle multiplicities are required to achieve the same value for $< N_{IMF} >$. Some fragments from the statistical decay of projectile-like residues are lost because they are emitted to angles smaller than 5.4°. This loss is most important for the higher two incident energies and leads to an unknown reduction in the fragment multiplicities at medium to low values of N_C. This problem is less important for central collisions. For measurements at E/A=35-200 MeV, the peak IMF multiplicity is observed for the most central collisions. In contrast to the data at lower incident energies, the data at E/A=400 MeV display a maximum at $N_C = 60$ and decline thereafter. Since comparable values of $< Z_{tot} >$ for the most central collisions are observed at the three highest incident energies, this decline in $< N_{IMF} >$ for $N_C > 60$ at the highest incident energy is not likely a trivial consequence of charge conservation or a loss in detection efficiency in the experimental setup.

The energy dependence of charged particle and fragment production in central collisions, $0 < \hat{b} < 0.25$ is shown as the solid points in the lower and upper panels,

MSU−94−027

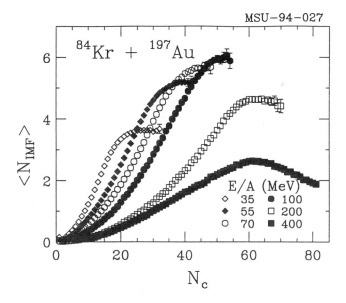

Fig. 2: The correlation between average detected IMF (Z=3-20) multiplicity, $< NIMF >$, and detected charged particle multiplicity, N_C, is shown for the six incident energies.

respectively, of Figure 3. The charged particle multiplicity increases monotonically with incident energy. The fragment multiplicity is observed to increase to a maximum at $E/A \approx 100$ MeV and decreases thereafter. The increase for $E/A < 100$ MeV is likely due to an increase in thermal excitation and in the collective expansion velocity with incident energy. Both are expected to cause an increase in fragment multiplicity for systems in which fragment production is excitation energy limited [9,25]. The relative importance of the two quantities for the present data set is unknown. A decrease at higher energies is expected from general arguments based upon entropy production; the wide incident energy range of the present data permits, for the first time, an approximate determination of the energy at which this decrease commences.

A similar maximum at $E/A = 100$ MeV has been predicted by microscopic molecular dynamics models [28] for Nb+Nb collisions. Thus it is interesting to explore whether such models can describe the present data. Predictions of the Quantum Molecular Dynamics (QMD) model [29] are shown as dashed lines in the left-hand-panels of the figures. (For this comparison, we assume $b_{max} = 10fm$.) After correcting for the experimental acceptance, one obtains the filtered QMD calculations shown by the solid lines. The general energy dependent trends of the data, i.e. a maximum

26

in $< N_{IMF} >$ at E/A=100 MeV and monotonically increasing values for $< N_C >$ are reproduced. However, the calculations significantly underestimate the number of charged particles and the number of intermediate mass fragments produced at $E/A < 100$ MeV.

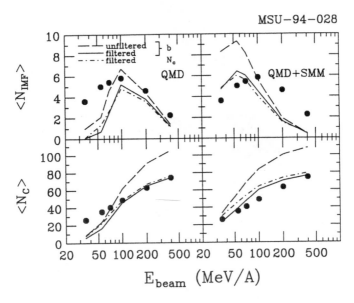

Fig. 3: The incident energy dependences of the detected IMF (Z=3-20) multiplicity, $< N_{IMF} >$ (upper panels), and the detected charged particle multiplicity, N_C (lower panels). Predictions of the QMD model are shown as dashed lines (left panels) and solid lines after the model calculations filtering through the experimental acceptance. The dotted-dashed lines depict the filtered calculations that were analyzed as data to assess impact parameter fluctuations. The right-hand panels show a similar presentation of the QMD+SMM model calculations.

The failure of QMD calculations to reproduce the large IMF multiplicities observed at low incident energies have been attributed to an inadequate treatment of the decay of highly excited heavy reaction residues produced in the QMD calculations [30]. To remedy this deficiency, the decays of all fragments with $A \geq 4$ were calculated via the statistical multifragmentation model (SMM) [31], which contains a "cracking" phase transition at low density. Input excitation energies and masses for the SMM calculations were taken from the QMD calculations at an elapsed reaction time of 200 fm/c. The results from these two stage calculations, shown by the dashed lines in the right-hand-panels, significantly overpredict the data at $E/A < 100$ MeV reflecting the additional contributions from heavy residue decay in SMM stage, but underpredict the data at higher energies because many fragments produced by the QMD stage are evaporated away in the later SMM stage. The numbers of IMF's produced in these same calculations, when filtered through the experimental acceptance (solid lines), remain very similar at $E/A > 100$ MeV. The efficiency of detection is reduced significantly at $E/A < 100$ MeV reflecting the fact that the calculated energy spectra are peaked at lower kinetic energies than the measured spectra and consequently many of the predicted fragments fall below the experimental thresholds.

All comparisons of data to calculations performed at fixed impact parameter implicitly assume a high precision for the experimental impact parameter constructed from the charged particle multiplicity via Eq. 1. We have tested this idealization by applying Eq. 1 to the calculated and filtered charged particle multiplicity from the QMD and QMD+SMM simulations to determine a corresponding value for N_C, which we define as N_{cut}, such that the cross section for events with a higher filtered multiplicity equals $\pi*(2.5fm)^2$. We then compute the calculated mean charged particle and IMF multiplicities for events with $N_C > N_{cut}$, i.e. we analyze the calculations as if they are data. The results, shown by the dotted-dashed lines in Fig. 3 do not differ from the calculations at fixed impact parameter significantly.

For further insight in the reaction mechanism it is useful to study the energy spectra. A quantity which allows to compare fragment energies of all incident energies is the transverse energy, here expressed classically since the transverse fragment energies are relatively small. For a thermal source with temperature T, an transverse energy of $< E_T >= T$ is independent of the fragment charge Z. Coulomb repulsion and collective motion, however, lead to increasing transverse energies; recoil effects to decreasing transverse energies with increasing fragment charge, respectively. The interplay between the contributions could lead to a complicated transverse energy distribution as function of mass or charge. Hence, we restricted ourselves in the following to the comparison of the transverse energies to QMD+SMM model predictions.

Transverse energies are shown as function of the fragment charge Z=2-4 for all incident energies in Fig. 4. $< E_T >$ for Li and Be fragments produced in mid-central collisions increases from 35 to 65 MeV when the beam energy is raised from 35 to 400 AMeV. Fragments emerging from central collisions show a larger increasing of the transverse energy from 35 to 100 MeV indicating a more violent collision. The low transverse energy of the ^{4}He fragments might be due to secondary decay where the ^{4}He energies are low. The transverse energies are higher than expected from a thermal source and connote preequilibrium emission of the fragments. The lines in Fig. 4 show the prediction of QMD+SMM. The predictions underestimate $< E_T >$ indicating that the fragments are produced from relatively explosive collisions. At low incident energies the model underpredicts the transverse energy since the fragments are produced by the SMM part of the hybrid model (see Fig. 3) which is a fairly gentle breakup. At 100-400 AMeV incident energy the most violent central collisions produce practically no fragments. The fragments come from more peripheral reactions from the hot spectators.

28

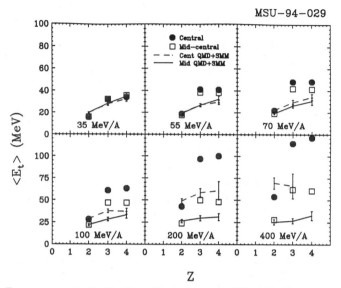

Fig. 4: Transverse energies for He, Li, and Be fragments for different incident energies. Fragments from central collisions and fragments from mid-central collisions are shown as solid circles and solid diamonds, respectively. The model calculations are shown as solid (central) and dashed (mid central) lines. The central collision correspond to an reduced impact parameter range of , the mid-central collision correspond to an reduced impact parameter range of , respectively.

In summary, we have presented the first comprehensive study of the multifragment emission over a broad range of beam energies, E/A = 35 - 400 MeV. For the $^{84}Kr+^{197}Au$ system, fragment multiplicities are greatest at $E/A \approx 100 MeV$. For central collisions, much of the energy dependence of fragment production agrees qualitatively with QMD molecular dynamics calculations, but the calculations significantly underpredict the data at low incident energies. Calculation of the statistical decay of residues via the SMM model improves the agreement between data and theory at the low incident energies but these calculations underpredict the fragment yields at higher incident energies and the peak fragment multiplicity is predicted at E/A=55 MeV, rather than at E/A=100 MeV as it is observed. The calculations also show a qualitative agreement with transverse energies for Li and Be fragments but fail to predict the data quantitatively. There is a need for an improved transport theory for the treatment of density fluctuations and fragment formation.

This work is supported by the National Science Foundation under Grant numbers PHY-90-15255 and PHY-92-14992, and the U.S. Department of Energy under Contract No. DE-FG02-87ER-40316. W.G. Lynch and L. G. Sobotka are pleased to acknowledge the receipt of U.S. Presidential Young Investigator Awards. N. Carlin and S. R. Souza gratefully acknowledge partial support by the CNPq, Brazil. We gratefully acknowledge the excellent support from the operations staff of the LNS and of the NSCL, and we wish to express our appreciation for their kind hospitality extended to us during our experiment at the LNS.

[a] Present address: Chalk River Laboratories, Chalk River, Ontario, K0J1J0 Canada.
[b] Present address: Lawrence Berkeley Laboratory, Berkeley CA 94720.
[c] Present address: Physics Department, Hope College, Holland MI 49423.

References

[1] J.W. Harris et al., Nucl. Phys. **A471**, 241 (1987).
[2] R. Bougault et al., Nucl. Phys. **A488**, 255 (1988).
[3] C.A. Ogilvie et al., Phys. Rev. Lett. **67**, 1214 (1991).
[4] Y. Blumenfeld et al., Phys. Rev. Lett. **66**, 576 (1991).
[5] D.R. Bowman et al., Phys. Rev. Lett. **67**, 1527 (1991).
[6] R.T. de Souza et al., Phys. Lett. **B268**, 6 (1991).
[7] K. Hagel et al., Phys. Rev. Lett. **68**, 2141 (1992).
[8] W.A. Friedman, Phys. Rev. Lett. **60**, 2125 (1988).
[9] W.A. Friedman, Phys. Rev. C **42**, 667 (1990).
[10] J. Hubele et al., Phys. Rev. C **46**, R1577 (1992)
[11] J.E. Finn et al., Phys. Rev. Lett. **49**, 1321 (1982).
[12] G. Bertsch and P.J. Siemens, Phys. Lett. **B126**, 9 (1983).
[13] L.P. Csernai and J. Kapusta, Phys. Reports **131**, 223 (1986).
[14] P.J. Siemens, Nature **305**, 410 (1983).
[15] T.J. Schlagel and V.R. Pandharipande, Phys. Rev. C **36**, 162 (1987).
[16] L.G. Moretto, Nucl. Phys. **A247**, 211 (1975).
[17] W.A. Friedman and W.G. Lynch, Phys. Rev. C **28**, 16 (1982).
[18] W.G. Lynch, Ann. Rev. Nucl. Part. Sci. **37**, 493 (1987).
[19] D.H.E. Gross, Rep. Prog. Phys. **53**, 605 (1990).
[20] L.G. Sobotka, Phys. Rev. Lett. **51**, 2187 (1983).
[21] J. Hubele et al., Z.Phys. **A340**, 263 (1991).
[22] M.B. Tsang et al, Phys. Rev. Lett. **71** (1993) 1502.
[23] R.T. de Souza et al., Nucl. Instr. Meth. **A295**, 109 (1990).
[24] R. de Souza et al., to be published
[25] D.R. Bowman et al., Phys. Rev. C **46**, 1834 (1992).
[26] L. Phair et al., Nucl. Phys. **A548**, 489 (1992).
[27] C. Cavata et al., Phys. Rev. C **42**, 1760 (1990).
[28] G. Peilert et al., Phys. Rev. C **39**, (1989) 1402.
[29] G. Peilert et al., Phys. Rev. C **46**, 1457(1992), and refs. therein.
[30] T.C. Sangster et al., Phys. Rev. C **46**, 1404 (1992).
[31] J.P. Bondorf et al., Nucl. Phys. **A444**, 460 (1985); A.S. Botvina et al., Nucl. Phys. **A475**, 663 (1987).

Exploring Multifragmentation with a 4π Detector: Xe-Induced Reactions at E_{beam}=60 MeV/nucleon[#]

W. Skulski,[*] K. Tso, N. Colonna, L. G. Moretto, G. J. Wozniak, D. R. Bowman, M.Chartier[+], C. K. Gelbke[+], W. C. Hsi[+], M. A. Lisa[+$], W. G. Lynch[+], G.F.Peaslee[+@], L. Phair[+$], C. Schwarz[+], and M. B. Tsang[+]

ABSTRACT

The reactions ^{129}Xe + ^{27}Al, ^{nat}Cu, ^{89}Y, ^{165}Ho and ^{197}Au at 60 MeV/nucleon were investigated with the combination of a high-resolution forward array of Si-Si-plastic telescopes and a high-efficiency phoswich ball. The projectile-like primary source was reconstructed. Its parallel velocity component is strongly correlated with the observed multiplicity and hence with dissipation. At high multiplicity, large fragments still show a projectile-like and target-like components in their parallel velocity distributions. A further sub-division of highly dissipative events into binary-like and non binary-like subclasses has been performed by applying an additional experimental selection of the orientation of the event in momentum space.

Introduction

The disintegration of highly excited nuclear systems is the subject of much current interest, both experimentally and theoretically (see (1) and references therein). The key quantities determining how the hot nuclear system decays are its size and excitation energy. In experimental studies these quantities are not directly measured. The energy dissipation in the collision can be inferred from the charged particle multiplicity.[2,3] The size of the hot primary "source" can be accessed by summing all the detected fragments over a suitable range of detection angles.[1] An important uncertainty still remains: how many hot primary fragments have been produced in the reaction? Theoretical calculations predict that two primary fragments are formed in peripheral collisions, one projectile-like and one target-like, whereas a single equilibrated nuclear system is formed for central events with small impact parameter. Due to fluctations these two different situations can happen at similar total excitation energy, and thus at similar multiplicity. Therefore, one cannot use particle multiplicity to cleanly separate binary from non-binary reaction events. This is exemplified by a study of the Xe+Bi reaction at 28 MeV/nucleon: a binary reaction mechanism persists even for high particle multiplicity and/or large intermediate mass fragment (IMF) multiplicity.[4] A clean non-binary reaction mechanism could not be separated at all.

The aim of the present paper is to characterize experimentally reactions of Xe with a range of targets to study to what extent a selection of binary from non-binary events can be achieved at 60 MeV/nucleon bombarding energy.

Experimental procedure

The experiment was performed at the K1200 Cyclotron of the National Superconducting Cyclotron Laboratory at Michigan State University. A 60 MeV/nucleon ^{129}Xe beam bombarded targets of ^{27}Al, ^{nat}Cu, ^{89}Y, ^{165}Ho and ^{197}Au of thickness 2.07, 2.0, 1.0, 2.0 and 1.3

mg/cm^2, respectively. The detection system subtended angles from 2° to 160° with respect to the beam axis and had a geometric acceptance of ~88% of 4π. At forward angles (2° - 16°), fragments (Z = 1 - 54) were detected in a 16-telescope Si(300 μm)-Si(5 mm)-Plastic(7.6 cm) array[5] with good energy and position resolution. The geometrical efficiency of the forward array was about 64%. Individual elements were resolved from Z=1 to 54, when counting statistics allowed. Representative detection thresholds in the forward array were 13, 21 and 27 MeV/u for fragments of Z = 8, 20 and 54, respectively. Energy and position calibrations were performed by utilizing analog beams[6] of q/A=1/6 [D^{4}He$^+$, ^{6}Li$^+$, ^{12}C^{2+}, ^{18}O^{3+}] at 22 MeV/nucleon, and two "cocktail" beams[6] at 60 MeV/nucleon: [^{30}Si, ^{60}Ni, ^{90}Zr] and [^{43}Ca, ^{86}Rb, ^{129}Xe]. (A "cocktail" beam consists of 3 different ion beams with nearly identical charge to mass ratios.) These low intensity beams were swept directly across each of the array telescopes. These data were used to measure the nonuniformity in the 300 μm ΔE detector thickness and corrections were made off-line. The pulse-height defect was measured and corrected for according to ref. (7). The overall energy calibration was accurate to about 1% and the position calibration to within 1.5 mm.

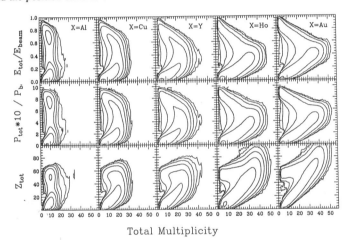

Total Multiplicity

Fig. 1. Logarithmic contour plots of the detected total energy (upper row), total detected parallel momentum (second row) and total detected charge (third row) vs. total charged particle multiplicity, for the 60 MeV/u ^{129}Xe + ^{27}Al, ^{nat}Cu, ^{89}Y, ^{165}Ho and ^{197}Au reactions.

At larger angles (16° - 160°), light charged particles and fragments (Z = 1 - 35) were detected in the MSU Miniball[8] consisting of 171 fast plastic (40 μm) - CsI(2 cm) phoswich detectors. The most forward-angle ring and four detectors of the second ring of the Miniball were removed to accommodate the forward-angle Si array. Representative detection thresholds were 2, 3, and 4 MeV/u for Z = 3, 10, and 18 fragments, respectively. For H and He, individual isotopes could be resolved. For heavier elements, only elemental resolution was achieved. The energy calibration was obtained by scaling the previous calibration[9] to hydrogen punch-through points in every detector. This procedure was checked for a subset of 8 Miniball detectors that were extensively calibrated by sweeping the q/A=1/6 beams across their face. For this limited subset of detectors, satisfactory agreement with the existing calibration was found.

Results

I. Selection of "completely measured" events

As the excitation energy of the primary reaction products is increased, one expects the number of evaporated neutrons and light charged particles to increase. (Recent measurements[4] have shown a strong correlation between the measured neutron and charged particle multiplicities.) In our measurements, we utilize the total charged particle multiplicity M, which consists of the measured number of light charged particles (LCPs) and intermediate mass fragments (IMFs). Low M values are characteristic of low-dissipation events, while high M values are associated with more violent collisions. For light targets (Al, Cu) the present experiment does not detect the most peripheral events with good efficiency due to the grazing angle falling below the minimum detection angle of the forward array. In this case a PLF can be detected in the forward array only if it is deflected to $\theta > 2^\circ$ due to sequential decay. This introduces a certain bias against large impact parameter events. For heavier targets (Y, Ho, Au), events from a large range of impact parameters can be observed.

In the present experiment the total detected energy can be smaller than the total available energy minus the Q-value because some charged particles may not be observed due to thresholds and dead regions between the detectors. Also, neutrons are not measured by the present experimental setup. Figure 1 shows contour plots of the total detected charge, kinetic energy and linear momentum versus the total number of detected particles M for five targets. For the lightest target ^{27}Al, the distributions have two peaks. For heavier targets, instead of two peaks one observes two branches (e.g., see the Au target). In both cases, the lower peak (or the lower branch) corresponds to the situation when the heavy projectile-like fragment (PLF) is not detected due either to it being emitted at an angle smaller than 2° or in the dead region between telescopes. When the heavy forward-going PLF is not detected, one sees only the light particles emitted from the target and projectile-like residues. (The slow moving target-like fragment (TLF) is below the detector thresholds.) We thus define "completely measured events" as the ones in which the PLF was *not* missed. Experimentally, they are defined as those with the total detected energy of at least 50% of the beam energy (see the first row in Figure 1). This 50% cutoff serves as a convenient discrimination against incompletely measured events, in which the PLF was not detected due to less than 100% detection efficiency.

II. Primary PLF fragments

When the energy dissipation increases, as reflected by the increasing multiplicity, the excitation of the forward-going primary PLF also increases. At large excitation energies, the primary PLF can decay into several secondary fragments. With our forward array we can detect most of them. During the off-line analysis, we reconstructed the primary PLF "source" defined as the sum of all fragments detected in the forward array ($Z_{fragment} > 4$).[1] When discussing "PLF sources", only "completely" measured events will be considered (see previous section).

One of the conclusions of the previous studies summarized in ref. (1) was that slower PLF sources correspond to increased energy dissipation. One can test this by looking at the total detected charge and/or multiplicity in coincidence with PLF sources moving with different velocities. In Figure 2 the parallel velocity of the reconstructed PLF source is presented versus its charge as a contour plot (upper-right panel). The triangular pattern of the charge-velocity contour plots is similar to those previously reported in the La-induced reactions between 35 and 55 MeV/nucleon.[1] This pattern was previously interpreted as a result of the incomplete fusion followed by extensive particle evaporation, see (1) for details.

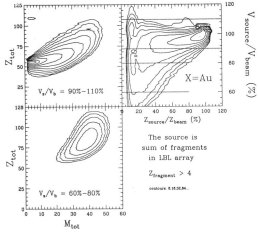

Fig. 2. *Upper right panel:* Contour plot of the mean parallel velocity of the PLF source versus the size of the source (see text). *Left-hand panels:* Distributions of the total detected charge Z_{total} versus multiplicity, gated on two ranges of the PLF source velocity: 90%-110% and 60%-80% of V_{beam}. The distributions are for the 60 MeV/nucleon $^{129}Xe + ^{197}Au$ reaction. The total detected charge Z_{total} includes Z_{source} as well as all charges of the light charged particles.

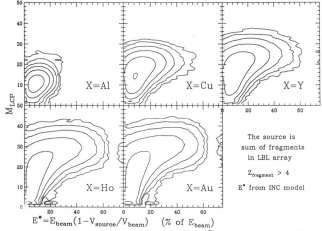

Fig. 3. Logarithmic contour plot of the excitation energy E^* generated in the collision versus LCP multiplicity. E^* was calculated from the PLF source velocity assuming the incomplete-fusion model.

Both total detected charge and total multiplicity are increased on the average, when the gate is set on the slower-moving PLF source (see the left-hand side of Figure 2). The increase is due to the higher degree of dissipation, which corroborates the conclusions of ref. (1). One notes a conspicuous absence of low-multiplicity collisions in the lower left panel of

Figure 2. Only high-dissipation events are selected by gating on the slow PLF source. The multiplicity of these events extends to the highest values observed in the experiment, as revealed by comparing Figures 1 and 2. However, these high-multiplicity events are still relatively peripheral, as indicated by the very existence of a PLF source of substantial size.

In Figure 3 the light-charged particle multiplicity M_{LCP} is plotted versus the excitation energy generated in the collision, as inferred from the PLF source velocity in the framework of the incomplete-fusion model.[1] For targets heavier than Al there is a correlation between both variables. It shows that the velocity of the PLF source can indeed be used to infer the energy dissipation. Very little correlation is present for Al target, since in this case only relatively central collisions are observed in the experiment, as discussed previously.

III. Event-by-event selection of non-binary events in high-multiplicity collisions

It has been shown in the previous section that sizeable sources are present at high multiplicity. It shows the existence of very dissipative binary-like collisions. In addition to these, also central collisions are expected to be detected as high-multiplicity events. Can one discriminate between these two types of events? Is it feasible to distinguish experimentally between the "true central" collisions and "highly dissipative binary" background on an event-by-event basis? Clearly, one has to use an additional experimental observable different from the multiplicity itself in order to achieve this goal. We propose an orientation of the event in the momentum space (defined below) as a such additional observable.

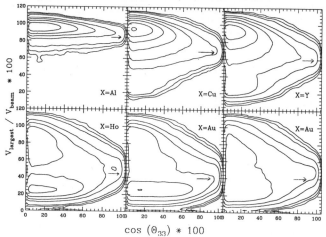

$$\cos \left(\Theta_{33} \right) * 100$$

Fig. 4. Logarithmic contour plots of the parallel velocity of the largest fragment vs. the orientation of the momentum tensor. Only "complete" high multiplicity events were included in the analysis. In the lower-rightmost panel, the largest detected fragment was dropped from analysis of the ^{129}Xe + ^{197}Au reaction to augment the effect of incomplete detection efficiency of the experimental setup. In each case, the compound nucleus velocity is marked by a horizontal arrow.

The orientation of an event in momentum space is defined by means of the sphericity tensor[11] constructed for each event according to:

$$T = \sum P_k{}^T P_k / 2 M_k, \qquad k = 1 \text{ to } M_{IMF}$$

The momentum tensor was determined using IMFs with Z>2, in order to minimize the effect of double-hits in the forward detectors. At least three IMFs were required to calculate the tensor. The tensor T is a real symmetric matrix and corresponds to an ellipsoid in momentum space.[11] Its eigenvectors are perpendicular to each other and oriented along the principal axes of the ellipsoid. For every event the tensor T was diagonalized to find its eigenvectors and their directions in space. The "orientation of the event" is defined as $\cos(\Theta_{33})$, where Θ_{33} is the angle between the shortest eigenvector and the beam axis.

In Fig. 4 the parallel velocity of the largest detected fragment is plotted as a function of $\cos(\Theta_{33})$. Only "completely measured" high-multiplicity events were selected to construct this Figure, to concentrate on the most dissipative events in each case. High multiplicity cuts were introduced at M= 10, 20, 25, 35 and 35 for Al, Cu, Y, Ho and Au, respectively; cf. Fig.1. If the momentum ellipsoid is oriented parallel to the beam, then Θ_{33} = 90 degrees, $\cos(\Theta_{33})$=0, which corresponds to the left side of Figure 4. The case of perpendicular orientation is to the right.

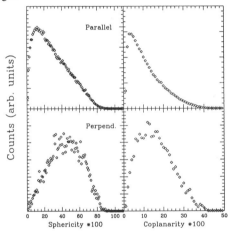

Fig. 5. Projected sphericity-coplanarity spectra for the parallel (upper row) and perpendicular (lower row) orientation of the momentum tensor, for the 60 MeV/u ^{129}Xe+^{197}Au reaction.

The parallel velocity of the largest detected fragment is used in Fig. 4 as an indicator of the binary character of the collision. As shown in the Figure, the largest fragment is likely to carry memory about the primary PLF for parallel event orientation (left side of the contour plots). For Au and Ho targets this is shown by the double-humped velocity distributions, when the $\cos(\Theta_{33})$ is close to zero. This conclusion is not sensitive to the efficiency of the forward array, as shown by the last box of the Figure, where the largest fragment was dropped from the analysis (also from the momentum tensor calculation). The *second* largest fragment in the event shows the same qualitative behaviour as the largest one. It shows, that for the most dissipative collisions the primary PLF breaks into a few fragments of substantial size, each one moving with a parallel velocity close to that of the projectile. The primarily binary character of the collision can be revealed by detecting any of these fragments.

As shown in Fig. 4, for highly dissipative high-M collisions one can distinguish binary from non-binary events by additionally selecting the orientation of the event in momentum space. Those events whose momentum ellipsoid is oriented parallel to the beam show binary PLF-TLF characteristics, namely the velocity spectra of the largest fragments show PLF and TLF components. For those events whose momentum ellipsoid is perpendicular to the beam, the largest detected fragment moves with a velocity similar to the CN velocity and the velocity distribution is no longer bimodal. It is important to note that the momentum ellipsoid is still non-spherical even for perpendicular event orientation. This is shown in more detail in the case of Au target in Figure 5 where the projected sphericity and coplanarity distributions are compared directly for parallel and perpendicular orientations of the momentum tensor. A transition from a very elongated shape (rod-like, longest:shortest axis ratio of 5:1) to a more spherical shape with the ratio 3.6:1 is observed in this case. The change of shape can presumably be interpreted as a transition to more equilibrated compound-like system, produced in more central collision events.

Summary and conclusions

We have established that in 60 MeV/nucleon ^{129}Xe + ^{27}Al, ^{nat}Cu, ^{89}Y, ^{165}Ho and ^{197}Au reactions the parallel velocity of the reconstructed PLF source is useful to select events with a varying degree of dissipation. More dissipative events are selected on the average in coincidence with slow sources, than the ones observed in coincidence with fast moving sources.

For the high-multiplicity events, large fragments still show target-like and projectile-like components in their parallel velocity distributions. A further sub-division of highly dissipative events into binary-like and non binary-like can be performed event-by-event by applying an additional experimental selection on the orientation of the event in momentum space. This selection is in addition to and independent of selections already performed according to multiplicity and total detected energy and/or momentum. It serves to distinguish binary from the "true-central" non-binary events in the class of high-multiplicity events (i.e., the most dissipative events) detected in the experiment.

Footnotes and References

#This work was supported by the Dir., Office of Energy Research, Div. of Nuclear Physics, Office of High Energy and Nuclear Physics of U.S. DOE.
*On leave of absence from Heavy Ion Laboratory, Warsaw University, Poland
+NSCL, Michigan State University, East Lansing, MI 48824
$Present address: Lawrence Berkeley Laboratory, Berkeley, CA 94720
@Present address: Hope College, Chemistry Dept., Holland, MI 49423
1) L.G.Moretto and G.J.Wozniak, Annual Review of Nucl. and Part. Physics **43** (1993) 379.
2) D. R. Bowman, et al., *Phys. Rev. Lett.* **67** (1991) 1527.
3) D. R. Bowman, et al., *Phys. Rev.* **C46** (1992) 1834.
4) B. Lott, et al., *Phys. Rev. Lett.* **68** (1992) 3141.
5) W. L. Kehoe, et al., *Nucl. Instr. Meth.* **A311** (1992) 258.
6) M. A. McMahan, et al., *Nucl. Instr. Meth.* **A253** (1986) 1.
7) J. B. Moulton, et al., *Nucl. Instr. Meth.* **157** (1978) 325.
8) R. T. d. Souza, et al., *Nucl. Instr. and Meth.* **A295** (1990) 109.
9) Y. D. Kim, et al., *Phys. Rev.* **C45** (1992) 338.
10) R. J. Charity, et al., *Nucl. Phys.* **A483** (1988) 371.
11) G. Fai and J.Randrup, *Nucl. Phys.* **A404** (1983) 551.

CRITICAL EXPONENTS ASSOCIATED WITH MULTIFRAGMENTATION

NORBERT T. PORILE
Department of Chemistry, Purdue University
W. Lafayette, IN 47907, USA

For the EOS Collaboration

ABSTRACT

A method for determining critical exponents from small systems is discussed. We apply this method to projectile fragmentation of gold nuclei in a recently completed experiment using the EOS-TPC.

1. Introduction

The Purdue group has had a long-standing interest in multifragmentation. In a previous study of 50-300 GeV proton-nucleus collisions using an internal gas jet target of heavy noble gases [1] we discovered that the fragment mass-yield distribution obeyed a power law with an exponent characteristic of a phase transition at the critical point [2, 3]. The energy dependence of the mass-yield distribution was examined in a similar experiment with 1-20 GeV protons [4] and the characteristic behavior seen at the higher energies was observed at ~10 GeV and above. Inclusive experiments such as our previous studies [2, 3] can only provide limited information. We report here preliminary results of an exclusive study of multifragmentation, in which all the charged fragments are detected on an event-by-event basis.

The experiment was performed at the Bevalac by the EOS Collaboration. We used reverse kinematics to study projectile fragmentation of 1 GeV/nucleon gold nuclei incident on a carbon target. The target resides just in front of the EOS time projection chamber (TPC). Behind the TPC sits a multiple-sampling ionization chamber (MUSIC II), followed by a 100 element two layer time- of-flight (TOF) wall. The TPC is capable of detecting charges one through eight, MUSIC II, seven through Z_{beam}, and TOF, charges one through Z_{beam}. For the preliminary analysis presented here, each detector covered an exclusive range: TPC, $1 \leq Z \leq 6$; MUSIC, $11 \leq Z \leq Z_{beam}$; TOF, $7 \leq Z \leq 10$. The total charge reconstructed from about 15,000 events is shown in Figure 1. The tail to the high multiplicity side of the peak is due to the lack of track matching between the TOF wall and the other two detectors. In what follows, we have accepted for analysis only those events with reconstructed charge 79±3 shown as the two vertical lines in Figure 1.

38

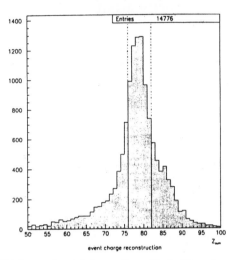

Fig. 1: Total reconstructed charge, Z_{sum}. Vertical lines denote 79 ± 3.

At small excitation energies, the excited projectile remnant decays via emission of a few nucleons or light clusters. At high excitation energies, total vaporization of the remnant is observed. Somewhere in between, a wide range of intermediate mass fragments are made via a simultaneous disassembly of the excited remnant. See Figure 2. (Charges 1-3 have been omitted from this figure so that the yield of the heavier fragments is apparent.) From the point of view of the liquid drop model of the nucleus we might describe the three regimes as follows. At low excitation energies, we observe evaporation. At the highest excitation energies, the nucleus is vaporized into a gas of nucleons and light clusters. At intermediate excitation energies, the nucleus is torn between the liquid and the gas phases. As already mentioned, the observation in inclusive studies of a power-law yield in the fragment mass distribution led us to suggest that perhaps nuclear fragmentation was a critical phenomenon occurring in a small system [4]. From this point of view, fragments result from the density fluctuations that are characteristic of second order phase transitions. Subsequent experiments have confirmed that fragmentation is a multi-body, simultaneous breakup that occurs when deposited energy is of the order of the total binding energy of the nucleus. Of course, the observation of a power law in the mass yield, with an exponent in the range of two to three, is suggestive but not conclusive of critical behavior. Critical behavior is characterized by a finite number of critical exponents, so if indeed nuclear multifragmentation is a critical phenomenon, then it too should possess critical exponents.

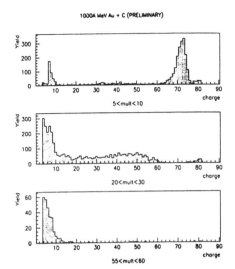

1000A MeV Au + C (PRELIMINARY)

Fig. 2: Fragment yields for different multiplicity regimes. Charges 1-3 have been omitted.

2. Signatures of Critical Behavior

Fluctuations are the hallmark of a phase transition occurring near a critical point. These can be fluctuations in density in a fluid system, magnetization in a magnetic system, or cluster sizes in a percolation system. It is the fluctuations that we must examine to see whether a system is exhibiting critical behavior. These fluctuations are largest in the critical region and persist even with infinite statistics. Far from the critical region, fluctuations exist, but these are fluctuations that are indicative of the 'mean field'. A second hallmark of second order phase transitions is a finite number of critical exponents. These govern the power law behavior of various quantities as the critical point is approached. For example, the rate at which the isothermal compressibility diverges as T_c is approached from either side is given by $K_T \propto |T\text{-}T_c|^{-\gamma}$. The density of clusters of size s at T_c, is given by $\rho(s) \propto s^{-\tau}$. Only two of these critical exponents are independent and are related to the others via a sum rule.

As a prototypical system, we will use bond percolation on a cubic lattice of linear dimension L. We will use percolation as a testing ground for our techniques for analyzing exclusive multifragmentation data. Campi has shown the usefulness of this approach in an important series of papers [5]. Quantities that diverge in macroscopic systems show peaking behavior as the constituent number of the system decreases. The critical point becomes a critical region and the location of the maximum is shifted somewhat from the location of the divergent behavior in the infinite system. In a fluid, we might examine the isothermal compressibility. The analogous percolation quantity is the second moment of the cluster distribution [6],

$$M_2 = \sum_s s^2 n_s(p) \tag{1}$$

where n_s is the number of clusters containing s sites and p is the probability that a bond is formed between neighboring sites. In the infinite system, percolation does indeed possess a critical 'point', in that one never makes the infinite or spanning cluster below a particular value of $p = p_c$ (for bond forming percolation). In Figure 3 we show the behavior of $\ln(M_2)$ versus reduced multiplicity, i.e. the number of clusters per lattice site formed, for three different sizes of percolation lattices. It is clear that a peak remains even as the system size grows quite small.

Making an analogy to a liquid-gas system, we will label the region above p_c as the 'liquid' region, and below p_c as the 'gas' region. This gives us a more physical picture. It is understood that in the summation of equation (1) the largest cluster, S_{max}, is to be omitted whenever we are on the liquid side of p_c, since it represents what would be the infinite cluster if $L \to \infty$. In Figure 3 we have removed the largest cluster everywhere for display purposes. As $L \to \infty$ the 'singular' part of the moments exhibit power law behavior:

$$\left[\sum_s s^k n_s \right]_{singular} \sim |\epsilon|^{\frac{-(1+k-\tau)}{\sigma}} \tag{2}$$

where $\epsilon = p - p_c$, and τ and σ are critical exponents. Other critical exponents are defined by

$$M_2 \sim |\epsilon|^{-\gamma} \tag{3}$$

$$S_{max} \sim |\epsilon|^{\beta} \tag{4}$$

$$\text{and } n_s \propto s^{-\tau} \text{ at } \epsilon = 0. \tag{5}$$

The behavior expressed in equations (3)-(5) holds only in the limit $L \to \infty$ and $s \to \infty$. The relations between exponents is:

$$\gamma + 2\beta = 2 - \alpha = \frac{\tau - 1}{\sigma} \tag{6}$$

$$\tau = 2 + \frac{\beta}{\beta + \gamma} \tag{7}$$

Of course, multifragmentation is not percolation on a lattice. If we wish to perform a cluster analysis in an effort to determine the exponents γ, β, and τ, we must find a suitable substitute for the percolation probability, p. Ideally we would like to use temperature, but since this is difficult to measure directly in our case, we will follow Campius suggestion [5] and use charged particle multiplicity, m, or the reduced multiplicity, $n = m/79$, where 79 represents the charge of the gold projectile nucleus.

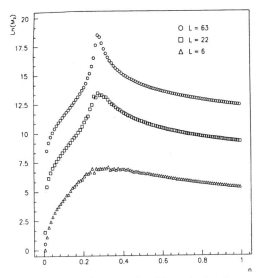

Fig. 3: ln(M_2) versus n for different lattice sizes.

Fluctuations indicative of power law behavior are indeed present *in a particular multiplicity* interval. These are observed in the deviations of the largest fragment produced when plotted as a function of multiplicity as shown in Figure 4. For each multiplicity bin we find the average size of the largest fragment. We then compute the standard deviation about this mean and divide this deviation by the mean for normalization. The number of entries in each multiplicity bin of Figure 4 is approximately constant, so the peaking is *not* simply a matter of counting statistics. These fluctuations represent the density fluctuations that occur on all length scales at the critical point in a macroscopic system. It is significant that these deviations are similar to those in a percolation lattice of comparable size.

3. Determination of Critical Exponents

Having now some evidence for critical fluctuations, one would like to pursue the question of critical behavior further. We will do this by:

- determining in an independent manner three critical exponents, γ, β, and τ.
- determining whether these exponents obey the sum rule, equation (7).

We have used small percolation lattices in order to develop a method for extracting critical exponents from small systems. The key to determining critical exponents using a finite system is to find a region in probability where the power law behavior of equation (2) holds. This means that one cannot expect power law behavior too

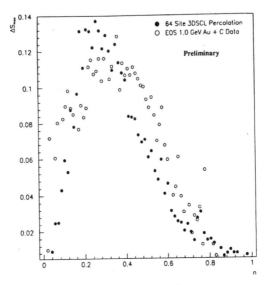

Fig. 4: Fluctuations in the average size of the largest fragment versus multiplicity.

close to the peak in M_2, for example, because finite size effects dominate there. On the other hand, one cannot go too far from the peak, since here, one is in the so-called mean field regime, where fluctuations are small relative to the mean.

We first discuss the determination of the exponent γ. It is well known in percolation theory that the critical percolation probability, p_c, in a small lattice is different from that for an infinite lattice [6]. One can account for finite size effects in part by choosing a p_c that is lattice size dependent. Thus, the methods consists of the following steps:

- choose a trial value of p_c
- on the 'gas' side of p_c, plot $\ln(M_2)$ versus $|p-p_c|$. Do the same for the 'liquid' side. This means choosing an appropriate fitting region for each side. This is determined by requiring the value of γ from the two regions to be the same. If one cannot find such a region, then adjust the value of p_c.

An example of this approach using a percolation lattice of $L = 63^3$ sites is shown in Figure 5. The extracted value is in excellent agreement with the canonical value of 1.80. We have also performed this same exercise on an $L = 6^3$ lattice with a similar result.

Having found both p_c and γ, we can now use the same p_c and fitting regions to determine β from equation (4).

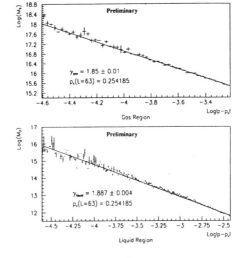

Fig. 5: Determination of γ for percolation.

As Campi pointed out [5], by plotting $\ln(M_3)$ versus $\ln(M_2)$, one can determine τ. This follows because in an infinite system, the slope of this correlation is:

$$\frac{\Delta[\ln(M_3)]}{\Delta[\ln(M_2)]} = \frac{\tau - 4}{\tau - 3}. \tag{8}$$

In a small system, one must be careful to avoid the 'liquid' side of this correlation, since here we must remove the largest piece. The effect of removing this largest cluster is a large perturbation on a small lattice. On the 'gas' side, however, the largest piece is small, and it makes little difference whether it is removed or not. Thus, we use only the gas side to determine τ when using this correlation. An alternative method is to fit the cluster distribution at p_c as a function of cluster size, since according to equation (5) it should be a power law with exponent τ. Finite size effects once again enter in if one tries to extend the fit to clusters of too large a size. In fact, even in large lattices, one typically restricts the region of the power law fit to be about 1% of the maximum possible cluster size. If one adheres to this rule of thumb, one can obtain good agreement between the two methods.

We now apply this procedure to multifragmentation of Au nuclei at 1 GeV/nucleon. We will use multiplicity, since we don't know, on an event-by-event basis, the temperature. We first make an estimate of the critical multiplicity, m_c. The critical fluctuations shown in Figure 4 give us a first guess. After some trial and error, we determine that $m_c = 27$ gives good agreement on both the liquid and gas sides of M_2. The fits to the data are shown in Figure 6. We then use this value of m_c in our

44

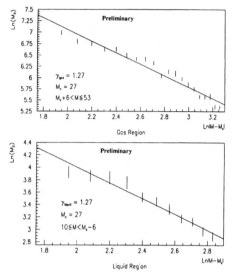

Fig. 6: Determination of γ for Au multifragmentation.

determination of β, Figure 7. Finally, using the correlation between M_3 and M_2 we can determine τ. This is shown in Figure 8.

Table 1: Critical Exponents for Various Systems

System	γ	β	τ	$2 + \beta/(\gamma + \beta)$
3D Percolation	1.80	0.45	2.20	2.20
Fluid	1.24	0.33	2.21	2.21
Mean Field Theory	1.0	0.5	2.33	2.33
Au+C fragmentation	1.27	0.31	2.18	2.20

We summarize our results in Table I. We list values for three-dimensional percolation lattices, real fluids, mean field theory and our fragmentation data. The uncertainty in the exponent values for the data are about 10-15%. These are due to the different values one can obtain by varying the choice of m_c and fitting region.

The significant point is that we were unable to find a set of values that was consistent with either the percolation or the mean field values. The similarity between nuclear matter values and those for fluid systems is very striking. Most remarkable is the fact that the sum rule is obeyed for the nuclear matter exponents. If this preliminary analysis is correct, then we have directly measured the approach of the equation of state of nuclear matter to the critical point using small samples (finite nuclei). The implication is that the fragmenting system has had sufficient time to approximate thermal equilibrium. In a small system, the approach occurs over a considerable range of excitation energy, temperature and density. It remains

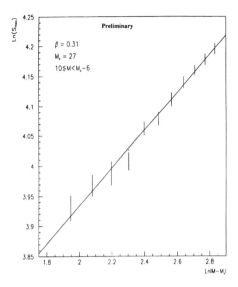

Fig. 7: Determination of β for Au multifragmentation.

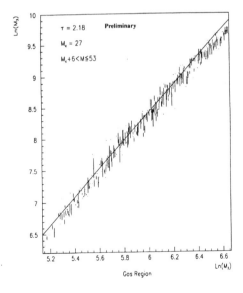

Fig. 8: Determination of τ for Au multifragmentation.

to determine these quantities from the data. This analysis is underway.
This work was supported by the U.S. Department of Energy.

4. References

[1] B. C. Stringfellow et al., Nucl Instrum. Methods A **251**, 242 (1986).

[2] J. E. Finn et al., Phys. Rev. Lett. **49**, 1321 (1982).

[3] R. W. Minich et al., Phys. Lett. B **118**, 458 (1982).

[4] N. T. Porile et al., Phys. Rev. C **39**, 1914 (1989).

[5] X. Campi, J. Phys. A **19**, L917 (1986); X. Campi, Phys. Lett. B **208**, 351 (1988).

[6] D. Stauffer and A. Aharony, Introduction to Percolation Theory, second edition (Taylor and Francis London, 1992).

The ATLAS Positron Experiment – APEX

I. Ahmad[a], S.M. Austin[b], B.B. Back[a], D. Bazin[c], R.R. Betts[a], F.P. Calaprice[d], K.C. Chan[e], A. Chisti[e], P. Chowhury[e], R. Dunford[a], J.D. Fox[f], S. Freedman[g], M. Freer[h], S. Gazes[i], J.S. Greenberg[e], A.L.Hallin[k], T. Happ[l], N. Kaloskamis[e], E. Kashy[b], W. Kutschera[a], C.J. Lister[e], M. Liu[k], M.R. Maier[g], A. Perera[m], M.D. Rhein[a], E. Roa[f], J.P. Schiffer[a], T. Trainor[n], P. Wilt[a], J.S. Winfield[b], M. Wolanski[a,i], F.L.H. Wolfs[m], A. Wuosmaa[a], A. Young[d] and J.E. Yurkon[b]

[a]Physics Division, Argonne National Laboratory, Argonne, IL 60439; [b]NSCL, Michigan State University, East Lansing, MI 48824; [c]GANIL, Caen 14021, France; [d]Physics Department, Princeton University, Princeton, NJ 08544; [e]Wright Nuclear Structure Laboratory, Yale University, New Haven, CT 06511; [f]Physics Department, Florida State University, Tallahassee, FL 32306; [g]Lawrence Berkeley Laboratory, Berkeley, CA 94720; [h]Department of Physics, University of Birmingham, Birmingham, B15 2TT, United Kingdom; [i]Department of Physics, University of Chicago, Chicago, IL 60637; [k]Physics Department, Queen's University, Kingston, Ontario, K7L 3N6, Canada; [l]GSI, Plankstrasse 1, 64291 Darmstadt, Germany; [m]NSRL, University of Rochester, Rochester, NY 14627; [n]Nuclear Physics Laboratory, University of Washington, Seattle, WA 98195

Presented by M.D. Rhein

ABSTRACT

APEX - the ATLAS Positron Experiment - is designed to measure electrons and positrons emitted in heavy-ion collisions. Its scientific goal is to gain insight into the puzzling positron-line phenomena observed at the GSI Darmstadt. It is in operation at the ATLAS accelerator at Argonne National Laboratory. The assembly of the apparatus is finished and beginning 1993 the first positrons produced in heavy-ion collisions were observed. The first full scale experiment was carried out in December 1993, and the data are currently being analysed. In this paper, the principles of operation are explained and a status report on the experiment is given.

1. Introduction

Extremely strong electromagnetic fields are produced in close collisions of high-Z atoms at energies close to the Coulomb barrier. This situation has been predicted to give rise to qualitatively new phenomena associated with the over-critical binding of the inner electron orbits, such as the spontaneous emission of positrons [1].
Experiments, originally motivated by these ideas and carried out at the GSI Darmstadt over the past decade, have produced some remarkable and unusual results, i.e. the observation of peak structures both in the positron spectra [2-5] and in the sum-energy spectra of coincident electrons and positrons [6-10]. If confirmed, these results would seem to signal the appearance of, certainly interesting and possible fundamental, new physics.
A number of different scenarios have been proposed to account for the observation of sharp lines, none of which is wholly consistent with the data or constraints imposed by other known physics. The most pressing questions then relate to a clarification of the experimental situation and, in particular, a precise determination of the kinematics of the coincident electron-positron pairs which give rise to the sharp sum-energy lines.

2. Experimental apparatus

The vast experience in positron spectroscopy gained by the GSI groups over the past decade and the numerous questions which arise with the observation of the peak structures helps to define the specific tasks a second generation experiment should fullfill [11,12]. Briefly, these are: (i) clear identification and measurement of positrons emitted in heavy-ion collisions, (ii) high detection efficiency for electrons and positrons in the energy range from $150 \text{ keV} \leq E \leq 600 \text{ keV}$, (iii) effective suppression of electrons with energies below 150 keV, (iv) ability to determine the emission angle of the leptons to perform a Doppler correction of the measured energies and to determine the opening angle between the lepton pairs, (vi) detection of the scattered heavy ions to determine the collision kinematics and the reaction Q-value. (v) high granularity of the detection devices to reduce the chance for multiple hits in one counter. The latter point is of major importance because high beam intensities have to be used to obtain statistically significant data sets within reasonable beam times.

Figure 1 shows a schematic view of the APEX apparatus and its major detector systems. It consists of a solenoid mounted transverse to the beam direction. Electrons and positrons produced at the target position spiral along the field lines of a homogeneous magnetic field (B = 0.03 T) and are transported to two silicon detector arrays mounted on the solenoid axis. These arrays each consist of 216 3×0.5 cm², 1 mm thick PIN diodes [13] arranged on a surface of a hexagonal cylinder 36 cm long with 1.5 cm sides. The detectors provide information on the energy, position and time of flight of each lepton, thus allowing a reconstruction of their vector momenta.

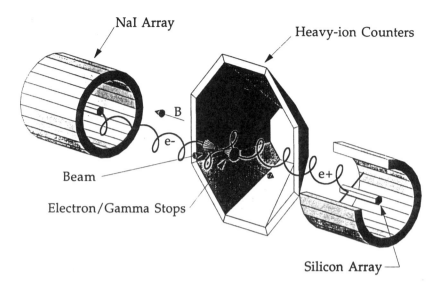

NaI Array

Heavy-ion Counters

B

e-

Beam

Electron/Gamma Stops

e+

Silicon Array

Fig. 1: Schematic drawing of the APEX setup and its major detectors.

The angle with respect to the solenoid is given by

$$cos\,\Theta \; = \; p_z \,/\, p\,, \quad p_z \; = \; m\,z\,/\,t,$$

where p_z, the component of momentum along the solenoid axis, is determined from the time of flight and the detection position. The total momentum is calculated from the measured energy. An important feature of this device is that the spiral orbits always have close to a complete number of turns and the time of flight is then an integer multiple of the cyclotron period

$$t \; = \; n\,T_{cyc} \; = \; n\,\tfrac{2\pi m}{eB}.$$

Providing the time resolution is sufficient to determine the number of turns a lepton has undergone, the time of flight can be determined to the precision of the energy measurement.

The energy calibration of the silicon detectors is obtained using radioactive electron conversion sources providing monoenergetic electrons with energies of 193.7 keV and 263.9 keV (^{203}Hg), 363.3 keV (^{113}Sn) and 481.7 keV (^{207}Bi). Figure 2 shows the energy spectrum obtained with the Sn and the Hg sources, exhibiting an energy resolution of 13.2 keV and 9.8 keV (FWHM) for 193 keV and 363 keV electrons, respectively.

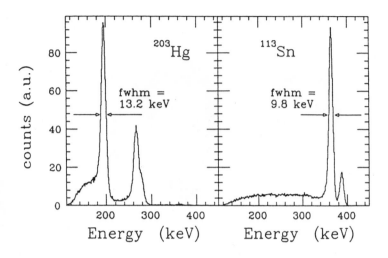

Fig. 2: Energy spectra obtained with the conversion electron sources ^{203}Hg (left part) and ^{113}Sn (right part).

An unique feature of APEX is the fact that both electrons and positrons are transported to the same detectors which results in a high degree of symmetry in their detection efficiencies. Positrons are distinguished from electrons by their annihilation radiation detected in two large position-sensitive NaI barrels surrounding the Si arrays [14]. Each barrel consists of 24 NaI bars with a thickness of 6 cm and a length of 55 cm. These crystals are specially made with a short attenuation length such that the pulse height information from both ends of the crystal can be used to determine the position of the annihilation photon. When both photons are detected the position on the silicon array where the annihilation took place thus can be reconstructed. The on-axis position resolution is $\Delta Z = 3$ cm (FWHM) and an energy resolution of 13% (FWHM) at 511 keV is achieved.

The crystals are equipped with *Hamamatsu H2611* phototube assemblies which work in the magnetic field without further shielding. Both barrels are enclosed in a cylindrical, Pb shielded cradle which protects the scintillators from ambient room background and γ-rays from the beam-dump. Near the target position, conical electron/gamma-stops (Fig.1) prevent direct radiation from the target hitting the inside of the array. For each barrel the outputs of the 96 photomultipliers are connected to a trigger processor which generates a trigger signal when a preloaded trigger pattern occurs. In normal operation two opposite NaI bars (or the neighbouring ones) are required to be hit to yield a trigger signal. Further conditions have to be satisfied

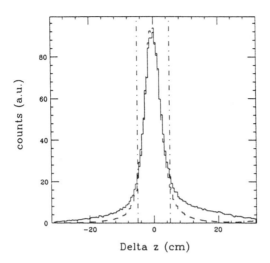

Delta z (cm)

Fig. 3: Difference of the reconstructed on-axis position and the position where the positron was detected (solid line: beam data, dashed line: positron source measurement). The dashed-dotted lines give the window limits for positron identification.

in the analysis to identify a hit on the silicon detector as a positron: (i) the two γ-rays detected in opposite NaI bars must be in prompt time coincidence ($\Delta t = 10$ ns), (ii) there should be prompt time coincidence between the γ-ray, the event in the silicon detector and the beam pulse ($\Delta t = 25$ ns), (iii) the energy deposited in at least one of the two NaI bars must be within the 511 keV full-energy peak region ($\Delta E = 200$ keV), (iv) the reconstructed on axis position of the annihilation γ-rays (z_{NaI}) must correspond to the z-position of the silicon detector which fired ($\Delta z = z_{NaI} - z_{Si} \leq 5$ cm), (v) no second hit on the silicon array within the Δz window.

The spectra displayed in Fig. 3 show the difference in the on-axis position as obtained from the reconstructed position where the annihilation took place and the z position of the silicon detector which detected a lepton in prompt time coincidence. The spectrum obtained from beam induced events is compared with the spectrum obtained from positrons emitted from a ^{68}Ge β^{+}-source. The similar shapes of the two spectra demonstrates that in beam positron events are well defined through the Δz condition (the gate is shown in Fig. 4, too).

During beam measurements the ratio of positrons to electrons produced is only of the order of about 10^{-4}. A source test was used to verify the ability to detect coincident electron-positron pairs above a huge background of electrons. For this test a ^{90}Y source was mounted in the target position. A small percentage of decays (0.011%)

52

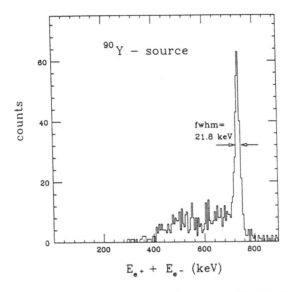

Fig. 4: Sum energy spectra of coincident positron-electron pairs originating from the E0 transition in ^{90}Zr.

populate the first excited 0^+ state in ^{90}Zr. This state can decay only by electron or pair conversion processes. When pair conversion takes place an electron-positron pair with a combined energy of $E_{e-} + E_{e+} = 739$ keV is emitted. The ratio of positrons to electrons emitted is $3.5 \cdot 10^{-5}$ per ^{90}Y decay. The background of electrons originating from decays to the ground state extends up to 2.28 MeV. Figure 4 shows the sum-energy spectrum of electron-positron pairs detected in coincidence. The sum energy can be reconstructed within 21.8 keV (FWHM). The tail originates from backscattered electrons.

An large solid-angle array of 24 Low Pressure Multi-Wire Proportional Counters serves to detect the scattered ions [15]. Eight three-element trapezoidal counters provide 360° coverage in the azimuth angle and 20° - 68° coverage in the polar angle. They are operated at a pressure of 5 Torr Isobutane and a typical anode voltage of 500 V. A plane of anode wires with 1 mm wire spacing yields the time-of-flight information. The scattering angle information is provided by a transmission-line delay cathode. The time-of-flight resolution is 1.1 ns (FWHM) which includes a 0.5 ns time resolution from the beam. The angular resolution is about 0.25° (FWHM) with systematic errors up to one degree. For asymmetric collision systems the projectile and the target like particles can be identified by their time of flight and scattering angle information. Figure 5 shows a two dimensional plot of these quantities for the system U + Ta at 6.1 MeV/nucleon. The U and the Ta branches are clearly separated. Thus a Doppler

53

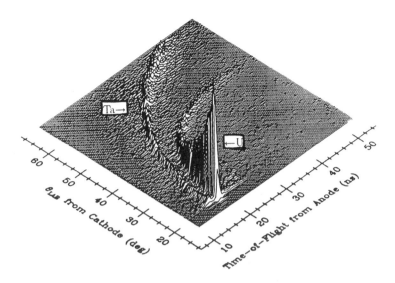

Fig. 5: Two dimensional representation of the time of flight versus the scattering angles as obtained from the heavy-ion arrays in U + Ta collisions at 6.1 MeV/nucleon.

correction for the leptons detected in coincidence to heavy ions can be performed.

3. Summary

The APEX spectrometer has recently been completed and, over the past year, has progressed from the testing phase to full operation. Initial experience has shown that, in essentially all respects, the apparatus performs as conceived and designed. Beams of ^{238}U of intensity up to 5 pnA from ATLAS have been used to bombard 1 mg/cm^2 ^{181}Ta targets and beam induced positrons have been measured with high resolution and extremly low background. The measured acceptance of the device is close to specifications.

In the first physics production run, which took place in December 1993, a total of over 600,000 positrons were measured with over 250,000 positron-electron coincidences. These data are currently being analyzed. The analysis of these data and of future experiments over the next several years should certainly lead to new insight into the origin of the line phenomenon.

54

4. References

[1] J. Rafelski, B. Mueller, W. Greiner, *Nucl. Phys.* **B68**, 585 (1974).

[2] T. Cowan *et al.*, *Phys. Rev. Lett.* **54**, 1761 (1985).

[3] W. Koenig *et al.*, *Z. Phys.* **A328**, 129 (1987).

[4] H. Tsertos *et al.*, *Z. Phys.* **A328**, 499 (1987).

[5] H. Tsertos *et al.*, *Z. Phys.* **A432**, 79 (1992).

[6] T. Cowan *et al.*, *Phys. Rev. Lett.* **56**, 444 (1986).

[7] E. Berdermann *et al.*, *Nucl. Phys.* **A488**, 683c (1988).

[8] W. Koenig *et al.*, *Phys. Lett.* **B218**, 12 (1989).

[9] P. Salabura *et al.*, *Phys. Lett.* **A245**, 153 (1990).

[10] I. Koenig *et al.*, *Z. Phys.* **A346**, 153 (1993).

[11] Proposal for an ATLAS Positron Experiment,
Argonne National Laboratory (1989), unpublished.

[12] R. Betts, *Nucl. Instr. Meth.* **B43**, 294 (1989).

[13] L. Evensen *et al.*, *Nucl. Instr. Meth.* **A326**, 136 (1993).

[14] N. Kaloskamis *et al.*, *Nucl. Instr. Meth.* **A330**, 447 (1993).

[15] D. Mercer *et al.*, *to be published.*

SHARP (e^+e^-) PAIRS from $(\beta^+ +$ ATOM)
and from HEAVY ION COLLISIONS

JAMES J. GRIFFIN

Department of Physics

University of Maryland, College Park, Maryland 20742-4111, USA

ABSTRACT

The Composite Particle $(Q_0?)$ Scenario provides a comprehensive phenomenology for the "(e^+e^-) Puzzle". It contradicts no data nor any established theory. It also implies a new "Supercomposite" molecule (the neutral $C(Q_0?)$particle bound to a nuclear ion), and predicts its pair decay by the special process of Sharp Annihilative Positron Emission (SAPosE): the electron is captured into a Bohr orbit and a single energetic positron emerges. The inverse process is Recoilless Resonant Positron Absorption (RRePosA). We show here that RRePosA can encompass the sharp electon lines observed in β^+ irradiation of heavy atoms as just another aspect of the same scenario. Thus the "(e^+e^-) Puzzle" expands into the broader "Sharp Lepton Problem", now accessible experimentally via both β^+ irradiation and high Z heavy ion collisions. The continued success of the scenario invites us to entertain the question of how Quadronium can be understood within (or beyond?) contemporary quantum electrodynamics.

1. INTRODUCTION: THE QUADRONIUM SCENARIO [1]

The Quadronium Scenario envisages[1] during a close high-Z heavy ion collision a spontaneous Landau-Zener process for creating from the vacuum a Quadronium (Q_o) particle tightly bound to the dinucleus in a strongly polarized state. As the dinucleus reseparates, the bound Q_0 particle depolarizes and its binding energy to the separating

---

[1]For the present data, this phenomenology does not specifically require the $(e^+e^+e^-e^-)$ composition, but only the decay to (e^+e^-). Therefore,we call the physical particle "$C(Q_0?)$" and the hypothetical Quadronium particle, "Q_o".

two-center Coulomb field decreases. If Q_0 does not dissolve adiabatically back into the vacuum, then it either becomes unbound and moves away from the nuclei, later to decay in isolation by Free Annihilative Pair Emission (FAPE), or it becomes bound to one of the departing nuclei, finally to decay, under the influence of its Coulomb field, by Bound Annihilative Pair Emission (BAPE). A non-zero average difference energy (as well as an enhanced difference energy width) therefore distinguishes the bound decays from the free[2].

This scenario naturally produces the following results:

- a good description of the EPOS $U + Th$ data[3, 4] as a FAPE decay[2, 5, 6]; and of the $U + Ta$ data, as a BAPE decay; the latter

- implies an extended supercomposite Q_0-Ion molecular binding state (with $R_0 \approx 9.4\,\hbar/mc \sim 4\times10^3\,Fm$), and a binding energy of only $\sim 1\,keV$ against separation of Q_0 from the nucleus[2].

- This small binding energy, in turn a) allows the decay energies of the bound and free decays to have nearly the same values, line by line, as observed, b) allows most of the bound Q_o decays to occur at such great distances from the nucleus that the pair momenta are nearly "back-to-back", but nevertheless have the unequal magnitudes required by the EPOS' $U + Ta$ energy difference data[3, 4, 5, 6], and c) nevertheless, still allows some decays close to the nucleus, whence the lepton opening angle is smaller and substantial momentum is transferred to the nuclear ion [2]

- The model also implies an attractive (Q_o, Ion) polarization force which
  1. is proportional to the square of the electric field at Q_o and, therefore, a) is some $100X$ weaker for (equi-)distant colinear $\{U(Z = 92), Th(Z = 90)\}$ ions than for $\{U(Z = 92), Ta(Z = 73)\}$, favoring capture in $U + Ta$, where the data do in fact exhibit the Coulombic signatures of the bound decay , and b) varies rapidly with Z, so that the weakly bound $[C(Q_0?), U]$ state i) should become unbound for Z-values only slightly smaller than 92, and

---

[2]In particular, under complete analysis, this BAPE process may accommodate the observations by the Orange group of nearly back-to-back opening angles in some pairs from some collisions under some circumstances[7], and larger opening angles under others[8], without additional hypotheses[8].

ii) may even be unbound by the ambient electrons of the neutral atom, as suggested below.

- Furthermore, the scenario restricts the properties of a model[9] Q_o Particle: If Q_o is to have a rest energy of $3\,mc^2$, is to be created spontaneously at $Z \gtreqless 160$, and is to become nearly unbound at $Z = 92$, then a) its radius must be of the order of its own Compton wavelength, $R_Q \sim \hbar/3\,mc \approx 135 Fm$, and b) its stiffness must be of the order of its rest energy divided by the square of its radius, $k_Q \sim M_Q^0 c^2/R_Q^2 \sim 27\,m^3 c^4/\hbar^2 \approx 8.5 \times 10^{-5}\ MeV/Fm^2$.

- Finally, a Q_o structure suggests that decay to $(e^+ e^- \gamma)$ should be more probable than the decay to a pair, as is consistent with the null results from Bhabha searches for this particle[10] and the preliminary report of $(e^+ e^- \gamma)$ decay[11] by Widman, et al.

2. BOUND PAIR DECAY and CREATION of Q_o in $(\beta^+ + \mathrm{AtOM})$

2.1. Special Decays Decays of the Bound $C(Q_o?)$ Particle: SAPosE

For bound Q_o decays, four new (so far unobserved) decay modes, forbidden by energy-momentum conservation for free isolated decays, become allowed[12]. Of these, consider Sharp Annihilative Positron Emission (SAPosE): it deposits the decay electron into a vacant Bohr-Dirac orbit and ejects a single energetic (800-950 kev) positron[2]. Its observation would imply its inverse process (Recoilless Resonant Positron Absorbtion, RREPosA), in which a positron colliding with an electron of a heavy atom creates $C(Q_0?)$ bound to the ion in a supercomposite molecule. We propose [13] that this process provides the key to understanding the e^+ lines and the very sharp e^- lines observed[14, 15, 16] in various β^+ irradiations of heavy elements.

2.2. Inverse SAPosE: Resonant Recoilless Positron Absorption (RRePosA)

In the RRePosA process, a positron of a properly resonant energy impinges upon a high-Z target ion which has an electron in some Bohr-Dirac orbit, creating a bound $[C(Q_o)$-Ion] supercomposite state, s. The large nuclear mass makes the final velocity and the kinetic energy of the molecule negligibly small ($v_s/c \sim 10^{-6}$; $K_s \sim 1eV$):

it is created essentially at rest in the laboratory, "Recoilless". Subsequent $C(Q_0?)$ decay yields the same (BAPE) pair distributions which successsfully describe[2, 5, 6] the EPOS $(U + Ta)$ pair data after Lorentz transformation to the laboratory frame.

2.3. Sharp Electrons Emerge from β^+ Irradiation of High Z Atoms

So far, this Recoilless Resonant Positron Absorption (RRePosA) process almost (but not quite! See below) describes the β^+ irradiations of neutral U and Ta atoms, which several investigators report to yield excess electrons and positron[14, 15] with kinetic energies of $\sim 330\,keV$ and, more recently, experiments[3] which produce[16, 19] *two* narrow electron lines with energies, $330 \pm 0.3keV$ and $410 \pm 0.3\,keV$, and widths, $\Gamma_{E^-} < 3\,keV$. No such sharp leptons follow irradiations of lower Z(=73) Ta atoms. These data suggest the creation of some object, with eigenstates at 1.68 MeV and 1.84MeV, which decays into pairs of leptons with sharp lab kinetic energies of $330keV$ and 410 keV, respectively. Could it be the particle of the "$C((Q_0?))$ Scenario"?

2.4. Narrow Width Requires Decay at Rest in the Lab

We note that the reported[16] upper bound, $(\Gamma_{E^-} \leq 3\,keV)$, upon the electron linewidths places a stringent condition upon such an description, by virtue of[4] the relation, $\Gamma_{E^-} = 4p\,_-^!\,[< \gamma^2 v_Q^2 > /3]^{1/2}$, between the width and the source velocity. For a 330 keV electron, this width requires a Q_o laboratory speed, $v_Q/c \leq 2 \times 10^{-3}$, and a kinetic energy $K_Q \leq 3\,eV$. Thus the β^+ data demands an isolated $C(Q_0?)$ particle *at rest* in the laboratory, decaying into a back-to-back pair of equal energy leptons.

Unfortunately, the Recoilless Resonant Positron Absorption process produces not the required isolated $C(Q_0?)$ particle at rest in the lab but the supercomposite molecular bound state, whose pair decays yield, not the separately sharp Free Annihilative

---

[3]Although others [17, 18] report no success in observing such features, we adopt the positive results reported as the basis for the present discussion. We note also that Sakai argues[16] that because of the experimental arrangements, these failures to observe the leptons should not be taken as evidence *against* the processes observed in other experiments.

[4]See Sec. 5.1 and 5.2 of Ref.[2], "Quadronium, Unravelling the (e^+e^-) Puzzle", which imply this result for the isotropic free (FAPE) decay of a $C(Q_0?)$ particle moving with velocity v_Q in the laboratory.

lepton energies, but Bound Annihilative Pair Emissions characterized by Coulomb broadened distributions of the separate lepton energies[5]. One therefore can never expect from these decays the separately sharp E^- lines observed[16, 19] in the β^+ irradiations.

2.5. β^+ (RReposA) Process May Produce Unbound Supercomposite States in Neutral Atoms

Then in the nearly neutral atom, perhaps due to the influence of the additional ambient electrons not present in the 65^+ U projectile ions of the EPOS experiment, the supercomposite molecule may shift from slightly bound to slightly unbound. Then it would break up spontaneously, releasing the $C(Q_0?)$ particle to escape from the ionic Coulomb field and to decay subsequently in a Free Annihilative Pair (FAPE) decay.

Remarkably, this assumption also is too simple: a two-body breakup process delivers most of the breakup energy to the light $C(Q_0?)$ particle. Then the break-up energy would have[16] to be less than 3 eV *(electron volts, not keV!)*, and the supercomposite states for both U and Th must have energy eigenvalues within 3 eV of zero binding energy[13]. For any single state to honor such a coincidence would strain credulity; for four states, two in each of two distinct ions, to do so without independent compelling cause is beyond belief.

2.6. Viscous Breakup Can Complete the Picture

Then the focus shifts to the slowing down of $C(Q_0?)$: How could $C(Q_0?)$ be brought to rest in the lab before its decay? Since the $C(Q_0?)$ is charge neutral and stiff against polarization, its Coulombic interaction with the electron cloud could be very weak. But if the composite particle is in fact a Q_0, with constituents $(e^+e^+e^-e^-)$, then it can interact with the ionic electron cloud by virtue of the Pauli principle [20], which forces a time-dependent readjustment of the electron cloud as the Q_0 particle moves through it, exciting the cloud and slowing down the Q_0.

We therefore suggest as a topic for further investigation the following hypothesis

---

[5]As noted above, these are the same distributions which, after Lorentz transformation to the laboratory, successfully describe the EPOS $U + Ta$ data in Refs. [2, 5, 6].

of "Viscous Breakup" of a slightly unbound supercomposite, $[C(Q_0?), Ion]$, molecule. During "Viscous Breakup" the $C(Q_0?)$ particle would deposit essentially all of its kinetic energy into the ambient electronic cloud of the ion as it travels from the interior ($r'_i \lesssim 4 \times 10^2 \, Fm$) to a location outside the cloud ($r'_f \gtrsim 5 \times 10^4 \, Fm$), reducing an initial kinetic energy of perhaps some keV to a final kinetic energy in the eV range, as required by the e^- width observed in the β^+ experiments.

"Viscous Breakup" suggests a mechanism for producing a free isolated $C(Q_0?)$ particle at rest in the laboratory frame, and promises therefore to explain the whole set of heavy element β^+ observations. The β^+ data are recognized as just another consequence of the $C(Q_0?)$Scenario. Then these β^+ data and the older data of the traditional heavy ion "(e^+e^-)" Puzzle" combine into the expanded "Sharp Lepton Problem".

2.7. Lifetime of a Supercomposite State

Thus, we view Sakai's experiments as measuring the energy-integrated cross section for Recoilless Resonant Positron Absorption (RRePosA), followed sequentially by Viscous Breakup and Free Annihilative Pair decay of the isolated $C(Q_0?)$ particle. A distinct resonance occurs for each target electron, with energies spread over the 132 keV range of electron binding energies in the U atom. The continuous energy range of the beta decay positrons provides a remarkably felicitous match to the process' preference for so many precisely resonant energies. In Ref.[13] we discuss this analysis in more detail and extract from Sakai's measured cross sections (for the 330 keV electron line from $^{82}Rb\beta^+$ particles on Uranium) the following upper bound upon the lifetime of the supercomposite molecular state

$$\tau^s = \hbar/\Gamma^s_{TOT} \leq \pi^2 < \lambda^2 > /4(\overline{\sigma\Delta})_{EXP} < \tau_{MAX} \sim 10^{-17} sec, \qquad (2.7.1)$$

a limit presumably set by the molecular break-up lifetime.

2.8. No Contradiction with Data or Theory

This lifetime is much shorter than the lower bounds frequently asserted for the $C(Q_0?)$ decay [10, 11, 21, 22, 23] from Bhabha scattering experiments on low-Z targets. But[24, 25] those bounds follow not from the data, but from the unverified assumption that the particle can decay only to (e^+e^-). In fact, the present upper bound contradicts no Bhabha data. Similar remarks apply[13, 24] to the claim[26] that Q_o violates high precision quantum electrodynamics.

3. SUMMARY and OUTLOOK: HOW to EXPLAIN $C(Q_o)$?

The "Composite Particle $(Q_0?)$ Scenario" for the (e^+e^-) puzzle has been reviewed, with special emphasis upon the Sharp Annihilative Positron Emission (SAPosE) process in bound pair (BAPE) decay, and upon the process inverse to it, Recoilless Resonant Positron Absorption (RRePosA). We propose here that this (RRePosA) process is plausibly the cause of the very sharp electron lines observed to follow the β^+ irradiation of heavy neutral U and Th atoms. But the description requires the (still to be analyzed) "viscous breakup" of the unbound supercomposite molecule, which deposits the breakup kinetic energy into the ambient ionic electrons, bringing $C(Q_0?)$ to rest in the laboratory, and thereby allowing the very narrow widths observed for the e^- lines following the β^+ irradiations. Thus can the previously unexplained β^+ data become an integral part of a newly generalized "Sharp Lepton Problem".

In the end the "Composite Particle $(Q_0?)$ Scenario" provides once again a specific detailed phenomenology which offers a place for every major feature of the data, even as its scope expands to include this β^+ data. If this phenomenological success persists, then one day we must come to recognize the Quadronium Conjecture and its "$C(Q_0?)$ Scenario" as no longer merely an efficient data-oriented phenomenology, but as a fundamental challenge for theoretical physics, and especially for quantum electrodynamics.

Acknowledgements

The author is grateful to C.E.N. (Bordeaux-Gradignan), I.S.N. (Grenoble), G.S.I. (Darmstadt), ORNL (Oak Ridge) and LANL (Los Alamos) for their kind hospitality during the early stages of this work, and to the U.S. Department of Energy which supported this research under Grant No. DE-FG02-93ER-40762; also, to his his colleagues and friends who have offered frequent and invaluable criticisms and comments, and especially to Dr. Thomas E. Cowan, who first asked the author (at a Gordon Conference in 1989), "What happens if the electron gets caught in a Bohr orbit?"

1. References

[1] James. J. Griffin. *J.Phys.Soc.Jpn.*, 58:S427, 1989. and earlier refs. cited there.

[2] James. J. Griffin. *Intl.J.Mod.Phys.*, A6:1985, 1991.

[3] T. E. Cowan. *Monoenergetic Positrons and Correlated Electrons from Superheavy Nuclear Collisions.* PhD thesis, Yale U., 1988.

[4] P. Salabura et al. *Phys.Lett.*, B245:153, 1990. See also references therein, and Ph. D. thesis, U.of Krakow, Poland, 1990.

[5] James. J. Griffin. *Phys.Rev.Lett.*, 66:1426, 1991.

[6] James. J. Griffin. *Phys.Rev.Lett.*, 68:1960, 1992.

[7] W. Koenig et al. *Phys.Lett.*, B218:12, 1989.

[8] W. Koenig. In *Vacuum Structure in Intense Fields, ed. H. M. Fried*, page 29 (Plenum, New York, 1991).

[9] Xuemin Jin. U.of Md. report in preparation, to be published.

[10] H. Tsertos et al. *Phys.Lett.*, B266:259, 1991. and earlier refs. cited there.

[11] E. Widmann et al. *Z.Phys.*, A340:2091, 1991.

[12] James J. Griffin. In *Proc. 6th Int. Conf. on Nuclear Reaction Mechanisms Varenna, Italy, June 1991, ed.E.Gadioli*, page 758 (Physics, U. of Milano, Italia).

[13] James J. Griffin. In *Topics in Atomic and Nuclear Collisions,Predeal Romania, Sept. 1992, ed. A. Calboreanu,V.Zoran* (Plenum Press, 1994).

[14] K.A. Erb et al. *Phys.Lett.*, B181:52, 1986.

[15] C. Bargholz et al. *Phys.Rev.*, C40:1188, 1989.

[16] M. Sakai et al. *Phys.Rev.*, C47:1595, 1993. and earlier refs. cited there.

[17] R. Peckhaus et al. *Phys.Rev.*, C36:83, 1987.

[18] T. F. Wang et al. *Phys.Rev.*, C36:2136, 1987.

[19] M. Sakai et al. In *Proc.Conf.on Nuclear Physics in Our Times,Sanibel Island, Florida, Nov. 1992* (World Scientific, Singapore).

[20] Brij L. Gambhir et al. *Phys.Rev.*, C7:590, 1973.

[21] S. M. Judge et al. *Phys.Rev.Lett.*, 65:972, 1990.

[22] X.Y.Wu et al. *Phys.Rev.Lett.*, 69:1729, 1992.

[23] S.D.Henderson et al. *Phys.Rev.Lett.*, 69:1733, 1992.

[24] James. J. Griffin. *Phys.Rev.C*, 47:351, 1993.

[25] James. J. Griffin. *Phys.Rev.Lett.*, 70:4158, 1993.

[26] J. Reinhardt et al. *Phys.Rev.*, C33:194, 1986. and earlier refs. cited there.

ISOSPIN EQUILIBRATION IN INTERMEDIATE ENERGY ION COLLISIONS

S.J. YENNELLO[1,2], B.M. YOUNG[2], J. YEE[2], J.A. WINGER[1,2],
J.S. WINFIELD[2], G.D. WESTFALL[2], A. VANDER MOLEN[2],
B.M. SHERRILL[2], D.J. MORRISSEY[2], W.J. LLope[2], T. LI[2],
H. JOHNSTON[1], E. GUALTIERI[2], D. CRAIG[2] and W. BENENSON[2],
1-Cyclotron Institute Texas A&M University, College Station, TX 77843
2-National Superconducting Cyclotron Laboratory
Michigan State University, East Lansing, MI 48824, USA

ABSTRACT

The effect of the neutron to proton ratio of the colliding system on the isotope ratio of the emitted fragments ($1 \leq Z \leq 5$) was studied for for a variety of constant mass beam-target combinations. To extend the N/Z ratio as far as possible, two of the projectiles employed, ^{40}Cl and ^{40}Sc, were radioactive secondary beams. The isotopic ratios depend on the N/Z of the target and beam in a way which is not consistent with N/Z equilibration on the timescale of the emission of intermediate-mass fragments at $E/A = 53$MeV.

1. Introduction

The isotopic composition of nuclear reaction products contains important information on the reaction mechanisms that govern the collision. Recently there has been much interest in the isotopic composition of intermediate-mass fragments (IMFs) emitted from intermediate energy light-ion and heavy-ion collisions [1,2,3,4,5,6]. Previously, the neutron to proton (N/Z) degree of freedom has been used to study deep inelastic collisions[7,8]. Interesting insights into the time scale for equilibration were obtained. If the composite system is equilibrated prior to the emission of the reaction products, then the N/Z ratio of these fragments should depend only on the N/Z ratio of the equilibrated system and not on the N/Z ratios of the entrance channel. If the emission of the IMFs takes place prior to equilibration of the N/Z degree of freedom, then the isotopic composition of fragments emitted can be used to set the timescale for the emission process. In the present measurement we have used a beams of radioactive ions and rare isotope targets to extend the N/Z available within a constant mass target and projectile. This permitted the collection of data points with overlapping values of N/Z. The results can not be explained by any simple model which assumes equilibration of the composite system at $E/A = 53$ MeV.

64

2. Experimental Details

The equilibration of the N/Z degree of freedom was studied in exclusive events in which the Michigan State University (MSU) 4π Array[9] provided event characterization from the reaction of E/A = 53 and E/A = 25 MeV A = 40 projectiles with A = 58 targets. The beams of radioactive ions were made available by the A1200 Fragment Mass Analyzer[10]. Details have been previously published [1]

3. Results and Discussion

3.1. Central Isotopic Ratios

The behavior of the observed inclusive isotopic ratios depends significantly on the Z of the fragment. The inclusive isotopic ratios at 40° for the most intense isotopes of helium through boron are shown in figure 1. The data points represent the ratio of the number of counts of the more neutron-rich isotope to that of the more proton-rich isotope. The isotopic ratios are clearly dependent on the N/Z of the composite projectile plus target system where

$$(N/Z)_{cs} = \frac{N_{projectile} + N_{target}}{Z_{projectile} + Z_{target}}.$$

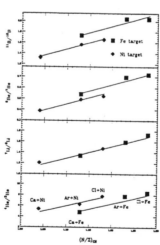

Fig. 1: Isotopic ratios plotted as a function of the N/Z of the combined target and projectile system. (a)^{4}He/^{3}He (b)^{7}Li/^{6}Li (c)^{9}Be/^{7}Be (d)^{11}B/^{10}B. Lines are linear fits to the data points on a given target.

The prominent feature in figure 1 is that no single line can fit all six data points of a given Z, except for the Li case. This difference implies that an equilibrated composite system cannot be responsible for these isotopic ratios, but that there must be some residual memory of the entrance channel. The solid lines in figure 1 are linear fits to the data for each target. Any effects of the entrance channel asymmetry present in the data should be present at $(N/Z)_{CS} = 1.13$ because the same composite system containing 46 protons and 52 neutrons was produced by two different target-projectile combinations. The comparison of the isotopic ratios for these two different constructions of the same composite system is a probe of entrance channel effects. Notice that for the He, Be, and B fragments the isotopic ratios have two different ratios for these identical composite systems, which indicates that these fragments are not being emitted from an equilibrated composite system of the target and projectile. The difference between the nickel target point and the iron target point is largest for B fragments and decreases with the Z of the fragment. At $Z = 3$ the two target points are nearly equal and for the lightest fragments the Fe target preferentially emits proton-rich fragments.

3.2. Equilibrated Subsystem ?

If one assumes that the emission comes from an equilibrated source the source must be some subset of the colliding system. Such a subset should be revealed by a

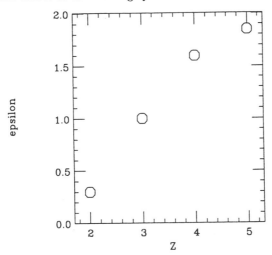

Fig. 2: Fit parameter epsilon versus charge of the fragment. Epsilon is derived from the measured isotopic ratios from the reaction of ^{40}Ca, ^{40}Ar, ^{40}Cl on ^{58}Fe and ^{58}Ni at E/A=53 MeV; $Ratio = ((N_p + \epsilon N_t)/(Z_p + \epsilon Z_t))$.

fitting procedure that assumes some weighting of the target and projectile to make a composite system. Hence the data were fitted with the following form

$$R_{subsystem} = \frac{N_p + \epsilon N_t}{Z_p + \epsilon Z_t}$$

Using this formalism we obtained a linear plot for the isotopic ratio as a function of $(N/Z)_{subsystem}$ for the Be fragments with $\epsilon = 1.6$ If this picture of an equilibrated subsystem resulting from a midcentral collision was to hold up the fragments would all appear to originate from the same subsystem, represented by the same ϵ. However this is not the case and in figure 2 we show how ϵ changes with the charge of the fragment. As we can see the lighter fragments come from a system with a much lower ϵ, i.e. less target and more projectile, than the heavier systems where a higher epsilon represents a system with more target than projectile. Note this is a smoothly varying function. Similar behavior has been reported for light charged particles[4] even though those systems were dominated by the target N/Z because they were very assymetric collisions, so that those studies were insensitive to this behavior in the intermediate mass fragments.

3.3. Target-like and Projectile-like sources at Forward and Backward Angles

Data at very forward and very backward angles also show isotopic ratios that are not simply a function of $(N/Z)_{CS}$. The isotopic ratios for the $Z = 2$ fragments at a forward, central, and backward angle are shown in figure 3.

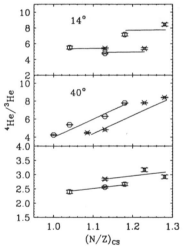

Fig. 3: Isotopic ratios plotted as a function of the N/Z of the combined target and projectile system for ^{4}He/^{3}He at 14° 40° and 140° in the lab. Lines are linear fits to the data points on a given target for the 40° and 140° angles and on a given projectile for the 14° data.

$Z = 2$ fragments at 140° show a much greater dependence on $(N/Z)_{target}$, while at 14° a much greater dependence on $(N/Z)_{projectile}$ is demonstrated. The forward detectors should be more sensitive to collisions that result in a projectile like source, the backward detectors should be more sensitive to a target like source, while the central detectors should be more sensitive to central collisions.

3.4. Centrality Cuts

In order to test the assumption that the fragments seen at 40° are originating from central collisions we can make a centrality condition and monitor the effect on the data. It is also interesting to look at isobaric fragments a well as isotopic fragments. Figure 4 shows the ratio of ^{10}Be to ^{10}B as measured at 40° for both "inclusive" and central collisions. The impact parameter of the event was determined by the mid-rapidity charge detected in the 4π Array[11]. We see two interesting features of this plot. First the isobaric ratios show the same behavior as the isotopic ratios. Second the centrality cut does not change the isobaric ratios.

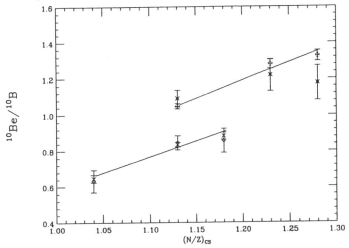

Fig. 4: Isobaric ratios plotted as a function of the N/Z of the combined target and projectile system for A = 10. Lines are linear fits to the data points on a given target.

In order to test the effectiveness of the centrality cuts we compared the multiplicity of events from a minimum bias trigger (the 4pi) with the multiplicity distribution when a valid fragment is detected at 40° . The minimum bias trigger had a peak multiplicity of charged particles of 1. With the requirement of a fragment at 40° the peak of the multiplicity distribution increased to 5.

3.5. Towards Equilibrium

Very similar system have been studied at much lower energies[7] At these energies there showed no signed of nonequilibration in the N/Z degree of freedom. The N/Z of the fragments was determined by the minimization of the potential energy for a given mass assymetry. Somewhere between these two energies there must be a transition. In figure 5 we show preliminary data for $E/A = 25 \text{Mev}$ that is in the same format as figure 2. We can see that initial indications show a much greater degree of equilibration at this lower energy, as a single line can be used to explain all six data points for the Be and B isotopes. While the data of Galin was looking predominately at the projectile-like and target-like fragments from deep inelastic collisions we are looking at lower Z intermediate mass fragments and still it appears we see the same equilibration at this lower energy.

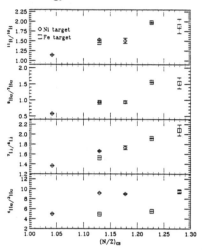

Fig. 5: Isotopic ratios plotted as a function of the N/Z of the combined target and projectile system. (a)^{4}He/^{3}He (b)^{7}Li/^{6}Li (c)^{9}Be/^{7}Be (d)^{11}B/^{10}B at $E/A=25 \text{MeV}$. Lines are linear fits to the data points on a given target.

4. Conclusions

In summary, secondary beams of exotic nuclei have been used to permit a systematic study, which includes a vital region of overlapping data concerning the effects of the N/Z ratio of the target and projectile on the isotopic composition of the final fragments. We have shown that complete equilibration of this degree of freedom does not occur in these reactions at $E/A = 53$ MeV. The same composite system results in different isotopic ratios depending on the initial projectile and target and on the choice of isotopic pairs. The reversal of the effect, with ^{11}B having a higher yield than ^{10}B for the Fe target, and ^{4}He having a lower yield than ^{3}He for the Fe target is puzzling. This difference indicates that the emission of IMFs occurs on a faster timescale than the equilibration of the N/Z degree of freedom. We conclude that the fragments from these near symmetric collisions at $E/A = 53$ MeV are predominately emitted from some subset of the colliding system that is dependent both on the character of the projectile and the character of the target. Preliminary results indicate that at $E/A = 25$ MeV for the same system at much greater degree of equilibration is achieved before fragmentation. The present work also shows the importance and utility of radioactive beams in future studies of N/Z equilibration.

The authors wish to thank the operations staff for the excellent beam quality. This work was supported by the NSF through grant #PHY-89-13815, the DOE through grant #DE-FG03-93er40773, and the Welch Foundation through grant #A-1266.

5. References

[1] S.J. Yennello et al., Phys. Lett. B 321 (1994) 15.

[2] J. Brzychczyk et al., IUCF preprint 1992.

[3] E. Renshaw et al., Phys. Rev. C 44 (1991) 2618.

[4] R. Wada et al., Phys. Rev. Lett. 58 (1987) 1829.

[5] J.L. Wile et al., Indiana University Nuclear Chemistry report# INC-40007-77 (1990).

[6] F. Deak et al., Phys. Rev. C 43 (1991) 2432.

[7] Galin et al Z. Phys. A278 (1976) 347, Chiang et al Phys. Rev. C20 (1979) 1408.

[8] W. U. Schroder and J. R. Huizenga in Treatise on Heavy Ion Science, ed D.A. Bromley Vol. (Plenum Press, 1984).

[9] G.D. Westfall et al., Nucl. Inst. Meth. 238 (1985) 347.

[10] B.M. Sherrill et al., Nucl. Inst. Meth. B 70 (1992) 298.

[11] Transverse Collective Motion in Intermediate Energy Heavy-Ion Collisions, C.A. Ogilvie et al.

NUCLEAR STRUCTURE AND REACTION STUDIES USING RADIOACTIVE BEAMS AT THE NSCL

B. M. SHERRILL

National Superconducting Cyclotron Laboratory
and Department of Physics and Astronomy
Michigan State University, East Lansing, MI 48824-1321, USA

ABSTRACT

A review of the radioactive beam program at the National Superconducting Cyclotron Laboratory will be presented. As an example of the program, experiments which study the parallel momentum distributions of fragments from ^{11}Li breakup will be presented in detail. It has been found that breakup on heavy targets can be described by the three body model of Esbensen and Bertsch. The breakup widths on lighter targets in not fully understood, but indicate an s-wave component of the halo neutrons. Other experiments on structure and reactions are also outlined. Finally, future upgrades to the radioactive beam capabilities will be presented. These include construction of the S800 spectrograph and a coupling project for the K500 and K1200 cyclotrons, which will increase secondary beam rates by up to a factor of 1000.

1. Introduction

Radioactive beams offer the possibility to study nuclei at the extreme limits of particle stability. These nuclei exhibit unusual structure and provide a new testing ground for nuclear models. Radioactive beams also provide a new tool for the study of nuclear reactions. Since the National Superconducting Cyclotron Laboratory Phase II project has been in operation approximately 50% of the beam time has been devoted to radioactive beam experiments. The studies have ranged from experiments to determine the structure of halo nuclei to reaction mechanism studies which have used the unusual ratio of neutrons in the radioactive projectiles to look at the degree to which target and projectile equilibrate [1]. The heart of the radioactive beam program is the A1200 fragment separator [2] which is used to filter the radioactive nuclei produced in projectile fragmentation like reactions (the exact nature of the production mechanism is in fully characterized). A schematic drawing of the A1200 is included in figure 4. A full review of the operation of fragment separators and the projectile fragmentation mechanism can be found in various reviews [3,4]. A significant advantage of the system at the NSCL is that the A1200 fragment separator is located at the start of the beamlines; and since operation of the A1200 facility began in late 1990, experiments have been done in all the experimental areas.

Table 1: Brief summary of some of the radioactive beam experiments done at the NSCL over the past two years.

Experiment	Beam	Rate [ions/s]	Reference
New Isotopes	^{78}Kr,^{92}Mo,^{106}Cd ‡	0.0001	5,6
Inelastic Scattering	^{11}Li,^{14}Be,^{11}Be	100-10^4	8,9
Parallel Momentum	^{11}Li,^{14}Be^{11}Be	100-10^4	10,11
Coulomb Excitation	^{11}Li	10^3	12
(^{13}O,^{12}O)	^{13}O	10^3	13
Mirror Charge Exchange	^{13}N	10^5	14
Reaction Mech. Studies	^{40}Sc,^{40}Cl	2×10^6	1

‡Rare isotope primary beams

Typical secondary beam rates range from a few ions per day to almost 10^7 ions/second. Table I gives a summary of some of the radioactive beam experiments performed at the NSCL, the beams involved, and the rates. Most of these will not be discussed in detail in this paper. Some of the experiments are described in other contributions, e.g. see the contributions of W. Benenson and S. Yennello. The first item in the table is not a radioactive beam experiment, but illustrates other uses for the facility. In these studies up to 15 new isotopes have been identified [5,6] and half-lives for some of them have been measured [7]. Section II will present the details of one of the experiments aimed at measuring properties of very weakly bound neutron rich nuclei. The weak binding of such nuclei leads to an extended neutron distribution, or halo.

Another use for the A1200 facility is been as a zero degree spectrometer. This mode has been used to make a precise measurement of the ^{11}Li mass via the measurement of the Q-value of the reaction ^{14}C(^{11}B,^{11}Li)^{14}O [15]. The binding energy of ^{11}Li is a key ingredient for attempts to calculate its structure. This mode has also been used to study the decays of GT resonances excited in the (^{6}Li,^{6}He) reaction [16], and in studies of double charge exchange.

2. Parallel Momentum Distribution from ^{11}Li Breakup

One of the key pieces of evidence in the halo picture of ^{11}Li is the very narrow transverse momentum distributions measured for ^{9}Li fragments [17]. However the original paper, ref. 17, where this result was reported also noted a very wide distribution for breakup on Pb. This measurement was difficult to reconcile with the halo picture. There is an inherent problem with transverse momentum distribution measurements. Effects such as multiple scattering in the target and nuclear diffraction in the breakup process and can spread the angular distributions. This is probably the case for the wide distribution measured on Lead. Recently, several authors have concluded that the widths measured on light targets may also be influenced [18,19].

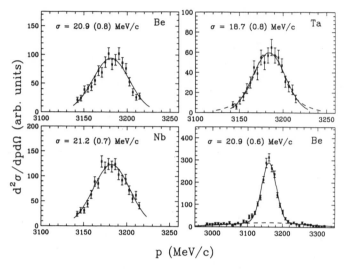

p (MeV/c)

Fig. 1: Parallel momentum distributions for ^{9}Li fragments from ^{11}Li breakup on various targets. The dashed line is from the Coulomb excitation model of Bertsch and Esbensen.

A solution to this problem would be to measure parallel momentum distributions. This is however difficult with secondary beams since the spread in beam momentum is typically on the order of 100 MeV/c and the width of the distributions to be measured are only on the order of 50 MeV/c. This problem has been overcome at the NSCL by use of the A1200 fragment separator in the energy loss mode [10]. We were able to measure narrow distribution for heavy targets as expected using 66 MeV/nucleon ^{11}Li beams. Figure 1 shows sample ^{9}Li distributions for a variety of targets. More recently we have extended the measurements to a U target [20]. The dashed line in the figure corresponds to the 3-body model of Bertsch and Esbensen for the Coulomb excitation and decay of ^{11}Li [21]. It very well reproduces the data. Figure 2 is a summary of the FWHMs measured for the various targets. The breakup on the lighter targets must be dominated by nuclear effects. The weak target dependence probably reflects that the mechanism for the higher Z targets is mostly Coulomb excitation and decay, while the lighter targets are dominated by nuclear processes. In the sudden approximation the width of this distribution indicates the Fermi momentum of the removed two neutrons. More correct reactions models which include final state interactions confirm that the simple picture is adequate and the ^{9}Li distribution is dominated by the 2n momentum. Also included in the figure are preliminary results from recent measurements done at 650 MeV/nucleon using the GSI fragment separator in the energy loss mode. The results at higher energy seem to agree well with the lower energy measurements. Recently, Thompson et al. have argued that the narrow momentum distributions indicate a strong s-wave component to the ^{11}Li structure [22].

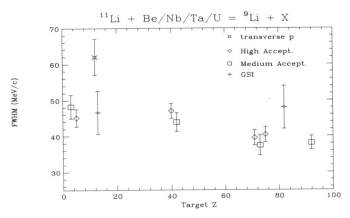

Fig. 2: Summary of the FWHM of the ^{9}Li parallel momentum distributions. The squares are from preliminary results from data taken at GSI at 650 MeV/nucleon and agree with the data taken at MSU at 66 MeV/nucleon. Medium and High acceptance refer to different collection efficiencies. Data for the same A are offset for clarity.

3. Future Developments in the NSCL Radioactive Beam Program

3.1. The S800 Spectrograph

A key piece of experimental equipment which has been missing from the NSCL is a large solid angle high resolution spectrograph. To solve this need we have begun construction of a spectrograph with a 20 msr solid angle and an energy resolving power of 1×10^{4}. The completion date is expected to be July, 1995. The 20 msr solid angle corresponds to opening angles of 7 degrees in the vertical plane and 10 degrees in the horizontal. This large angular acceptance would match well with experiments which are now done with radioactive beams. For example in the elastic scattering studies 10 degrees in the lab is 20 degrees in the CM and most of the interesting information is contained within this angular range [8]. This large angular acceptance would also eliminate and uncertainties fragment widths mentioned in section II due to the finite acceptance of the A1200. In experiments with stable beams the spectrograph will also be used to look for multiphonon excitation in nuclei, spin and isospin resonances, and general nuclear spectroscopic investigations.

3.2. The Coupled Cyclotron Project

By coupling the K500-K1200 cyclotrons it should be possible to significantly increase the intensity of light beams of high energy and for heavier beams make a large advance in beam energy. For example it should be possible to accelerate Uranium

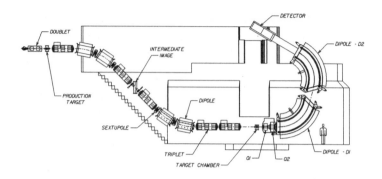

Fig. 3: Schematic drawing of the S800 Spectrograph.

ions to near 100 MeV/nucleon. This would be of interest to reactions mechanism studies. The increase in intensity for the lighter beams would be a great improvement for radioactive beams. It would provide a factor of up to 1000 increase in the secondary beam rates over what we now have. For example, the current [11]Li rate is about 5000/second, while with the coupled project this would increase to 5×10^6 ions/second. At this rate is would be possible to search for subthreshold pion production and study the effect of the low Fermi momentum of the halo neutrons [23]. It will also be possible to perform single and double nucleon exchange reactions to look at the detailed nuclear structure of these nuclei.

Besides the coupling of the cyclotrons we also hope to upgrade our fragment separator capabilities. The current A1200 separator has a limited momentum and angular acceptance, especially for lighter beams and lower production energies. Also to match the higher energies we would like to increase the maximum rigidity of the separator. A final consideration is to provide sufficient space for shielding of the high radiation present at the secondary beam production target and primary beam dump. Figure 4 illustrates the current A1200 separator and the proposed A1900 separator. The A1900 will have a 10 msr solid angle, and a 6 % momentum acceptance compared to the 1 msr and 3% of the current A1200.

The net result of the improvements will be a significant advance in the number of nuclei which can be studied. This includes several more cases of halo nuclei. Further, there is a chance to be able to produce and study nuclei near 6 new doubly magic nuclei, e.g. [78]Ni, [100]Sn, and [56]Ni. Figure 5 shows a sample of the possibilities with the new facility. It illustrates the number of new masses which could be measured using the TOF/Bρ technique [24]. There are 100 to 200 new masses, just in this region, which could be measured.

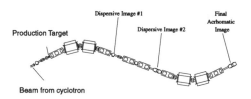

A1200 Radioactive Beam Facility

Dispersive Image #1

Final
Achromatic
Image

Production Target

Dispersive Image #2

Beam from cyclotron

A1900 Radioactive Beam Faciltiy

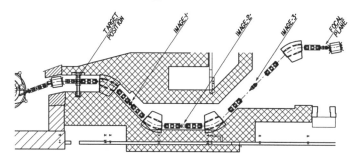

Fig. 4: Comparison of the present A1200 fragment separator and the proposed A1900 separator. The new separator would have a factor of 20 greater acceptance and bend up to 6.0 T-m rigidity beams.

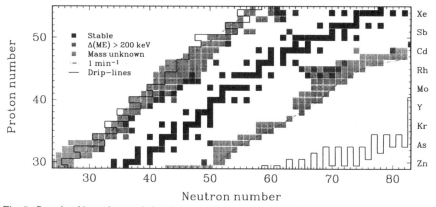

Fig. 5: Sample of how the coupled cyclotron project would extend the range of nuclei which could be studied. This figure shows the nuclei in the mid mass range which have unknown masses which could be measured with the upgrade. The close to 200 new masses.

4. Conclusions

A large number of unique nuclear structure studies have been made possible with the radioactive secondary beams at the NSCL. Much of these studies have centered on the study of light halo nuclei. A few radioactive beam experiments have pursed information on nuclear reactions. The contribution by Sherry Yennello and Walter Benenson at this meeting illustrate those experiments. The upcoming addition of the S800 spectrograph will greatly enhance the available experimental equipment and allow breakup and elastic and inelastic scattering experiments with good resolution. The Coupled Cyclotron Project could increase the secondary beam rates by up to a factor of 1000, and the construction of a new separator, the A1900 could improve the fragment collection efficiency from a few percent to 20 percent. The new facility would open hundreds of new nuclei for study including many more light halo nuclei and proton dripline nuclei up to mass 100.

Acknolwdgements

The work described here represents the contributions of many people. I would like to acknowledge the hard work of the whole A1200 group in these studies. In particular Dave Morrissey, Maggie Helstrom, Bob Kryger, Danial Bazin, and former members Nigel Orr and Jeff Winger. I would like to thank H. Geissel and W. Schwab for providing the resent results from the work at GSI.

5. References

[1] S.J. Yennello et al., accepted in Phys. Lett. B.
[2] B.M. Sherrill et al., Nucl. Instr. Meth. B70, 289 (1992).
[3] B.M. Sherrill, **Radioactive Nuclear Beams**, ed. Th. Delbar (Adam Hilger, London, 1992) p. 1.
[4] G. Münzenberg, Nucl. Instr. Meth. B70, 265 (1992).
[5] M.F. Mohar et al., Phys. Rev. Lett. 66, 1571 (1991).
[6] S.J. Yennello et al., Phys. Rev. C 46, 2620 (1992).
[7] J.A. Winger et al., Phys. Lett. B299, 214 (1993).
[8] J.J. Kolata et al., Phys. Rev. Lett. 69, 2631 (1992).
[9] J.J. Kolata, et al., accepted for publication in Phys. Rev. C Rapid Comm.
[10] N.A. Orr et al., Phys. Rev. Lett. 69, 2050 (1992).
[11] J. Kelley et al., Proc. Third Int. Conf. on Radioactive Nuclear Beams, Ed. D.J. Morrissey, **Editions Frontieres** (Gif-sur-Yvette, 1993) to be published.
[12] K. Ieki et al., Phys. Rev. Lett. 70, 730 (1993).
[13] M. Thoennessen, B. Kryger et al., private communication.
[14] M. Steiner et al., Proc. Third Int. Conf. on Radioactive Nuclear Beams, Ed. D.J. Morrissey, **Editions Frontieres** (Gif-sur-Yvette, 1993) to be published.
[15] B.M. Young et al., Phys. Rev. Lett. 71, 4124 (1993).

[16] S. Gales, private communication.

[17] T. Kobayashi *et al.*, *Phys. Rev. Lett.* **60**, 2599 (1988).

[18] C.A. Bertulani and K.W. McVoy, *Phys. Rev. C* **46**, 2638 (1992).

[19] P. Roussel, Ch.O. Bacri, and F. Clapier, *Nucl. Phys.* **A559**, 646 (1993).

[20] N.A. Orr *et al.*, Proc. Third Int. Conf. on Radioactive Nuclear Beams, Ed. D.J. Morrissey, **Editions Frontieres** (Gif-sur-Yvette, 1993) to be published.

[21] H. Esbensen and G.F. Bertsch, *Nucl. Phys.* **A542**, 310 (1992).

[22] I.J. Thompson *et al.*, Proc. Third Int. Conf. on Radioactive Nuclear Beams, Ed. D.J. Morrissey, **Editions Frontieres** (Gif-sur-Yvette, 1993) to be published; and submitted to Phys. Rev. Lett.

[23] X. Li, M. Hussein, and W. Bauer, *Phys. Lett.* **B258**, 749 (1991).

[24] N.A. Orr *et al.*, *Nucl. Phys.* **B258**, 29 (1991).

PARTICLE PRODUCTION AT AGS ENERGIES

S.G. STEADMAN, P.J. ROTHSCHILD, T.W. SUNG, and D. ZACHARY

Laboratory for Nuclear Science and Physics Department
Massachusetts Institute of Technology
Cambridge, MA 02139-4307, USA

for the E802 Collaboration

ABSTRACT

We discuss particle production from 14.6 A·GeV/c Si and 11.6 A·GeV/c Au projectiles on Al and Au targets. The second-level trigger utilized by E859 allows high precision measurements of K^-, $\bar{p}$, Λ and $\bar{\Lambda}$. The $\bar{\Lambda}$ yield is larger than expected, and a surprisingly large fraction of the $\bar{p}$'s are observed to arise from the decay of $\bar{\Lambda}$.

1. Introduction

At AGS energies of 11-15 A·GeV/c one achieves a high degree of stopping, thereby reaching the highest baryon densities achievable in heavy-ion reactions. Indeed, cascade Monte Carlo calculations, such as ARC[1], indicate that densities exceeding 9 times the ground state nuclear matter density may be achievable in Au+Au collisions. It is generally believed that densities exceeding 5-6 times normal density are required to make the transition to the quark-gluon plasma (QGP) phase. If such new physics is to be observed, it is important to provide a broad base of data for comparison to such Monte Carlo models in order to test their validity.

This work concentrates on the systematics of particle production from the series E802/E859/E866 of experiments at the Brookhaven AGS. The very successful implementation of the second level trigger in E859 with Si beams allows the E802 spectrometer to perform on-line particle identification within 40 μs. This allows an event selectivity that increases the data taking rate by up to an order of magnitude, providing a far better data sample for the detection of particles such as K^- and $\bar{p}$.

2. Experiment

The E802 spectrometer[2] consists of a 25 msr acceptance rotatable single arm spectrometer that can be used together with event characterization devices to obtain semi-inclusive spectra and measure two-particle correlations. For asymmetric collisions, such as 14.6 GeV/c Si+Au, a target multiplicity array (TMA) sensitive to a large fraction of the total charged particles emanating from the target, is used to select the collision geometry, specifically the upper 7% of the interaction cross-section (central collisions) or lower 50% (peripheral collisions). The central trigger corresponds roughly to a range of impact parameters for which the Si projectile is totally occluded by the Au target. For the symmetric Si+Al collisions at 14.6 GeV/c, as well as Au+Au collisions at 11.6 A·GeV/c incident momentum, a zero-degree calorimeter is used to measure the forward going energy in a cone of $\approx$ 1.5 degrees relative to the beam axis. By dividing this measured forward energy by the kinetic energy per projectile nucleon we obtain the mean number of projectile participants.

3. Pion production

Of particular interest is the determination of the number of produced pions per participant. At AGS energies the dominant part of the inelastic N-N cross section is excitation of the nucleon resonances Δ and N*.[1] Indeed, ARC calculations[3] indicate that a large fraction of the participants in central Au+Au collisions become these excited baryonic resonances. One might expect, then, that in the subsequent decay one would observe about one produced pion per participant. The charged pion distributions are observed to have an exponential dependence on $p_\perp$, with inverse slopes typically about 160 MeV/c at mid-rapidity. By integrating over $p_\perp$ one obtains the rapidity distribution, for which our acceptance covers approximately 80% of the total yield. Assuming a Gaussian form for dN/dy, we show in Fig. 1 the phase space integrated π^+ multiplicity as a function of measured projectile participants, as determined by ZCAL for 11.6 A·GeV/c Au+Au collisions and 14.6 A·GeV/c Si+Al collisions. (One should note that the energy of the Au beam is somewhat lower than for the Si beam, which from p-p systematics would lead one to expect about a 14% reduction in the π^+ yield.) The obvious linear dependence is seen. From a similar observed dependence for π^- and assuming that the yield for π^0 is the mean of the π^+ and π^- yield, one obtains an overall multiplicity of 1.2 produced pions per participant.

4. Kaon and lambda production

We now turn to strangeness production, whose enhancement has long been considered a signal for QGP production[4] and whose observation by E802[5] and others has stimulated much theoretical interest. With the combined use of the second level trigger and extended particle ID using a segmented threshold gas Čerenkov detector, we have far more precise data available for both K^+ and K^- production. The invariant cross sections per trigger (TMA central or peripheral) are found to be well described by exponentials in the transverse mass, $m_\perp = \sqrt{p_\perp^2 + m^2}$. The fitted inverse slope parameters, as a function of rapidity, are shown in Fig. 2 for both central and peripheral Si+Al and Si+Au collisions. Unlike pion production, the inverse slope parameters depend on centrality and generally these more central collisions have higher inverse

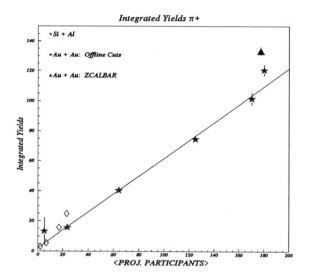

Fig. 1: Integrated yields of π^+ as a function of projectile participants for ^{28}Si and ^{197}Au projectiles.

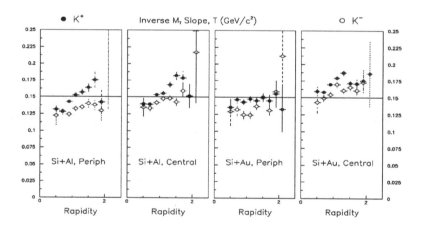

Fig. 2: E859 $K^{\pm}$ inverse $m_{\perp}$ slopes versus rapidity for peripheral and central Si+Al and Si+Au collisions.

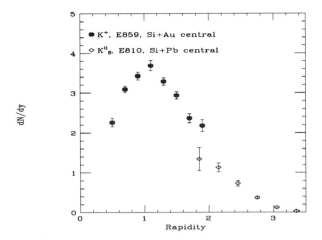

Fig. 3: E859 K^+ yield versus rapidity for central Si+Au collisions. The E810 K^0_S yield versus rapidity is also shown for central Si+Pb collisions.

slope parameters at a given rapidity. The K^+ tend to have larger inverse slope parameters than K^- for a given target and centrality. These transverse mass distributions can be integrated to yield rapidity distributions, dN/dy. Fig. 3 shows our results for K^+'s from central Si+Au collisions, together with other AGS measurements from E810 for K^0_S production in central Si+Pb collisions (albeit with a somewhat different centrality trigger).[6] It is noteworthy that the dN/dy for K^+ peaks at a rapidity of 1.1, behind the participant center-of-mass rapidity of 1.3 for central Si+Au collisions. Rescattering mechanisms, such as $\rho N \to K\Lambda$ have been suggested[7] as likely mechanisms for producing the observed enhancement. Such secondary collisions have peak yields at a lower rapidity than the participant rapidity. Thus, the observed rapidity distribution may be consistent with this mechanism.

To provide a measure of how the different mechanisms contribute to K^+ production, we have also measured the Λ production at y about 1.4. This was done by observing the Λ decay products, (p,π^-), for which the invariant mass distribution is shown in Fig. 4. Background distributions were obtained using (p,π^-) pairs with the protons and pions coming from different events. Invariant mass distributions for four $p_\perp$ intervals were generated for central collisions (upper 20% of the TMA multiplicity distribution). Background subtraction was done for each interval to obtain the number of Λ's for that interval. No vertex cuts were required on the Λ candidates. The large, momentum dependent acceptance correction was obtained by a Monte Carlo analysis. This was checked by using the acceptance to determine the Λ decay constant from the data. Our determined value for $c\tau = 7.98 \pm 0.65$ cm is in agreement with the accepted value[8] of 7.89 cm, giving us confidence in the validity of this acceptance correction. From an exponential $m_\perp$ fit to the four bins (Fig. 5), we obtain an inverse

82

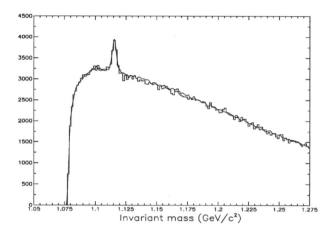

Invariant mass (GeV/c^2)

Fig. 4: The invariant mass distribution of (p,π^-) pairs from central Si+Au collisions.

slope parameter of 171 ± 13 MeV/c^2. Integrating over $m_\perp$ gives a rapidity density at this rapidity of 3.85 ± 0.58. A surprising result is the K^+/K^- ratio, as shown in the left panel of Fig. 6, for both central and peripheral Si+Al and Si+Au collisions. The rapidity dependence of the ratio is found to be independent of target and centrality. A similar rapidity dependence is also found for Au+Au collisions (right panel) though the statistics are poorer.

5. Antiproton and antilambda production

Because of the large energy required to produce a $\bar{p}$ in NN collisions at the AGS, one is very near threshold ($E_{th} = 4m_{proton} = 3.8$GeV, $E_{cm} = 5.4$GeV). Thus, one expects that the $\bar{p}$ production is particularly sensitive to the earlier stage of the collision, and therefore to possible QGP production. The on-line second level triggering capability of E859 allowed an order of magnitude improvement in statistics compared with E802.[9] Fig. 7 shows the extracted inverse $m_\perp$ slopes for central and minimum bias Si+Al and Si+Au collisions. The new data are consistent with the earlier E802 data, but with much greater precision. Overall, these inverse slope parameters are lower than for protons, but are similar to those obtained for the protons for central Si+Au. This is somewhat surprising. One proposed explanation for the large proton inverse slope parameter is the increased boost in $p_\perp$ given by multiple N-N collisions. The decay of these excited baryons yields an additional boost to the proton $p_\perp$. For $\bar{p}$, however, one does not expect significant rescattering, due to the high annihilation cross section. Thus, one would expect the mean $p_\perp$ for the $\bar{p}$ to be similar to the produced π's and K's, as observed in the minimum bias data.

Even though the cross-section for $\overline{\Lambda}$ production is very small, we have been able to measure the $\overline{\Lambda}/\Lambda$ ratio (integrated over the spectrometer acceptance) by switching

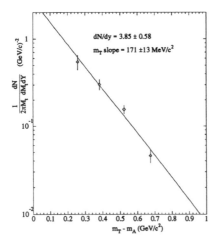

Fig. 5: E859 Λ $m_\perp$ distribution for central Si+Au collisions.

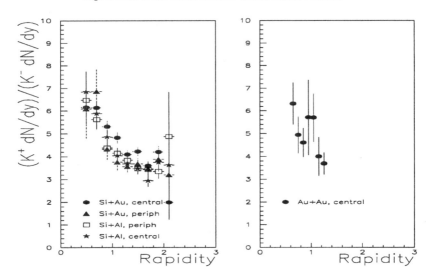

Fig. 6: E859 K^+/K^- ratio versus rapidity for central and peripheral Si+Al and Si+Au collisions (left panel). The right panel shows the same ratio for central Au+Au collisions.

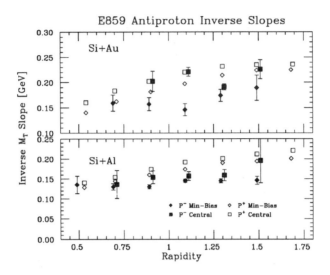

Fig. 7: Inverse $m_\perp$ slope parameters for antiprotons in Si+Al and Si+Au collisions.

the direction of the spectrometer magnetic field at the same angle setting of the spectrometer for which the Λ measurement was made and recording the opposite charged particles $(\bar{p}, \pi^+)$. The lower $\bar{p}$ yield results in a much reduced combinatoric background. This is shown by the $(\bar{p}, \pi^+)$ invariant mass distribution in Fig. 8. The only correction required in order to obtain the ratio is the simple correction for the $\bar{p}$'s that annihilate within the spectrometer ($\approx 5\%$). The spectrometer-integrated $\overline{\Lambda}/\Lambda$ ratio within the same rapidity interval about y=1.4 is $(3.7 \pm 0.5 \pm 0.3) \times 10^{-3}$, where the errors are statistical and systematic, respectively. With the correction for the decay branch to $\bar{p}$'s one obtains that $(63 \pm 13 \pm 11)\%$ of the $\bar{p}$'s within the rapidity interval would arise from $\overline{\Lambda}$ decay. It should be noted that this result depends on the assumption that the Λ and $\overline{\Lambda}$ have similar inverse slope parameters (similar slopes have been observed by NA36 for S+W at 200 A·GeV/c) [10] and that the $\bar{p}$'s from the $\overline{\Lambda}$ decay have the same rapidity distribution as the $\overline{\Lambda}$'s (confirmed by a Monte Carlo study). This large value for $\overline{\Lambda}$ production could explain two striking features of the $\bar{p}$ production, which otherwise are hard to understand, namely the large inverse slope parameters and the large yield for central Si+Au collisions. The latter may result from the relatively small annihilation cross section of $\overline{\Lambda}$'s compared with $\bar{p}$'s in the nuclear medium.

6. Conclusions

Even though calculations such as ARC predict high baryon densities, the volume of this high density region is small compared with the total interaction volume. Thus,

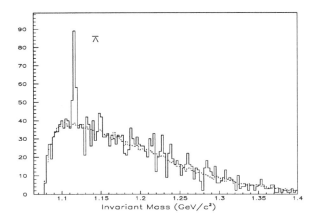

Fig. 8: Invariant mass spectrum of $(\bar{p}, \pi^+)$ pairs from central Si+Au collisions. The scaled background is shown by the dashed curve. Both the $\bar{p}$ and π^+ were required to be fully in the spectrometer acceptance.

it seems likely that any signal for QGP formation will be found in weak signals, such as strangeness production, and more likely in extremely rare processes such as $\overline{\Lambda}$ production. The preliminary indications from the data indicate that the $\overline{\Lambda}/\Lambda$ ratio is considerably larger than what one expects from a cascade code calculation.[11] Further measurement will be required to extend the rapidity range. More importantly, the recently commissioned Au beam offers the opportunity to create a much larger volume of high density nuclear matter. Under these conditions it will be particularly interesting to see if such large $\overline{\Lambda}$ yields are observed.

7. References

[1] Y. Pang et al., Phys. Rev. Lett. **68**, 18 (1992).

[2] T. Abbott et al., Nucl. Inst. Meth. A **290**, 41 (1990).

[3] S. Kahana et al., in Heavy-Ion Physics at the AGS: HIPAGS'93, edited by G.S.F. Stephans, S.G. Steadman, and W.L. Kehoe, MITLNS-2158 (1993), p.263.

[4] P. Koch, B. Müller, and J. Rafelski, Phys. Rep. **142**, 167 (1986).

[5] T. Abbott et al., E802 collaboration, Phys. Rev. Lett. **64**, 847 (1990).

[6] A.C. Saulys et al., E810 collaboration, in HIPAGS'93, op cit., p. 196.

[7] H. Sorge, in HIPAGS'93, op cit., p. 283.

[8] Review of Particle Properties, Phys. Lett. B **239** (1990).

[9] T. Abbott et al., E802 collaboration, Phys. Lett. B **271**, 4477 (1991).

[10] R. Zybert and E. Judd, Particle Production in Highly Excited Matter, edited by H.H Gutbrod and J. Rafelski (Plenum Press, NY, 1993), p. 545.

[11] A. Jahns, private communication

Pion Interferometry in E814 – Toward Equilibrium at the AGS

Nu Xu for the E814 Collaboration[1]
Department of Physics
State University of New York at Stony Brook
Stony Brook, New York 11794 - 3800

ABSTRACT

We report recent pion interferometry results from AGS experiment E814. By comparing to the results of RQMD calculations, a freeze-out size $R_{rms} = 8.3$ fm is found in Si + Pb central collisions. A consistent thermal equilibration picture is established by comparing experimental data with the results of both hydrodynamic and cascade model calculations.

1. Introduction

The ultimate goal of ultra-relativistic heavy-ion program at the BNL AGS and CERN SPS is to study highly excited nuclear matter and the transition from hadronic matter to quark gluon plasma. In the QGP state, quarks and gluons are no longer confined but they move freely in a fairly large volume. According to our knowledge, the de-confined state can be approached via either compression or heating of nuclear matter in heavy-ion collisions.

At AGS energies, the most relevant process is believed to be compression. However, before we try to identify any exotic events such as, for example QGP phase transition or chiral symmetry restoration, we have to answer the following questions: (1) Were the densities(due to the compression) high enough? (2) Was the high density region large enough? (3) Was the system thermalized?

The first question has been addressed in depth in reference [1], where the main conclusion is: a high degree of stopping has been reached in the Si + Pb central collisions. The baryon density is as high as $5\rho_0$ over a period of about 5 fm/c[2]. In order to address questions (2) and (3), we will organize the paper as follows: in section 2, we will discuss recent pion interferometry results from the E814 collaboration. The pion source size as a function of time will also be studied. In section 3, we will discuss the thermal equilibration issue at AGS energies using two different approaches, namely, a hydrodynamical type calculation and a cascade model, RQMD [2], calculation. Finally, we will make some concluding remarks.

---

[1]BNL, GSI, McGill University, University of Pittsburgh, SUNY Stony Brook, University of Sao Paulo, Wayne State University, Yale University

2. Recent HBT results from E814

For the data presented here we used the E814 apparatus which provides a 4π event characterization with a target calorimeter and a charged particle multiplicity detector. The forward spectrometer acceptance was determined by a lead collimator to $-115 < \theta_x < 14$ mr (bending plane) and $-21 < \theta_y < 21$ mr, both with respect to the beam direction. Particle identification is obtained by combination of time-of-flight and momentum measurement [1]. The acceptance in transverse momentum p_t and rapidity y of identified π^+ and π^- is shown in Fig. 1. Note that pions of both charges are detected at $p_t \geq 0$ with a mean rapidity of about 3.

Experimentally the two-particle correlation function C_2 is defined as:

$$C_2(q) = \frac{N_{tr}(q)}{N_{bk}(q)} \quad (1)$$

where $q = \sqrt{-(p_1 - p_2)^2}$ is the relative 4–momentum between the two identical particles. As done frequently, the numerator $N_{tr}(q)$ is obtained by taking two particles from the same event, while the denominator $N_{bk}(q)$ is constructed using two pions from different events (mixed event technique). This way one ensures that the statistical errors in the correlation function are determined by the statistics of the true pion pairs only.

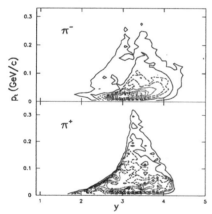

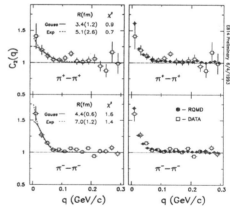

Figure 1. E814 acceptance.

Figure 2.   Correlation functions: (left) measured data with Gaussian and exponential fits; (right) data with RQMD results.

The pion correlation functions corrected for Coulomb effects and two-particle acceptance are shown in Fig. 2 (open symbols). The total number of selected π^- (π^+) pairs is 23.4k (4.3k) and about 80% of the pairs are in the relative momentum range of $0 \leq q \leq 0.3$ GeV/c. Error bars are statistical only. The background distribution is normalized to the total number of the entries in the true distributions within the range of $0 \leq q \leq 1$ GeV/c. It can be seen from this figure that, for $q \geq 0.1$ GeV/c, the distribution is consistent with unity. The Bose-Einstein enhancement is clearly visible in the low relative momentum region for both like-sign pion pairs.

The measured correlation functions have been fitted by two commonly used parameterization, namely, a Gaussian and an exponential function:

$$C_2^g(q) = 1 + \lambda_g \cdot exp(-q^2 R_g^2); \qquad C_2^e(q) = 1 + \lambda_e \cdot exp(-q R_e). \qquad (2)$$

where λ is the chaoticity parameter and R is determined by the space-time extent of the pion source. In Fig. 2 (left), the solid and dashed lines represent the Gaussian and exponential fits, respectively. The extracted fit parameters are also summarized in the figure. While the reduced χ^2 is slightly smaller for the exponential fit, both functional forms are consistent with the data. However, the source parameters are quite different.

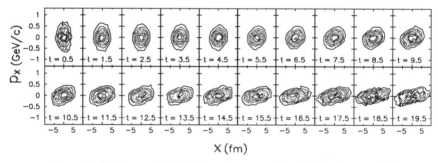

Figure 3. Contours of RQMD pion phase-space distribution for central Si + Pb collisions. Each window represents a time interval of $\Delta t_{cm} = 1.0$ fm/c.

Non-Gaussian shapes of the pion correlation function have been observed in heavy-ion collisions as well as in hadron-hadron collisions [3,4,5]. The shape of correlation function depends, besides the source distribution, on the experimental acceptance, resonance decays, dynamics of the emitting source, and other effects[6]. To overcome these ambiguities, rather than extracting the source size from fitting a certain functional form to the data, we use dynamical models with a known space-time characteristics of the source, impose the Bose-Einstein effect and evaluate the two particle correlation functions to compare them with the experimental results.

The event generator RQMD [2] has been successful in describing many aspects of pion, proton, kaon, and other measured single-particle spectra [1,7,8,9] at both AGS and SPS energies. It is natural to use the model to also calculate the two-particle correlation functions.

Pion phase-space distributions from RQMD for a dimension x transverse to the beam direction are shown in Fig. 3 for 14.6 A·GeV/c ^{28}Si + Pb collisions (for $b \leq 1.0$

fm). These distributions were generated for pions at different times in the collision. As shown in Fig. 3, as time elapses, (1) the spatial distributions become wider implying that the source expands; and (2) a correlation between momentum and spatial coordinates also developes. An immediate consequence of the momentum-spatial correlation is that a spectrometer will not be able to 'see' particles emitted from all locations of the source. Hence the source size extracted by fitting a functional form to the measured C_2 will be reduced. Indeed, with a coverage $p_t \geq 150\text{MeV/c}$, our π^+ correlation functions [10] have shown a smaller size parameter $R_g = 2.2$ fm.

In order to generate a two particle correlation function, the Koonin-Pratt method [11] was used to construct a symmetrized pion wave function from the RQMD generated single-particle distributions. The experimental conditions were imposed on the RQMD event before feeding them into the calculation. Finally the correlation function was corrected by the Gamow factor as has been done for the experimental data. Results of the calculations, for both π^+ and π^- within the E814 spectrometer, are shown in Fig. 2 (right) as filled circles. The agreement between the experiment and the model is excellent. Inspecting the RQMD phase-space distribution at freeze-out stage, we obtain a source size of $R_{rms} = 8.3$ fm(in a frame with $y_{cm} \approx 1.3$).

3. Thermal equilibrium at the AGS[2]

Using a hydrodynamical model one can calculate the bulk properties of the relativistic heavy-ion collisions which will provide a macroscopic description of such collisions. On the other hand, a cascade type calculation which is basically the superposition of elementary collisions including resonance and mean-field effects provides a microscopic picture of the collisions. However, in the limit of thermal equilibrium, these two approaches should converge. It is important to realize that the thermal characteristics in the measured hadron spectra provides a necessary[12], but not sufficient, proof of thermal equilibration at the early stage of the collision.

3.1. The hydrodynamical approach

Let us start with a isotropic stationary thermal source:

$$E\frac{d^3N}{d\vec{p}} \propto Ee^{-E/T} \propto m_t cosh(y)e^{-m_t cosh(y)/T} \tag{3}$$

where $m_t = \sqrt{p_t^2 + m^2}$ and T is the Boltzmann temperature. All kinematic variables are in the nucleon-nucleon center of mass system.

Integrating over the transverse variable m_t, one obtains the rapidity distribution:

$$\frac{dN'}{dy} \propto m^2 T(1 + 2\chi + 2\chi^2)e^{-1/\chi} \tag{4}$$

---

[2]In collaboration with P. Braun-Munzinger and J. Stachel

90

with the parameter $\chi = T/(mcosh(y))$. It is clear that the distribution is rather sensitive to the ratio of m/T. While the distribution for pions is close to that for massless particles, protons from an isotropic source are much narrower.

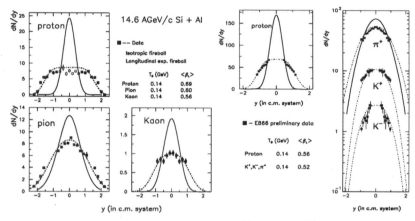

Figure 4. Rapidity distributions for 14.6A·GeV/c Si+Al collisions. Data are taken from E814 and E802 [1,13].

Figure 5. Rapidity distributions for 11.6A·GeV/c Au+Au collisions. Data are from E866 [14].

Figs. 4 and 5 represent rapidity distributions for 14.6A·GeV/c Si + Al and 11.6A·GeV/c Au + Au central collisions, respectively. Experimental data are shown as square symbols and the isotropic thermal calculations (for $T = 0.14$ GeV) are shown as solid lines. Obviously the agreement between the calculation and data is poor. From previous studies [1,2] we learned that a high degree of stopping has been reached at AGS energies and a density gradient is building up in the collision zone. Such collision induced pressure will eventually lead to collective flow. In order to include such effects into our model, following [15,16], the final rapidity distribution is calculated by superposition of individual isotropic thermal sources within a rapidity interval $[-\eta_{max}, \eta_{max}]$:

$$\frac{dN}{dy} = \int_{-\eta_{max}}^{\eta_{max}} d\eta \frac{dN'(y-\eta)}{dy}. \tag{5}$$

Here the integration limit η_{max} is treated as a fit parameter. The results of a calculation for a longitudinal expanded fireball are shown as dash-dotted lines in Figs. 4 and 5 and the agreement with the data is excellent. While the extracted mean longitudinal velocity[3] β_z for protons is higher than that of the produced particles (pions and kaons) for the Si + Al collisions, the extracted values for protons, pions

---

[3]The mean value of the β_l is defined as: $\beta_l = tanh(\eta_l)$ with $\eta_l = \eta_{max}/\sqrt{3}$.

and kaons are found to be rather close to each other for the heavy collision system Au + Au. Similar agreement is also obtained for the transverse spectra in Si + Au collisions (Fig. 6) where a common freeze-out temperature and the maximum transverse velocity are found to be $T = 0.14 - 0.15$ GeV and $\beta_t^{max} \approx 0.5$, respectively. A velocity profile $\beta_t = \beta_t^{max} \cdot (r/R)^2$ ($R = 7.0$ fm) has been used in the calculation[15].

3.2. The cascade approach

As we already mentioned, in the limit of thermal equilibration, a cascade type calculation should also show hydrodynamic effects. In the following we will discuss mean transverse outward[4] velocity β_{out} and longitudinal velocity β_z from RQMD events. It is important to perform such tests since they can demonstrate, that even starting from purely binary collisions, thermal equilibrium and collective motion can be reached in heavy-ion collisions at AGS energies.

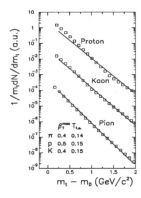

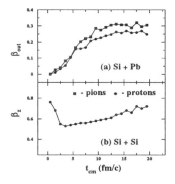

Figure 6. m_t spectra (E802 [13]) for central Si + Au collisions. Thermal calculations are shown as open symbols.

Figure 7. RQMD (a) transverse outward, (b) longitudinal velocity distributions for central collisions.

Fig. 7(a) shows the mean outward velocity β_{out} in 14.6A·GeV/c Si + Pb central collisions. Firstly we find a rapid increase in β_{out} for the first 6 fm/c, indicating a strong push from the high density region; secondly, when the acceleration is large, pions and protons share a similar velocity, implying transverse matter flow. This behaviour is indeed what one would expect from hydrodynamics; thirdly, for $t_{cm} > 10$ fm/c, pions and protons are separated slightly and β_{out} reaches a value of about 0.3 and 0.25 for pions and protons, respectively.

In the NN c.m. system, mean longitudinal velocities β_z are calculated for Si + Si central collisions at 14.6A·GeV/c (Fig. 7(b)). There are clearly two distinct regions:

---

[4]β_{out} is the mean outward velocity: $\beta_{out} = \vec{\beta}_t \cdot \vec{r}_t$. In this case, β_{out} is the only relevant variable in order to compare to hydrodynamic calculations.

for $t_{cm} \leq 4$ fm/c, when the two incoming nuclei collide violently, the system is in a process of compression and β_z decreases. At $t_{cm} \approx 4$ fm/c, β_z reaches its minimum value of 0.5. After this point the system experiences expansion and β_z gradually approaches 0.7. The initial decrease and later on rise of the flow velocity was also predicted [17] by Brown *et al.* in a thermal model calculation. Remember that if a transparency scenario was suited for such collisions, no increase of β_z should occur.

4. Conclusions

We demonstrated, from analysing two-pion correlation functions, that the pion freeze-out source is large. The RMS value of the size is consistent with $R_{rms} = 8.3$ fm in central Si + Pb collisions. We also demonstrated, from comparing the experimentally measured spectra with the results of thermal and cascade calculations, that the thermal equilibrium scenario provides a consistence picture for collisions at the AGS energies. The large final stage source size is, in this picture, due to collective flow which expands the system from an initial transverse size $R_T = 2.5$ fm to freeze-out size of $R_T = 6.7$ fm. Taking the maximum transverse velocity as 0.5 and assuming a linear dependence of the flow velocity on the radius, this implies a minimum expansion time of $t \geq 15$ fm/c.

5. Acknowledgements

The author is indebted to Drs. S. Pratt and H. Sorge for valuable discussions and their kindness to provide computer codes. We would also like to thank Drs. G. Brown, V. Koch, Y. Pang, and E.V. Shuryak for exciting and valuable discussions. We are grateful for support received from the U.S. DOE, U.S. NSF, Canadian NSERC, and CNPq Brazil.

6. References

[1] J. Barrette, *et al.*, E814 Coll., Z. Phys. **C59**, 211(1993).
[2] H. Sorge, A. von Keitz, R. Mattiello, H. Stöcker, and W. Greiner, Phys. Lett. **B243**, 7(1990) and H. Sorge, R. Mattiello, H.Stöcker, and W. Greiner, Phys. Lett. **B271**, 37(1991).
[3] A. Bamberger, *et al.*, NA35 Coll., Phys. Lett. **B203**, 320(1988).
[4] T. Åkesson, *et al.*, AFS Coll., Z. Phys. **C36**, 517(1987).
[5] R. Albrecht, *et al.*, Z. Phys. **C53**, 225(1992).
[6] W.A. Zajc, in *"Particle Production in Highly Excited Matter"*, H.Gutbrod and J.Rafelski, eds. (Plenum, New York,1993)p435.
[7] T. Hemmick, E814 Coll., Nucl. Phys. **A566**, (1994).
[8] J.P. Sullivan, *et al.*, Phys. Rev. Lett. **70**, 3000(1993).
[9] Th. Schönfeld, *et al.*, Nucl. Phys. **A544**, 439c(1991).
[10] N. Xu E814 Coll., Proceedings of HIPAGS'93, Workshop, MIT LNS-2158.
[11] S. Pratt, Phys. Rev. Lett. **53**, 1219(1984); S. Pratt, Phys. Rev. **D33**, 72(1986).
[12] E.V. Shuryak, Phys. Lett. **B42**, 357(1972); E.V. Shuryak, *"The QCD Vacuum, Hadrons and the Superdense Matter"* (World Scientific, Singapore, 1988).
[13] C. Parsons E802 Coll., Proceedings of HIPAGS'93, Workshop, MIT LNS-2158.
[14] M. Gonin E802/E866 Coll., Proceedings of HIPAGS'93, Workshop, MIT LNS-2158.
[15] E. Schnedermann, J. Sollfrank, and U. Heinz, Phys. Rev. **C48**, 2462(1993).
[16] A.N. Makhlin, private communication.
[17] G. Brown, C.M. Ko, Z.G. Wu, and L.H. Xia, Phys. Rev. **C43**, 1881(1991).

PION INTERFEROMETRY IN HADRONIC MODELS

G. M. WELKE
Department of Physics and Astronomy
Wayne State University
Detroit, MI 48202, USA

H.W. Barz,[a] G.F. Bertsch,[b] P. Danielewicz,[c] and H. Schulz[d]

ABSTRACT

Pion–pion correlations are calculated for a meson source that models the central rapidity region of central ^{16}O on Au collisions at 200 GeV/nucleon, via a Boltzmann equation with bosonic phase space enhancement, and an in–medium π–π scattering–amplitude. Extracted source sizes are small because of strong momentum–position correlations.

1. Introduction

In ultra–relativistic heavy ion collisions, the transient central region produces an abundance of pions which evolve and eventually cease to interact. To understand the collision dynamics, it is of interest to infer the freeze–out surface from experimental data. One way to do this is via the pion–pion correlation function:[1]

$$C_2(\vec{p}_1, \vec{p}_2) \equiv \frac{d^6 N_{\pi\pi}}{dp_1^3 dp_2^3} \bigg/ \frac{d^3 N_\pi}{dp_1^3} \frac{d^3 N_\pi}{dp_2^3} \quad . \tag{1}$$

If the pion source function contains no dynamical momentum–space correlations, and is suitably parametrized, its size may easily be inferred from the width of the correlation function. However, the interpretation of C_2 becomes far less clear once a more realistic source is postulated. For example, source geometry, expansion dynamics, resonance decays, and final state interactions will cloud the determination of the source size.

In this contribution these problems are examined by calculating the π–π correlation function from a realistic hadronic model. As discussed in section 2, the dynamics is to be described by a set of Boltzmann equations with Bose statistics for the mesons,[2,3] and an in–medium cross section for the pions.[4] The initial (hadronization) conditions approximate the mid–rapidity region of central ^{16}O + Au collisions at 200 GeV/nucleon.[5] Section 3 discusses the known collision geometry, which is compared in section 4 with results from the pion–pion correlation functions, for a pure pion gas. It is shown explicitly how the strong momentum–position correlations lead to small apparent source sizes.[6] We also discuss briefly the role of resonances, and compare the present results to data.[7] Concluding remarks are presented in the last section.

2. Evolution model including resonances

The time evolution of the pion phase space density $f_1(\vec{x}, \vec{p})$ may be modeled approximately by the Boltzmann equation

$$[p^0 \partial_t + \vec{p} \cdot \vec{\nabla}_x] f_1(\vec{x}, \vec{p}) = -\frac{1}{4} \sum \int d\Gamma_2 d\Gamma_3 d\Gamma_4 (2\pi)^4 \delta^{(4)}(p_1 + p_2 - p_3 - p_4)$$
$$\times |T|^2 \{f_1 f_2 \bar{f}_3 \bar{f}_4 - f_3 f_4 \bar{f}_1 \bar{f}_2\} , \qquad (2)$$

where $d\Gamma_i = d^3 p_i / [(2\pi)^3 2\omega(p_i)]$, $\omega(p_i) = \sqrt{m_\pi^2 + p_i^2}$, $\bar{f}_i \equiv 1 + f(\vec{x}_i, \vec{p}_i)$, m_π is the mass of the pion, with spatial coordinates $\vec{x}_i$ and momentum $\vec{p}_i$. For clarity, additional source terms from resonances have been dropped in Eq. (2). Note that baryonic degrees of freedom and mean field interactions have not been included in the present calculation. The T–matrix is of the Bethe–Goldstone form and describes the interaction of a pair of pions embedded in a pure pionic medium.[4] For scattering with and between other resonances a Hauser-Feshbach formalism is assumed.

The initial distribution of particle species i is defined on the hyperbola $t^2 - z^2 = \tau_{eff}^2$ to be of the form[2]

$$\frac{d^6 N_i}{dy_b \, dr_\perp^2 \, dy \, dp_\perp^2} = \mathcal{N}_i \left[e^{\beta m_\perp \cosh(y - y_b)} - 1 \right]^{-1} e^{-y_b^2 / 2\sigma_b^2} \int_{-\infty}^{\infty} dz' \, \rho_{\mathrm{Oxy}}(r_\perp^2 + z'^2) , \qquad (3)$$

where y is the particle rapidity, $p_\perp^2 = m_\perp^2 - m_i^2$ the transverse momentum, y_b is the boost rapidity that fixes the longitudinal position $z = \tau_{eff} \sinh y_b$, $r_\perp$ is the transverse position, and ρ_{Oxy} is the empirical projectile baryon density. The parameter τ_{eff} is the proper time needed for particle formation, and determines the spatial density of the initial distribution via its relation to the longitudinal position. The width of the boost rapidity $\sigma_b = 1.3$ and the parameter $\beta^{-1} = 130$ MeV are fixed from the corresponding experimental pp distributions. We further suppose that the mesons are initially produced in the ratio $\pi : \rho : \omega : \eta \sim 3 : 9 : 3 : 1 \sim 38 : 114 : 38 : 13$, which fixes the normalization constants $\mathcal{N}_i$. String breaking models yield such ratios, but are probably somewhat over–predicting the ρ abundance.

The initial local momentum distribution does not correspond to local thermal equilibrium, but collisions will partly thermalize the system. The phase space factors $\bar{f}$ in Eq. (2) increase the low momentum state occupation if the initial pion phase space density is high enough.[2,3] The experimental transverse momentum distribution of the pions may be explained in terms of Bose enhancement and resonance decays, if a small proper hadronization time of $\tau_{eff} \sim 1$ fm/c is chosen.[3] Here we shall also consider longer hadronization times of 4.5 and 8 fm/c.

3. Collision geometry

To get a rough orientation for the collision geometry we show in Fig. 1(a) the collision rate (solid line) and the total number of collisions (dashes) as a function of the time t, for $\tau_{eff} = 1$ and 8 fm/c. Initially, the collision rate rises rapidly as particles are created in the hot central zone, but then decreases as the hadron gas expands. The integrated collision rate suggests that there is considerable thermalization of the source, especially for shorter hadronization times, but kinetic pressure will ensure a superposed collective motion of the particles.

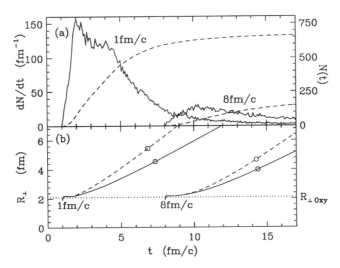

Fig. 1: (a) The collision rate (solid line) and collision number (dashed curve), and (b) the transverse rms radius of the system (solid: all rapidities; dashes: $|y| < 0.5$), as a function time. The dotted line in (b) indicates the initial size, while the open circles correspond the last points of interaction. The indicated times give τ_{eff}.

Fig. 1(b) exhibits the time dependence of transverse rms radius $R_\perp(t) \equiv \langle |\vec{r}_{i\perp}(t)|^2 \rangle^{1/2}$ of the expanding cloud of mesons, for the rapidity interval $|y| < 0.5$ (dashes) and all particles (solid line). Results for two different hadronization times are shown. Also shown in Fig. 1(b) are the transverse rms radii obtained from the points of last interaction. They decrease weakly with increasing τ_{eff}. For both τ_{eff}, the mean fireball freeze–out time is $t \sim 7$ fm/c, at which point $\sim 80\%$ of the asymptotic number of collisions has been reached. The duration of the expansion phase does not change with increasing τ_{eff}, and the expansion of the system closely resembles the behavior of a relativistic hydrodynamic fireball.[8]

4. The phase space distribution at freeze–out

In principle, the above geometric freeze-out information is obtainable from a study of correlation functions. However, in solving the kinetic equation (2) the correlation function contains not only information on the source geometry, but also its collective expansion, and resonance decays. To disentangle these effects, we begin by considering only those pions whose last interaction was an elastic collision. We assume that Coulomb corrections (Gamov factor) have been applied to experimental data.

Eq. (1) may be written[6] in terms of the invariant rate of production of pions in the final state, g :

$$C_2(p_1, p_2) = 1 + \frac{\int d^4x d^4x' \, g(x, P)g(x', P) \, \cos[q(x - x')]}{\int d^4x d^4x' \, g(x, p_1)g(x', p_2)} , \qquad (4)$$

where $q = (p_1 - p_2)$ and $P = (p_1 + p_2)/2$. The g's in the numerator should be evaluated off–shell, but we use on–shell values. For typical temperatures $\beta^{-1} \sim 150$ MeV, and correlation function widths of $|\vec{q}| \sim 100$ MeV/c, the energy shift results in a maximum drop of 11 % in the correlation function. We shall only consider C_2 as a function of the sideways momentum difference, $\vec{q}_{side} \cdot \vec{P}_\perp = 0$ and $\vec{q}_{side} \cdot \hat{z}_{beam} = 0$. Then particle emission times play no role, and one directly obtains the transverse spatial size of the source $R_{side} \equiv \langle [(\vec{x}_i - \vec{X}) \cdot \hat{q}_{side}]^2 \rangle^{1/2}$. Here, $\vec{x}_i$ are the spatial coordinates of pion emission, and $\vec{X}$ are the center of mass coordinates of particles contributing to g in Eq. (4). Note that we use here the one–dimensional average; in a cylindrically symmetric system, $R_\perp$ as used in Fig. 1 is $\sqrt{2} R_{side}$.

The test particle solution of the bosonic Boltzmann equation (2) yields a source function $g(x, p)$ that is too singular to display interference effects, so we smooth it in momentum space :

$$g(x, p) = \sum_{i=1}^{N} \exp(-|\vec{p}^*_{(i)}|^2 / 2\sigma_s^2) \, \delta^{(4)}(x - x_i) , \qquad (5)$$

where x_i is the test–pion's last point of interaction, and $\vec{p}^*_{(i)}$ is the spatial component of the momentum p in the rest–frame of test–pion i at emission. The choice of σ_s is important, as illustrated in Fig. 2. As will become clear, the σ_s–dependence of R_{side}^{cor} results from the destruction of dynamical momentum–coordinate correlation for values of σ_s that are too large. For small smearing, the size extraction becomes increasingly uncertain; of course, we have made use of the cylindrical and reflection symmetries to increase statistics. This allows one to choose sufficiently small σ_s for each $P_\perp$.

The solid line in Fig. 3(a) shows the source radius inferred from the correlation function versus $|\vec{P}_\perp|$, with fixed $\hat{P}_\perp$, $P_\parallel = 0$, and $\tau_{eff} = 1$ fm/c. One sees that $R_{side}^{cor} \sim 1.4$ fm is practically independent of the transverse momentum. This curve agrees with values R_{side} obtained by computing directly the sideways size of the last interaction coordinates, after the particle positions have been rotated by the angle required to align the particle transverse momentum with the $\hat{P}_\perp$ direction (dashed line). The parametrization (5) thus seems reliable in predicting the extension of the emitting source, despite rather involved expansion dynamics.

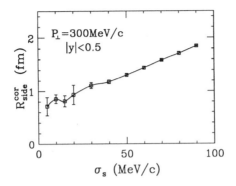

Fig. 2: The sideways source radius as extracted from the correlation function versus smearing width.

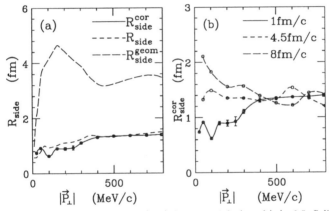

Fig. 3: (a) The sideways source size versus $|P_\perp|$, for $\tau_{eff} = 1$ fm/c and $|y|< 0.5$. Solid: correlation function measurements; short dashes: actual transverse size from all pairs with a fixed $P_\perp$-direction; long dashes: full geometric source size. (b) Correlation source sizes for different τ_{eff}.

The correlation measurement of our non–chaotic source provides only restricted access to the full transverse geometry, R_{side}^{geom}, illustrated by the long dashes.[6] Note that this curve is also the result of the calculation of the correlation function, if one artificially randomizes the direction of the transverse momenta of the test–particles. The structure in R_{side}^{geom} at intermediate $|\vec{P}_\perp|$ results because fast particles leave the reaction zone at an early stage of the evolution, while slower particles suffer many collisions as a result of large Bose enhancement factors, and are emitted at a later stage. This structure is washed out in R_{side}^{cor}.

Fig. 3(b) shows the correlation radii for the three hadronization times. The correlation and geometric source sizes still differ by a factor $\gtrsim 2$ at larger τ_{eff}, but the discrepancy becomes smaller for larger τ_{eff} (see also Fig. 1) at low momenta. This results because at lower densities (long τ_{eff}) collisions are less effective in correlating momenta with coordinates, particularly where boson $\bar{f}$-factors in Eq. (2) play a role.

The discrepancy in Fig. 3(a) is graphically illustrated in Fig. 4. Shown are the transverse positions of mid-rapidity particles that contribute to the correlation function (4) (upper panel: $|\vec{P}_{\perp}| < 50$ MeV/c; lower panel: $|\vec{P}_{\perp}| \sim 150$ MeV/c). Low momentum particles leave the reaction zone at a relatively late stage, and are located centrally with a small sideways dimension. The smaller circle indicates the transverse rms radius $\sqrt{2} \times R_{side}^{cor}$. For pairs with transverse momentum about the median value, (lower panel), the shift in the $\hat{P}_{\perp}$-, i.e., $\hat{x}$-direction is largest. For comparison, the dimensions of the corresponding quasi-chaotic sources, representing the actual source geometries, are shown as the larger circles.

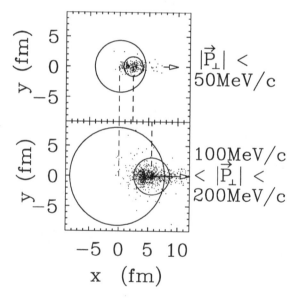

Fig. 4: Particle location in the transverse plane, for $\vec{P}_{\perp}$ pointing in the $\hat{x}$-direction, and $|y| < 0.5$. The range of $|\vec{P}_{\perp}|$ is indicated for each panel. The smaller circles in each part have a radius equal to the transverse rms radius of particles emitted in the $\hat{x}$-direction, while the larger circles show the full source size in each case. This illustration is for $\tau_{eff} = 1$ fm/c.

Freeze-out thus reflects a complicated interplay between thermalization and expansion inherent the bosonic Boltzmann equation. The values of R_{side}^{geom} and R_{side}^{cor} agree only for $\vec{P}_{\perp} = 0$, since there is no preferred direction then. For large transverse momenta the reduction in observed radii is constant, while at intermediate $|\vec{P}_{\perp}|$

the source structure is completely washed out. For a hydrodynamically expanding spherical shell[6] one expects $R_{\text{side}}^{\text{cor}}$ to decrease for increasing transverse momenta. Here we see that the dynamics lead to a less simple, model dependent relation between the observed and geometric sizes; a flat function $R_{\text{side}}^{\text{cor}}(|\vec{P}_\perp|)$ can also point to an expanding system with an involved space–time structure.

Let us briefly examine the role of the resonances. In Fig. 5, we show the pion correlation function for $\tau_{eff} = 1$ fm/c, at a fixed $|\vec{P}| = 250$ MeV/c, as a function of the sideways momentum difference of the pair. The dashed line indicates C_2 for pions that emerge from the source after an elastic collision or a ρ–decay. The solid line is obtained by taking into account the decays of all resonances within the calculation. Roughly 10% of the pions stem from long–lived η mesons; these are not correlated, and result in a reduction of the intercept of C_2. A hint of a peak structure emerges for $q_{\text{side}} \lesssim 10$ MeV/c, and may be attributed to the decay of ω–mesons with a lifetime of ~ 20 fm/c. Some further reduction of the C_2–intercept may result from an averaging of inclusive yields; here, we have chosen q_{out} and $q_\parallel$ to be zero.

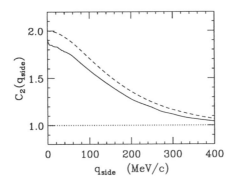

Fig. 5: Sideways correlation function at $P_\perp = 250$ MeV/c, and $P_\parallel = 0$. The solid line is for elastic and ρ–decay pions only, the dashed line for all pions.

The NA35 collaboration reports[7] $200A \cdot$ GeV ^{16}O + Au sizes of $R_{\text{side}}^{\text{cor}} \sim \sqrt{24.1}$ fm $\sim$ 2.9 fm $\sim 2R_{\text{side Oxy}}$, approximately a factor 2 times greater than what is found here. This suggests that the present model produces a too much collectivity, or is missing vital physics. Larger initial proper times increase the calculated radii, but not enough to account for the discrepancy (see Fig. 3(b)).

5. Conclusion

Using pion interferometry, we have studied the spatial freeze–out surface of a gas of π, η, ρ, and ω mesons, modelling mid–rapidity central ^{16}O + Au collisions at 200 GeV/nucleon. We point out the importance of smoothing the test–particle solution of the Boltzmann equation correctly. The resulting source has a non–trivial space–time structure caused by (a) the overall expansion due the thermal pressure; (b) thermalization processes via in-medium interactions with correct Bose statistics; and (c) resonance decays. The momentum–coordinate correlations implied by (a) mean that actual source sizes are reduced by a factor of 2–4 over the size of the emission region. The dynamics of the expansion implies that this region does not shrink as a function of the total momentum of the pair, as one would expect from a hydrodynamic scenario.[6] Freeze–out sizes deduced from experimental two-particle correlations – which are approximately twice the initial nuclear radii[7] – may be expected to underestimate considerably the full source dimensions.

Our model does not reproduce the width of a measured two-pion correlation function,[7] suggesting that a scenario with only pions and heavier mesons is missing some vital physics. Larger initial proper times reduce this discrepancy, since they reduce the degree of collectivity at freeze–out. The explanation of the low transverse-momentum pions in terms of a Bose enhancement generated during expansion – which requires high initial phase-space densities – must be re–examined in light of this evidence. A recent RQMD calculation[9] does yield widths that reproduce the data, pointing to the possible importance of baryonic degrees of freedom.

G.W. and P.D. acknowledge support by US DOE grant DE-FG02-94ER40831 and NSF PHY–9017077, respectively.

[a]FZ-Rossendorf, Dresden, Germany; [b]Department of Physics, University of Washington, Seattle, USA; [c]NSCL, Michigan State University, East Lansing, USA; and [d]Niels Bohr Institute, Copenhagen, Denmark.

6. References

[1] D.H. Boal, B.K. Jennings and C.K. Gelbke, *Rev. Mod. Phys.* **62**, 553 (1990).

[2] G.M. Welke and G.F. Bertsch, *Phys. Rev. C* **45**, 1403 (1992).

[3] H.W. Barz, P. Danielewicz, H. Schulz, and G. Welke, *Phys. Letts.* **B287**, 40 (1992).

[4] Z. Aouissat, G. Chanfray, P. Schuck and G. Welke, *Z. Phys. A* **340**, 347 (1991); H.W. Barz, G.F. Bertsch, P. Danielewicz and H. Schulz, *Phys. Letts.* **B275**, 19 (1992).

[5] The NA35 collaboration: H. Ströbele *et al.*, *Z. Phys. C* **38**, 89 (1988).

[6] S. Pratt, *Phys. Rev. Lett.* **53**, 1219 (1984).

[7] The NA35 collaboration: D. Ferenc *et al.*, *Nucl. Phys. A* **544**, 531c (1992).

[8] H.W. Barz, B. Friman, J. Knoll and H. Schulz, *Nucl. Phys. A* **519**, 831 (1990).

[9] The NA44 collaboration: J.P. Sullivan *et al.*, *Phys. Rev. Lett.* **70**, 3000 (1993).

Antiproton production at 0 degrees for Si + A and Au + A Collisions at the AGS

R. Debbe (for the E878 collaboration)
Brookhaven National Laboratory
Upton NY 11973, USA

D. Beavis[1], M. Bennett[2], J.B. Carroll[3], J. Chiba[4], A. Chikanian[2], H. Crawford[5], M. Cronqvist[5], Y. Dardene[5], R. Debbe[1], T. Doke[6], J. Engelage[5], L. Greiner[5], R.S. Hayano[7], T.J. Hallman[3], H.H. Heckman[8], T. Kashiwagi[6], J. Kikuchi[6], B.S. Kumar[2], C. Kuo[5], P.J. Lindstrom[8], J.W. Mitchell[9], J. Nagle[2], S. Nagamiya[10], K. Pope[2], P. Stankus[10], K.H. Tanaka[4], R.C. Welsh[11], and W. Zhan[10]

1. Brookhaven National Laboratory, Upton, NY
2. Yale University, A.W. Wright Nuclear Structure Laboratory, New Haven, CT 06511
3. University of California at Los Angeles, Los Angeles, CA
4. KEK, Japan
5. University of California Space Sciences Laboratory, Berkeley, CA
6. Waseda, Japan
7. University of Tokyo
8. Lawrence Berkeley Laboratory, Berkeley, CA
9. USRA/GSFC
10. Nevis Laboratory, Columbia University, Irvington, NY
11. Johns Hopkins University, Baltimore, MD

ABSTRACT

We present results on antiproton production obtained in the AGS experiments E858 and preliminary results of E878. The yields of antiprotons in Si + A collisions are shown to scale with the number of first collisions. The rapidity distributions for all targets (Au, Cu and Al) and both beams (Si at 14.6 GeV/c and Au at 11. GeV/c) have gaussian shapes peaking at y_{NN} and with similar standard deviations. From E878 we report a difference in the shape of the antiproton rapidity distributions obtained from two samples of the data populated with central and peripheral events respectively. In Au induced reactions the A dependence of the antiproton yields is small.

1. Introduction

The AGS experiment, E858, and its subsequent upgrade, E878, were proposed as high rate studies of the production of long-lived negative secondaries in reactions of heavy ion beams and nuclear targets. Both experiments were implemented in existing AGS beam lines, transforming them into double focusing spectrometers that measured negative or positive secondaries at zero degrees in the laboratory frame within a small momentum bite.

The beam fluence exceded 10^8 ions/spill for the Si beam runs. A redundant system of two ion-chambers upstream of the target, and a secondary particle telescope (SPT) made of three scintillator paddles aligned so as to view the target within a small solid angle, were used to monitor the beam intensity. For E878, several improvements were implemented at the target area; a multiwire proportional chamber upstream of the target to monitor the position of the beam as it impinges on the target. A quartz fiber hodoscope was also placed upstream and close to the target to provide another measurement of the beam position on the target, and finally, a second SPT set was placed upstream of the target making the same angle with the beam as the first one.

Figure 1 is a schematic of the instrumented AGS A3 line used for the E878 experiment. The E858 was implemented in the A1 line but the detector setup is similar. The spectrometer has $\Delta\Omega \approx 200\mu sr$ centered around zero degrees and a $\pm 3\%$ momentum bite.

Fig. 1: Experimental layout showing the instrumented A3 AGS beam line.

The best technique for particle identification in such a long spectrometer is the measure of time of flight. Both experiments had redundant systems to measure the time of flight. The first TOF station was placed at the first focus F1 30 meters downstream of the target and just after a collimator that defines the momemtum bite of the spectrometer. This first TOF counter was a scintillator paddle viewed by two photomultipliers. A second focus is formed 25 meters downstream of the first one, and there, another time of flight counter made of a single scintillator paddle viewed by two photomultipliers was installed. Two other TOF counters were placed close to the second focus in order to make redundant measurements. During E858 these additional time of flight stations were made of a single scintillator paddle viewed by two PMT's, and later in E878, they were upgraded to five scintillator paddles covering the same area in order to remove ambiguities in cases where more than one particle was present in the line.

All the PMT's from these TOF counters had their pulse height as well as their time information recorded, providing information about the time of flight as well as the charge of the detected particle. From that information, and the value of the tuned rigidity of the beam line, the mass of the particles can be infered. Figure 2 shows the mass distribution at 6.4 GeV/c measured during the E858 run. The entries to this histogram are tracks in the spectrometer that had consistent time of flight measured with two pairs of TOF stations and information from the tracking system indicating that these particles were produced in the target and were transported by the beam line all the way to the second focus region.

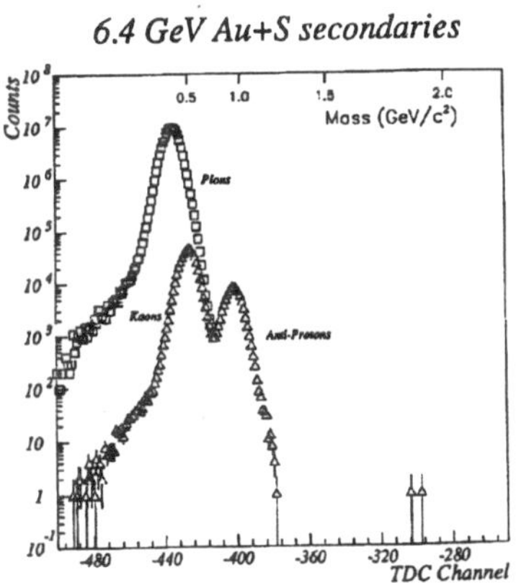

6.4 GeV Au+S secondaries

Fig. 2: TOF spectra for the negative secondaries at 6.4 GeV/c produced in 14.6 AGeV Si+Au collisions during the E858 run. Information from the Cerenkov counters was used to identify the pions

2. Data Analysis

A complete description of the data reduction can be found in ref. 1 and ref. 2. The total number of triggers recorded during the E858 run in the Spring of 1989 was 10^6 events, and taking into acount the scaledown pion signal, that represents some $5 \times 10^8 \pi^-$, a million K^-, some 3×10^5 antiprotons, and two antideuteron candidates.

A prominent result from E858 were the two antideuteron candidates at p=6.1 GeV/c ref. 2. This translates into a measured invariant cross section of 40 nb/GeV^2. In the same reference an attempt was made to estimate the production rate of antideuterons using a coalescence model from ref. 3. The measured antiproton cross sections were used in that calculation in order to be consistent, and the result at p=6.1 GeV/c turned out to be $4 \times 10^{-4} mb/GeV^2$, which is a factor of 10 larger than the measured value.

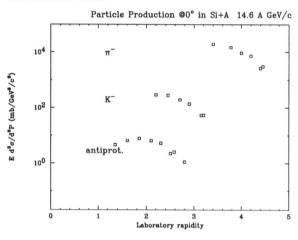

Fig. 3: Invariant cross sections for negative pions, kaons and antiprotons as functions of the laboratory rapidity. The negative secondaries were produced in 14.6 AGeV Si+Au collisions during the E858 run.

Figure 3 shows a sample of the invariant cross sections measured in E858, These cross sections are shown as functions of the laboratory rapidity, for π^-, K^- and antiprotons produced with the silicon beam incident on a gold target. As can be seen the acceptance of this experiment is quite forward for pions and kaons, but the measured antiprotons produced at midrapidity can be compared to the other heavy ion experiments at the AGS. Such comparison was made and is summarized in the following table where the quantity compared is the number of antiprotons per event at y=1.6 and $p_T = 0$:

Comparison to other AGS experiments			
E814	E859 Central	Estimated E859 Min. Bias.	E858
0.0035 ± 0.0006	0.0068 ± 0.002	0.0027 ± 0.0008	0.0027 ± 0.0003

The number of antiprotons per collision from E814 was obtained from ref. 4 after dividing their result by the total Si + Au cross section equal to $\sigma = 3750$ mb. The comparison to the E859 used data from ref. 6 and was made by extrapolating the invariant cross sections as functions of the transverse mass at y=1.6 to $p_T = 0$. These cross sections were extracted from a sample of central events. In order to make the comparison to minimum biased events from E858 the result of the extrapolation of E859 data was multiplied by 0.4 This conversion factor of 0.4 was extracted from ref. 5 as the ratio of antiproton yield in minimum biased events to the one from central events. More recent results from E859 indicate that this ratio can be equal to 0.5. And finally the E858 numbers were extracted from ref. 1 As can be seen in the table above, all three experiments agree within the error of their measurements.

The invariant cross sections as functions of the laboratory rapidity corresponding to each of the targets (Au, Al and Cu) are well described by gaussian shapes (ref. 7). For all three targets the distributions peak at the nucleon-nucleon center of mass rapidity, y_{NN}, and they all have standard deviations around $\sigma = 0.51 \pm 0.03$. These distributions are much wider than a simple proton-antiproton pair production from nucleon-nucleon interactions. A simple phase space calculation for this mechanism produces distributions with a width of some $\sigma \approx 0.35$. It was investigated if Fermi motion could account for the widening of the distribution, and found that, to do so, the Fermi momentum distribution would have to have a Fermi level P_F greater than 500 MeV/c (ref. 7).

The yield of antiprotons at $p_T = 0$, (calculated in this case as the area under the gaussian fit to the dN/dy distributions obtained with different targets) scales with the number of first collisions in an unbiased Si + A reaction.

3. The E878 Au + Au data

E878 has an integrated flux on target of some 5×10^{12} ions. The number of detected antiprotons was 1.2×10^5. Among the positive secondaries the sample of heavy particles is considerable; 2.2×10^5 deuterons, 6×10^3 tritons, 6×10^3 ^3He, and some 300 alpha particles.

Besides the data collected with the Silicon beam, E878 has a good data sample obtained with the gold beam at 11. GeV/c during the fall of 1993. In what follows, some preliminary results are presented for proton distributions collected during the portion of the run when the magnets of the beam line had their polarities set so as to transport positive secondaries produced at the target. The sample of positive particles ranges from pions, kaons and protons to heavier fragments like ^3He and alpha particles. A more detailed presentation of antiproton distributions is also included.

Figure 4 shows the invariant cross section of protons as a function of the laboratory rapidity. The sample of data used to extract this cross section does not have any condition on the centrality of the reaction and should be considered as minimum bias. In the same graph a cross section calculated from simulated protons using the ARC model (see ref. 8) is presented. The impact parameter of the simulated events ranges from zero to some 20 fm making it a minimum biased sample. As can be seen in the figure, the model agrees well with the measured data.

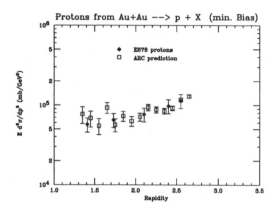

Fig. 4: Invariant cross section of protons produced by the Au beam at 11. GeV/c incident on a Au target. This cross section is shown as a function of the laboratory rapidity. No conditions on the centrality of the event were imposed on the data, these are minimum biased events

Figure 5 shows the minimum biased invariant cross section for antiprotons produced by the gold beam incident on a gold target as function of the laboratory rapidity. The statistical errors are smaller than the symbols used to display the results. The E878 collaboration is presently analysing these data and the present results have an assigned systematic error of some 15 to 20 % because there are still some uncertainties in the normalization of each measurement; the acceptance of beam line has been fixed as a constant for all the used tunes, the effects of multiple scattering, beam energy loss and absorption of secondaries in the target continue to be worked. A more thorough analysis will certainly smooth the results.

A fit was made to the measured points using a gaussian distribution and the result gave a peak position at 1.56 ± 0.1 units of rapidity or practically the NN center of mass rapidity. The width of this distribution obtained from the fit is equal to 0.5 ± 0.1 very similar to the width of the rapidity distributions of antiprotons in Si + A measured during the E858 runs. Eventhough these results are similar, it has to be pointed out that the available energy in the nucleon-nucleon center of mass is different for these systems. (E878 has reported the same measurement obtained during a short run in 1992 ref. 11. The wider rapidity distribution of antiprotons obtained from that run may be related to a higher momentum of the beam (11.7 GeV/c)) In the same figure the measured points were reflected around y_{NN} because the Au + Au is a symmetric system. The reflected points are displayed as open symbols, and it can be seen that wherever there is an overlap the discrepancy falls within the assigned systematic error.

An attempt was made to characterize the A dependence of the antiproton yields in the Au + A reactions. In order to reduce the effects of corrections not yet completed, ratios of yields are presented as functions of the laboratory rapidity. The cross sections

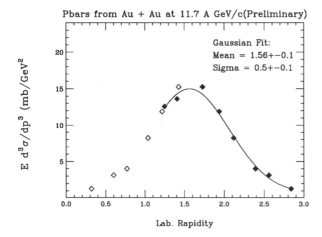

Fig. 5: Minimum biased invariant cross section for antiprotons produced in Au + Au reactions at 11. GeV/c.

used to calculate these ratios did not have a correction for beam energy loss in the targets. Figure 6 shows the following ratios: $\frac{Au}{Al}$ and $\frac{Au}{Cu}$. Within most of the range of measured rapidities, these two ratios are very similar and equal to roughly 1.2 indicating a weak dependence in the A of the target.

Among the new detectors added to E878, the so called Multiplicity Detector constitutes a major improvement because it characterizes the centrality of the event. This counter was designed to detect gamma rays produced by the decay of neutral pions. Each cell of this counter consists of 1/4 of an inch of Pb as converter, the same thickness of quartz radiator and finally high rate, square face photomultipliers (Hamamatsu 2248). The thickness of the converter was chosen so that in average each count correspond to one neutral pion. These detectors were arranged in two rings of 14 modules each making 40 and 60 degrees with the beam. The time information from each module is recorder in a multihit TDC system and the extracted information is the number of modules that have hits with time related to events recorded in the main spectrometer. This detector is also sensitive to charged particles with $\beta > 0.68$. Monte Carlo simulations indicate that this contribution accounts for 30% of the total multiplicity.

Two sets of events can be defined by requiring that the number of modules with hits in the Multiplicity Detector be greater than 20 (this set will be called the central events set) and less than 5 (these events are called peripheral)

It is estimated that the central events set represents a 10 % of the total cross section and the peripheral ones correspond to the 30 % most peripheral collisions.

Figure 7 shows the invariant yield of antiprotons from the Au + Au collisions

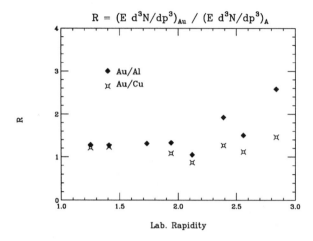

$$R = (E \ d^3N/dp^3)_{Au} \ / \ (E \ d^3N/dp^3)_A$$

Lab. Rapidity

Fig. 6: Ratios of yields of antiproton per event as a function of the laboratory rapidity in Au + A collisions at 11. GeV/c. The statistical errors are smaller that the display symbols.

as functions of the laboratory rapidity. The upper panel shows this distribution for the set of central events, and the lower panel the one corresponding to the peripheral events. It should be made clear that the units of the y axis in these plots are arbitrary because the analysis of these data is still in its preliminary stages. The remarkable feature in these distributions is the fact that they have different widths. The peripheral events, with a $\sigma = 0.35$ seem to be indicating a simple proton-antiproton pair production mechanism, and the central events point to some additional process that would have widened the distribution. Some models of antiproton production that are in wide use in the field, like RQMD ref. 9 and ARC ref. 10, contend that baryon resonances as intermediate states or the screening of the antiprotons by copious number of pions that render annihilation less probable, are important in shaping the observed antiproton spectra.

4. Summary

AGS experiments E858 and E878 were designed as high rate measurements of antiproton production in heavy ion reactions, as well as a search for antideuterons and stable negatively charged exotic particles. The present work reports the following results obtained in the E858 experiment: The yield of antiprotons scales with the number of first collisions and the rapidity dsitributions of antiprotons with $p_T = 0$ for all targets used have the same gaussian shape peaking at y_{NN} and having a standard deviation of $\sigma = 0.51 \pm 0.03$. An agreement between E858 and the other two AGS experiments that measured antiprotons is reported.

Preliminary results from experiment E878 were presented. A very interesting

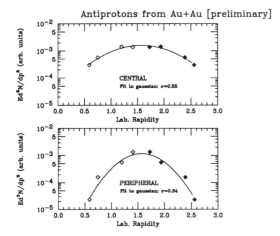

Fig. 7: Invariant yields of antiprotons per event as functions of rapidity. The upper panel correspond to the central events selected by a high number of E_T modules with hits. The lower panel is the same distribution but this time, it was extracted from the sample of peripheral events. The statistical are smaller than the symbols used to display tha data.

difference in the shape of the rapidity distribution of antiprotons appears between samples of events defined as central and peripheral. These samples of data are defined by the multiplicity of hits measured with the new Multiplicity Detector.

5. References

[1] P. Stankus Production of Antinuclei and Rare Particles in Si+A Collisions at 14.6 GeV/c at the BNL-AGS *PhD Thesis, Columbia University*

[2] Aoki. *et al.*, Measuremnt at 0 degrees of Negatively Charged Particles and Antin-uclei Produced in Collision of 14.6 GeV/c Si on Al, Cu, and Au targets. *Phys. Rev. Lett.* **69**, 2345 (1992).

[3] C. Dover, U. Heinz, E. Schnedermann, and J Zimanyi, *Phys. Rev. C* **44**, 1636 (1991).

[4] Barrette *et al.*, *Phys. Rev. Lett.* **70**, 1763 (1993).

[5] T. Abbot *et al.*, *Phys. Lett.* **271**, 447 (1991).

[6] P. Rothschild Antiproton production in Central Si+Au Collisions at 14.6 GeV/c for Experiment E859 (the E802 Collaboration) *Proceedings of HIPAGS ;93* , 153 (1993).

[7] P. Stankus Antiproton production in Si+A Collisions a detailed look at the results of BNL-E858 for Experiment the E858 Collaboration *Proceedings of HIPAGS ;93* , 134 (1993).

[8] Y. Pang, T. J. Schlagel and S. H. Kahana, *Phys. Rev. Lett.* **68**, 2743 (1992).

[9] A. Jahns *et al.*, Antiproton production and absorption in Nucleus-Nucleus collisions. *Proceedings of HIPAGS ;93* , 120 (1993).

[10] S. H. Kahana, Y. Pang and T. Schlagel Physics at the AGS with A Relativistic
Cascade *Proceedings of HIPAGS ;93* , 263 (1993).

[11] B. Shiva Kumar *et al.*, Rapidity Distributions of Antiprotons in Si+A and Au+A
Collisions. *Proceedings of the Quark Matter '93, Borlange, Sweden, June 1993* ,
To appear in Nucl. Phys. A.

FREEZE-OUT and HADRONIZATION

L. P. CSERNAI, and, T. CSÖRGŐ

Section for Theoretical Physics, Physics Department,
University of Bergen, Allegaten 55, N-5007 Bergen, Norway
and KFKI RMKI, H-1525 Budapest 114, POB 49, Hungary

ABSTRACT

Time-scales of first-order deconfinement phase transition in high energy heavy ion collisions at RHIC and LHC energies are considered when the system must supercool below T_c before the nucleation of hadronic bubbles becomes sufficiently rapid to overcome the expansion rate. It is shown that the expected time-scales of high energy heavy ion reactions are sufficiently short to prevent the reheating of the system via homogeneous nucleation near T_c. If quark-gluon plasma is produced in these collisions, it may have to hadronize from a supercooled state and the hadrons produced during rehadronization may freeze-out almost immediately.

1. Introduction

In the search for Quark-gluon plasma (QGP)[1] the most frequently used simple fluid dynamical model is the Bjorken model. [2] According to this model the colliding nuclei pass through each other at high energies, leaving behind a highly excited volume filled with gluons and quarks, which then expands mainly along the beam axis. Numerical simulations confirmed this picture predicting a nearly equilibrated and baryon-free plasma of about 150 fm^3 with an initial temperature of 300-350 MeV.[3]

The dynamics of the rehadronization of the expanding and cooling plasma phase is very sensitive to the formation rate of hadronic bubbles inside the plasma. The characteristic nucleation time as a function of the temperature was found to be of the order of 100 fm/c. [4] The rate for the nucleation of the hadronic phase out of the QGP can be written as

$$I = I_0 \exp(-\Delta F_*/T), \qquad (1)$$

where ΔF_* is the change in the free energy of the system with the formation of a critical size hadronic bubble, T is the temperature and I_0 is the prefactor. The prefactor has recently been calculated in a coarse-grained effective field theory approximation to QCD.[5]

The nucleation time and the time needed to complete the transition,[6] are different because supercritical bubbles grow during hadronization. In ref.[4] the temperature as a function of (proper)time was presented based on the integration of coupled dynamical equations describing bubble formation and growth in a longitudinally expanding

Bjorken tube. According to these calculations the matter continues to cool below T_c until T falls to about 0.95 T_c when noticeable nucleation begins. When the temperature has fallen to a "bottom" temperature, $T_b = 0.8T_c$, bubble formation and growth is sufficient to begin the reheating of the system.

Detailed calculations including dilution factor for the bubble formation, spherical expansion, bubble fusion or varying the values for the surface tension do not change the qualitative behaviour of the rehadronization process. According to the calculations in,[7] the time-scales become somewhat shorter due to the fusion of the bubbles which increases the speed of the transition. Thus bubble fusion brings the temperature versus time curve closer to the Maxwell idealization. A spherically expanding system cools faster than a longitudinally expanding one.

2. Freeze-out shown by experiments

Present experiments indicate early freeze-out: i) HBT results, ii) strange antibaryon enhancement, iii) high effective temperatures and iv) unchanged hadronic masses.

We may estimate the time-scales available for the rehadronization process at RHIC and LHC using data taken at present energies, and extrapolating them to higher energies.

i) The rapidity density for CERN SPS Pb+Pb collisions increases by about a factor of 3.5 when compared to the $S + U$ reactions at the same energies. Extrapolating this finding, at RHIC Au+Au reactions the rapidity density increases by a factor of about 7, at LHC Pb+Pb collisions by a factor of 13, when compared to the $S + U$ reactions at CERN SPS. This in turn implies scaling factors of 1.51, 1.91 and 2.35 in estimating the increase of the freeze-out proper times obtained from HBT analysis. As a result we obtain for the freeze-out times 6-10 fm/c at CERN SPS with Pb+Pb, 8-13 fm/c at RHIC Au+Au and 11-16 fm/c at LHC Pb+Pb collisions.

Comparing the time-scales of the QCD phase transition with the time-scales obtained from extrapolating present interferometry data to RHIC and LHC, we observed very interesting coincidences. If one starts with an initial state as suggested by the parton cascade simulations in ref.[3] the critical temperature is reached by 3 fm/c after the collision. By 10 fm/c time, which is about the freeze-out time according to the interferometry estimate, the system is far from being completely rehadronized, according to the calculations in.[4,7] At this time, the system is still very close to the bottom of the temperature curve.

ii) The idea that the quark-gluon plasma has to hadronize suddenly in a deeply supercooled state leads to the consequence that the strange particle composition and especially the production rate of strange antibaryons as suggested by [8,9,10] could become a clean signature of the quark-gluon plasma formation at RHIC and LHC energies as well as at the present CERN SPS energy. The WA85 collaboration found large production rates of strange antibaryons at CERN SPS S+W interactions. [9,10] The ratio for Ξ^-/Λ observed by WA85 was found to be compatible with those from other interactions. However, the ratio $\overline{\Xi}^-/\overline{\Lambda}$ was found to be about five times greater than those obtained by the AFS collaboration, corresponding to a two standard deviation effect. Rafelski was able to reproduce this enhancement only by assuming sudden rehadronization from QGP near equilibrium, which would not change the strangeness abundance.[8] Really, the long time-scale of the nucleation compared to the short time-scales of the pion freeze-out times at CERN SPS energies supports the coincidence of the maximal supercooling of the QGP with freeze-out time of 4.5-6.5 fm/c, leaving very short time for the strange antibaryons for reinteraction in the hadronic gas already at CERN SPS energies. Spacelike detonations and spacelike deflagrations from a supercooled baryon rich quark-gluon plasma were related to strangeness enhancement at CERN SPS energies in ref.[11]

iii) During such a sudden breakup the latent heat might be released as high kinetic energy of the hadrons in a timelike deflagration. [15] This is in qualitative agreement with the observation that the multistrange antibaryons observed by the WA85 collaboration are all at transverse momenta above 1.2 GeV/c, and show an effective $T_{slope} >$ 200 MeV. At lower energies it is known that if one produces a hot fireball where resonances (deltas) are in thermal equilibrium, after the freeze-out and the resonance decays the effective temperature for the protons (baryons) will be larger than those of pions. [17] Further, the effective slope of the baryons will be about 10% lower, than the freeze-out temperature. Thus we may expect that the slope parameters of the multistrange antibaryons provide more information about the freeze-out temperature than those of the pions (which at the present CERN experiments come out with more moderate slope parameters).

iv) Hadronic masses are expected to decrease considerably [16] in dense and hot hadronic matter. Nevertheless, in the dilepton spectra the observed masses of hadronic resonances (e.g. ϕ) were identical to their free masses in heavy ion reactions at CERN SPS energies. This also can be attributed to simultaneous hadronization and freeze-out, where the medium effects are ceased to exist, when hadrons are formed.

Thus from trends in interferometry data the *freeze-out* time scale is short enough to prevent reheating and the completion of the *rehadronization* of the quark-gluon plasma through bubble formation in the supercooled state. This in turn implies that other mechanisms must dominate the final stages of the hadronization.

3. Freeze-out and hadronization dynamics

Using the bag constant is $B^{1/4} = 235$ MeV, the critical temperature is given by $T_c = 169$ MeV and the pressure of the supercooled QGP vanishes at $T = 0.98T_c$ already. According to the above considerations and ref.[7] the temperature of the system in the supercooled phase reaches $T = 0.7 - 0.9T_c \simeq 120 - 150$ MeV. One can observe, that at such a low temperatures the pressure of the QGP phase takes large negative values in the bag model. Systems with negative pressure are *mechanically unstable*, either they don't fill the available volume or they spontaneously clusterize.

In the abstract many dimensional coordinate space corresponding to possible instabilities, after crossing the borderline of stability on the phase diagram there is always one channel which opens first. This usually corresponds to a spherical configuration of instability. A deeper penetration into the supercooled region may lead to the opening of other channels of instability. These other channels may include string-like, cylindrical and later layered instabilities or the spinodal decomposition. Thus the calculated nucleation rate gives an accurate description of the initial hadronization at small supercooling. Furthermore, the calculated nucleation rate calculated was dominated by thermal near - equilibrium processes and by the thermal interaction of the neighboring particles, or thermal damping. This is a valid assumption when the critical temperature is reached, but after further expansion and a considerable supercooling, e.g. 30% or more, the matter is not so dense any more and the collective near-equilibrium interaction with the surrounding matter may not be the dominant process. Instead of these quantum mechanical processes including very few particles may dominate the transition.

This mechanical instability of the QGP phase below 0.98 T_c on one hand and the typical 100 fm/c nucleation times on the other hand are the basic reasons for the sudden rehadronization which we propose. It is understood, that the expansion in an ultrarelativistic heavy ion collision is so fast, that the temperature drops below T_c by 20-30 % before nucleation becomes efficient enough to start reheating the system. By that time, the QGP phase is far in the mechanically unstable region. The transition proceeds from a mechanically unstable phase to a mechanically stable and thermodynamically (meta)stable phase, the (superheated) hadron gas state.

Let us consider the sudden freeze out from supercooled QGP.[12,13,14] The baryon free case was discussed in[13] in detail including the possibility of converting latent heat to final kinetic energy locally and instantly [in timelike deflagration], while[14] did not include this possibility. Although in ref.[13] it is argued that a superheated hadronic state is not realizable as final state [because one has to pass the mixed state on the way], this restriction does not apply for sudden freeze-out which we consider as a discontinuity across a hypersurface with normal Λ^μ ($\Lambda^\mu \Lambda_\mu = +1$). We can satisfy the energy and momentum conservation across this discontinuity expressed via the energy momentum tensors of the two phases, $T^{\mu\nu}$, $(T_H^{\mu\nu} - T_Q^{\mu\nu})\Lambda_\nu = 0$, with entropy production in the $Q \rightarrow H$ process.

Relativistic timelike deflagrations are governed [13] by the Taub adiabat,

$$\frac{p_1 - p_0}{X_1 - X_0} = \frac{\omega_1 X_1 - \omega_0 X_0}{X_1^2 - X_0^2}, \tag{2}$$

the Rayleigh-line

$$\frac{p_1 - p_0}{X_1 - X_0} = \omega_0 \tag{3}$$

and the Poisson-adiabat,

$$\frac{\sigma_1^2}{\omega_1} = \frac{R^2}{X_1} \frac{\sigma_0^2}{\omega_0}, \tag{4}$$

where $\omega_i = \varepsilon_i + p_i$ denotes the enthalpy density and X_i is defined as $X_i = \omega_i/\omega_0$. The index 0 refers to the quantity before the timelike deflagration, while 1 refers to after deflagration. If we suppose that the flow will be given by a scaling Bjorken-expansion before and after the timelike deflagration, one may simplify the equations governing the relativistic timelike shocks, introduced in ref.[13]. For the scaling 1d expansion the Taub adiabat reduces to the equality of the energy densities on the two sides of the timelike hypersurface

$$\varepsilon_1 = \varepsilon_0, \tag{5}$$

the Rayleigh-line becomes an identity and the Poisson-adiabat simplifies to the requirement that the entropy density should not decrease during the transition

$$R = \frac{\sigma_1}{\sigma_0} \geq 1. \tag{6}$$

If we start the timelike deflagration from a 30% supercooled state the initial state is a mixture including already 15-25% hadronic phase. Indicating the volume fraction of hadrons in the initial state by h, the initial energy density is given by $\varepsilon_0(T_0) = h\varepsilon_H(T_0) + (1 - h)\varepsilon_Q(T_0)$ and the expression for the entropy densities is similar. In the bag model, the energy densities are given as $\varepsilon_Q = 3\,a_Q\,T_Q^4 + B$, $\varepsilon_H = 3\,a_H\,T_H^4$, $\sigma_Q = 4\,a_Q\,T_Q^3$, $\sigma_H = 4\,a_H\,T_H^3$ with coefficients $a_Q = (16 + 21n_F/2)\pi^2/90$ and $a_H = 3\pi^2/90$. The quantity $r = a_Q/a_H$ gives the ratio of the degrees of freedom of the phases. The temperatures of the initial and final state can be determined from the Taub and Poisson adiabats as

$$T_1 = T_H = T_c \left[\frac{x - 1}{3\left(1 - (R^4 x)^{-1/3}\right)} \right]^{1/4},$$

$$T_0 = T_Q = T_c \left[\frac{x - 1}{3\left(1 - (R^4 x)^{-1/3}\right)} \right]^{1/4} (xR)^{-1/3}, \tag{7}$$

where the ratio of the effective number of degrees of freedom is given by

$$x = h + r\,(1 - h). \tag{8}$$

These equations provide a range for possible values of T_H and T_Q for a given initial hadronic fraction h. These bounds are given as

$$\left[\frac{x - 1}{3}\right]^{1/4} \leq \frac{T_H}{T_c} \leq \left[\frac{x - 1}{3(1 - x^{-1/3})}\right]^{1/4},$$

$$0 \leq \frac{T_Q}{T_c} \leq \left[\frac{x - 1}{3(x^{4/3} - x)}\right]^{1/4}. \tag{9}$$

Note, that the largest possible values for the temperature of the hadronic phase as well as the minimum degree of supercooling in the initial phase-mixture corresponds to the adiabatic ($R = 1$) timelike deflagrations, while entropy production in the transition decreases both the final and the initial temperatures at a given hadronic fraction h. If the transition starts from a pure quark phase, the initial temperature must be at least $0.7\,T_C$. However, as the initial hadronic fraction approaches unity, the maximum of the possible initial temperatures for timelike deflagrations approaches T_C the critical temperature. Another interesting feature is that for large initial hadronic fraction, $h \geq 0.9$, timelike deflagrations to final hadronic state with $T_H \leq 0.8\,T_C$ become possible. At $T_H = 0.8\,T_C$ the hadronic phase is already frozen out.

It was pointed out earlier that timelike deflagrations to a pion gas at freeze-out temperature are in principle possible even in the Bjorken model with bag equation of state, and the initial temperatures necessary for such transitions are rather low. However, up to this point only the effects coming from the admixture of hadrons to the initial state and the effects related to possible entropy production in the timelike deflagrations were taken into account. The following mechanisms may make such a sudden freeze-out more feasible:

- Quark degrees of freedom equilibrate much slower than the gluons during the first 3 fm/c at RHIC or LHC energies. A hot glue scenario[18] is also discussed where a hot gluonic plasma develops from the preequilibrium parton collisions. In such a plasma the number of degrees of freedom is less than in a quark-gluon plasma, and so the released latent heat in a timelike deflagration is also less. For the same initial supercooling, cooler hadronic gas states are reached.
- The completion of the microscopic quantummechanical processes that lead to the sudden timelike deflagration take some time. Thus, the deflagration front will have a timelike thickness of about 1-2 fm/c, which leads to a further dilution of the matter.
- The timelike deflagration may convert part of the latent heat into the collective kinetic energy of expansion and not into the internal thermal energy of the hadronic phase,[15] leading to the development of a collective transverse flow. The effective temperature of the hadrons, as measured by the transverse momentum distribution, will be larger than the temperature because of the transverse boost by the flow. For example, a freeze-out temperature of $T_f \approx 140$ MeV with a transverse flow of $\beta_T = 0.4$ results in an effective temperature $T_{eff} = 210$ MeV.
- Inclusion of more realistic equation of state in the hadronic and in the partonic phases might further change the amount of the latent heat which together with the transport coefficients and the surface tension are playing a major role in determining the nucleation dynamics. The amount of the latent heat strongly influences how large initial supercooling is necessary to reach a given final hadronic state. Inclusion of higher hadronic resonances further decreases the ratio of the number of degrees of freedom, r.
- The sudden freeze-out will lead to a baryon excess and particularly to a strange baryon excess compared to thermal and chemical equilibrium in the hadronic phase. This reduces the number of light mesons with high thermal velocities

and thus, advances freeze-out in the hadronic phase also.

4. Conclusions

Time-scales of rehadronization for a baryon-free QGP at RHIC and LHC energies were considered. We found that the time-scale for reaching the bottom of the temperature curve during the cooling process via homogeneous nucleation[4] is surprisingly close to the time-scale of the freeze out. We argued that the QGP has to complete rehadronization in a 10 - 30 % supercooled phase quite suddenly, and we have shown that such a sudden process is possible and satisfies energy and momentum conservation with non-decreasing entropy.

This rehadronization mechanism is signalled by a vanishing difference between the sidewards and outwards components of Bose-Einstein correlation functions, in the observation of the free masses of the resonances in the dilepton spectra, and in a clean strangeness signal of the QGP. Detailed microscopic calculations have to be performed for making more quantitative predictions.

5. Acknowledgments

We thank J. Kapusta, D. Boyanovsky, M. Gyulassy, P. Lévai and G. Wilk for stimulating discussions. We would like to thank the Organizers for invitation and support. This work was supported in part by the Norwegian Research Council (NFR) under Grant No. NFR/NAVF-422.94/008, /011, by the NFR and the Hungarian Academy of Sciences exchange grant, by the Hungarian Science Foundation under Grants No. OTKA-F4019 and OTKA-2973.

6. References

[1] Proceedings of the Quark Matter conferences, especially Nucl. Phys. **A498**, (1989), Nucl. Phys. **A525**, (1991) and Nucl. Phys. **A544**, (1992).
[2] J. D. Bjorken, Phys. Rev. **D 27**, 140 (1983).
[3] K. Geiger and B. Müller, Nucl. Phys. **B369**, 600 (1992); K. Geiger, Phys. Rev. **D 46**, 4965, 4986 (1992); ibid. **47** 133, (1993).
[4] L. P. Csernai and J. I. Kapusta, Phys. Rev. Lett. **69**, 737 (1992).
[5] L. P. Csernai and J. I. Kapusta, Phys. Rev. **D 46**, 1379 (1992).
[6] J. S. Langer and A. J. Schwartz, Phys. Rev. **A 21**, 948 (1980).
[7] L. P. Csernai, J. I. Kapusta, Gy. Kluge and E. E. Zabrodin, Z. Phys. **C 58**, 453 (1993).
[8] J. Rafelski, Phys. Lett. **B262**, 333 (1991); Nucl. Phys. **A544**, 279c (1992).
[9] S. Abatzis, et al., WA85 Collaboration, Phys. Lett. **B259**, 508 (1991).
[10] S. Abatzis, et al., WA85 Collaboration, Phys. Lett. **B270**, 123 (1991).

[11] N. Bilić, J. Cleymans, E. Suhonen and R. W. von Oertzen, Phys. Lett. **B311** (1993) 266; N. Bilić, J. Cleymans, K. Reidlich and E. Suhonen, preprint CERN-TH 6923/93;. J. Cleymans, K. Reidlich, H. Satz and E. Suhonen, Z. Phys. C **58**, 347 (1993).

[12] A. K. Holme, et al., Phys. Rev. **D40**, (1989) 3735.

[13] L.P. Csernai and M. Gong, Phys. Rev. **D37**, (1988) 3231.

[14] J. Kapusta and A. Mekjian, Phys. Rev. **D33**, (1986) 1304.

[15] P. Lévai, G. Papp, E. Staubo, A.K. Holme, D. Strottman and L.P. Csernai, In Proc. of the Int. Workshop on **Gross Properties of Nuclei and Nuclear Excitations XVIII, Hirschegg,** Kleinwalsertal, Austria, Jan. 15-20, 1990 (TH Darmstadt, 1990) p. 90.

[16] T. Hatsuda, Nucl. Phys. **A544** (1992) 27c, NA38 collaboration, contribution to the Quark Matter'93 conference (to appear in Nucl. Phys.)

[17] J. Harris in "Particle Production in Highly Excited Matter", H. Gutbrod and J. Rafelski eds, **NATO ASI series, B303** (Plenum, 1993) p. 89

[18] E. V. Shuryak, Phys. Rev. Lett. **70**, 2241-2244 (1993)

RESULTS FROM CERN EXPERIMENT WA80

WA80 COLLABORATION

F. Plasil,[a] R. Albrecht,[b] T. C. Awes,[a] C. Barlag,[c] F. Berger,[c] M. Bloomer,[d]
C. Blume,[c] D. Bock[c] R. Bock,[b] E. Bohne,[c] D. Bucher,[c] G. Claesson,[b]
A. Claussen,[c] G. Clewing,[c] R. Debbe,[e] L. Dragon,[c] A. Eklund,[f] S. Fokin,[g]
S. Garpman,[f] R. Glasow,[c] H. A. Gustafsson,[f] H. H. Gutbrod,[b] O. Hansen,[e]
G. Hölker,[c] J. Idh,[f] M. Ippolitov,[g] P. Jacobs,[d] K. H. Kampert,[c]
K. Karadjev,[g] B. W. Kolb,[b] A. Lebedev,[g] H. Löhner,[c] I. Lund,[b]
V. Manko,[g] B. Moskowitz,[e] F. E. Obenshain,[a] A. Oskarsson,[f]
I. Otterlund,[f] T. Peitzmann,[c] A. M. Poskanzer,[d] M. Purschke,[c] B. Roters,[c]
S. Saini,[a] R. Santo,[c] H. R. Schmidt,[b] K. Soderstrom,[f] S. P. Sorensen,[a,h]
K. Steffens,[c] P. Steinhaeuser,[b] E. Stenlund,[f] D. Stüken,[c]
A. Vinogradov,[g] H. Wegner,[e] and G. R. Young[a]

a. Oak Ridge National Laboratory,* Oak Ridge, Tennessee 37831
b. Gesellschaft für Schwerionenforschung, D-6100 Darmstadt, Germany
c. University of Münster, D-4400 Münster, Germany
d. Lawrence Berkeley Laboratory, Berkeley, California 94720
e. Brookhaven National Laboratory, Upton, New York 11973
f. University of Lund, S-22362 Lund, Sweden
g. Kurchatov Institute of Atomic Energy, Moscow 123182, Russia
h. University of Tennessee, Knoxville, Tennessee 37996
* Managed by Martin Marietta Energy Systems, Inc., under contract
DE-AC05-84OR21400 with the U.S. Department of Energy

ABSTRACT

Two-photon invariant mass distributions have been obtained, and reconstructed π^0 and η results have been deduced with signal/noise ratios as low as 0.002. The production of η mesons was found to be consistent with m_T scaling observed in hadron-hadron collisions. "Direct," single-photon yields were extracted from high-statistics S+Au photon data obtained at 200 GeV/nucleon. A slight excess of photons over the number of photons that can be accounted for by hadronic decays was observed for central events. Two-proton correlations were obtained in the target fragmentation region in p+A, ^{16}O+A and ^{32}S+A reactions at 200 GeV/nucleon. Source radii extracted from the correlations were found to follow an $A^{1/3}$ dependence. This indicates that entire target nuclei are involved in the collisions and that a simple target-spectator picture does not apply at these energies.

1. Introduction

In previous presentations at these conferences, we have emphasized global characteristics of nucleus–nucleus collisions at energies (60 and 200 GeV/nucleon) available from CERN's Super Proton Synchrotron, SPS. It was concluded that the degree of nuclear stopping is high and that the energy densities attained may be in the threshold region for possible quark-gluon plasma (QGP) formation.[1] In this paper, we focus on data that are unique to the WA80 collaboration: results from photon measurements and from measurements in the target fragmentation region. Photons, being color-neutral, are potentially good probes of the QGP (or of dense hadronic matter). However, measurements of photons from nucleus–nucleus collisions must cope with a huge combinatorial background from the decay of π^0's (and of other mesons) and also with the background resulting from hadronic showers produced by a large number of charged pions impinging on the photon detectors. Original WA80 photon data, obtained with the SAPHIR detector, suffered from limited acceptance and from low statistics. As a result, extracted yields of single photons were subject to errors in the 15% range. In order to reduce these errors, SAPHIR was expanded to a total of 3800 lead-glass modules, and high-statistics photon data were obtained in subsequent runs with ^{32}S beams at E/A = 200 GeV. We present here data from reactions between ^{32}S and ^{197}Au nuclei.[2]

Plastic Ball data in the target fragmentation region were obtained with beams of protons and of ^{16}O and ^{32}S ions in the two initial SPS runs. Several different targets were studied, ranging from ^{12}C to ^{197}Au; and azimuthal correlations, as well as two-particle correlations for pions and protons, were obtained. We discuss here source radii extracted from the proton–proton correlation data. These are particularly interesting since they have implications for the role of the target nucleus in nucleus–nucleus collisions at SPS energies.

2. Production of Neutral Mesons

The data presented here were measured with the expanded WA80 lead-glass spectrometer which had a 60% azimuthal coverage in a pseudorapidity range from 2.1 to 2.9. Through the application of the event-mixing technique,[3] it is possible to reconstruct π^0 and η mesons down to signal-to-background ratios of 0.002, as illustrated in Fig. 1. Absolute π^0 transverse momentum spectra are shown in Fig. 2 for central and peripheral events from S+S and S+Au reactions at 200 GeV/nucleon. It is noteworthy that the pion yields have been measured over a range of seven orders of magnitude. Exponential fits are indicated in Fig. 2, together with the corresponding slope parameters. However, as can be seen, these fits are not valid over the entire p_T range. Consequently, we have extracted slopes in limited p_T regions as shown in Fig. 3. Monotonic increases in the slope parameters are observed for both central and peripheral events with the slopes from peripheral events lying consistently about

20 MeV below those from central events. Minimum-bias proton–proton results are shown for comparison.

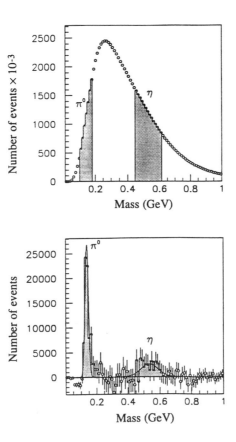

Fig. 1: Two-photon invariant mass distributions for very central events ($E_T = 120 - 140$ MeV) for ^{32}S+Au in the p_T range $0.2 - 1.2$ GeV/c. The upper and lower parts of the figure depict data before and after combinatorial background subtraction, respectively.

The higher-statistics data and improved combinatorial background evaluations make it possible, for the first time, to extract η-meson yields in nucleus–nucleus collisions. In Fig. 4 η/π^0 ratios are shown as a function of p_T for sulfur- and oxygen-induced reactions, together with results from similar hadron–hadron and hadron–nucleus measurements.[4] The solid line is the result of a fit to all data shown in Fig. 4

following the m_T-scaling parametrization of Bourquin and Gaillard.[5] We conclude that m_T scaling of meson production in nucleus–nucleus collisions is consistent with the phenomenological m_T scaling previously established in hadron–hadron and hadron–nucleus reactions. We make use of this conclusion to estimate the small contribution that decays of hadrons, other than the measured π^0's and η mesons, make to the total observed yield of photons.

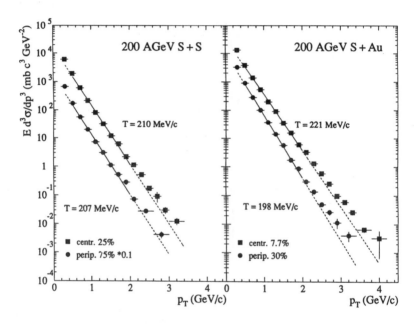

Fig. 2: π^0 transverse momentum spectra for peripheral and central collisions of S nuclei with S and Au nuclei at 200 GeV/nucleon. The curves represent exponential fits in the transverse momentum range from 0.5 to 2.0 GeV/c, with slope parameters, T, as indicated.

3. Yields of Single Photons

The determination of the yield of single photons is the primary goal of the WA80 collaboration, since this yield may be related to the possible production of the QGP. Single "direct" photons are expected from a variety of different sources such as hard QCD processes and thermal radiation from the QGP and/or from a "hadron gas."[4] Specific processes include quark-quark annihilation and the quark equivalent of the Compton process. However, as is indicated in the introduction, the extraction of single-photon yields from the data is difficult due to very large combinatorial and other

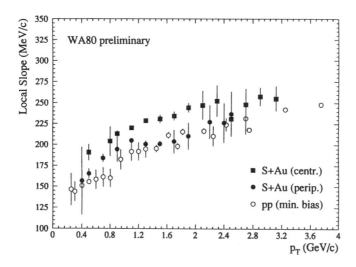

Fig. 3: "Local" slopes of π^0 transverse momentum spectra as a function of p_T for central and peripheral events from S+Au reactions at 200 GeV/nucleon and for minimum bias events from p-p reactions at 200 GeV.

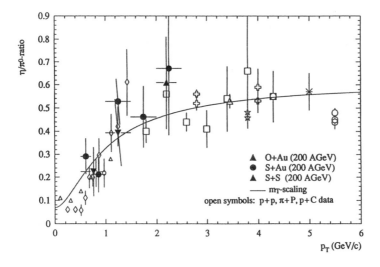

Fig. 4: η/π^0 ratios as a function of p_T for oxygen- and sulfur-induced reactions. Hadron–hadron and hadron–nucleus data are shown for comparison. The solid curve represents a fit to an m_T-scaling parametrization (see text).

124

backgrounds. A multivariate procedure has been implemented to obtain single-photon/π^0 ratios as a function of transverse momentum. Inputs to this procedure include measured total gamma yields, yields of reconstructed π^0's and η's, shower-reconstruction algorithms (for electromagnetic and hadronic showers) based on both measurements and simulations, geometrical factors, and estimates (based on m_T scaling) of the contributions to the total gamma yield from the decay of nonreconstructed mesons, such as ω, η', and K_0. (The total contribution from the decay of nonreconstructed mesons is estimated to be less than 2% of the contribution from reconstructed hadrons.) The improved data-reduction procedure outlined above, together with the accurate determination of the combinatorial background and the availability of higher-statistics data, resulted in a considerable reduction of errors compared to earlier WA80 results. This is indicated in Fig. 5 where the errors for the current S+Au data at 200 GeV/nucleon are shown for peripheral, central, and minimum-bias events. The approximate error range for earlier O+Au data is shown for comparison. Note that the range or errors has been lowered from $9 - 12\%$ to $6 - 7\%$ in the p_T range $0.3 - 1.5$ GeV/c.

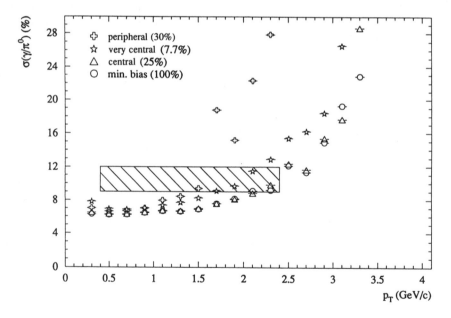

Fig. 5: Total errors of background-subtracted γ/π^0 ratios as a function of p_T for different regions of centrality from S+Au reactions at 200 GeV/nucleon. The shaded region depicts the range of errors in earlier WA80 O+Au data.

Preliminary γ/π^0 ratios are shown in Fig. 6 as a function of p_T for central and peripheral events from S+Au reactions at 200 GeV/nucleon. The circles indicate the inclusive photon yield, and the histograms show the yield that can be accounted for by hadronic decays. Note that a small, but statistically-significant, excess of photons (over those due to hadronic decays) is observed in central collisions, but not in peripheral collisions. The results shown here are preliminary, and work continues via independent evaluations in order to confirm the tantalizing effect depicted in Fig. 6.

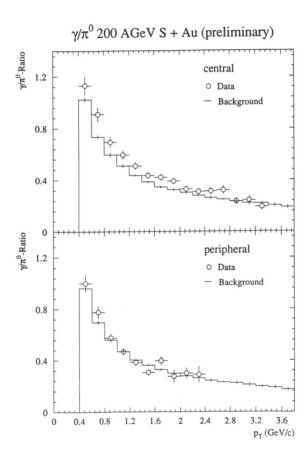

γ/π^0 200 AGeV S + Au (preliminary)

Fig. 6: Preliminary γ/π^0 ratios as a function of p_T for central and peripheral collision data from S+Au reactions at 200 GeV/nucleon. The open circles give the inclusive photon yield, and the histograms indicate the photon yield that can be accounted for by the decay of hadrons.

126

4. Two-Proton Correlations in the Target Fragmentation Region

Perhaps the most interesting results that have been obtained in the target fragmentation region ($-1.7 < \eta < 1.3$) by means of the Plastic Ball are source radii extracted from measured two-proton correlations.[6] The general principles which make it possible to extract these radii from correlation data are well-known, and are not described here. Following Pratt,[7] our specific approach involves comparisons of numerical simulations to data, in the absence of reliable analytical expressions that can be fitted directly. We have assumed a spherical Gaussian-shaped source distribution with a lifetime of $\tau = 0$ fm/c, which results in upper limits for spatial extents of extracted sources. Detector effects are included in the calculations, and care was taken to correct for misidentified particles and for other experimental effects. Extracted radii obtained by this procedure are shown in Fig. 7 for reactions of protons and ^{16}O ions with C, Cu, Ag, and Au targets, as well as for ^{32}S interactions with Al and Au nuclei at 200 GeV/nucleon. Remarkably, the radii follow closely a monotonic increase with the mass of the target, proportional to $A_{Target}^{1/3}$. Geometric radii of target nuclei are indicated in the figure by means of the hatched band. The radii extracted from heavy-ion reactions lie somewhat above the target geometric radii, while those obtained from p-induced reaction are very close to the target radii. This suggests that, even in proton-induced reactions, the whole target is involved in the reaction and that the simple participant-spectator picture which is often used to describe nucleus–nucleus collisions at relativistic energies may have to be modified.

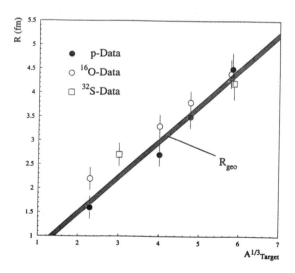

Fig. 7: Target-dependence of radii, R, extracted from proton–proton correlation data, as a function of target $A^{1/3}$. The hatched band labeled R_{geo} represents geometric radii of target nuclei.

5. Conclusion

The unique aspects of WA80 are measurements of photons and measurements in the target fragmentation region. As a result of improvements in experimental equipment as well as in data reduction techniques, yields of η mesons have been obtained, for the first time, in nucleus–nucleus collisions. In addition, errors associated with determinations of single-photon yields have been reduced by more than a factor of two. A small excess of photons over those that can be accounted for by the decay of mesons has been observed in central collisions of S nuclei with Au nuclei. Radii of emission sources deduced from proton–proton correlation data are found to be consistent with geometric radii of target nuclei, indicating that the whole target nucleus takes part in any given collision.

6. References

[1] R. Albrecht et al., Phys. Rev. C 44, 2736 (1991).

[2] R. Santo et al., Nucl. Phys. A566, 61c (1994).

[3] A. Lebedev et al., Nucl. Phys. A566, 355c (1994).

[4] Reference 2 above gives a complete list of references for the hadron–hadron and hadron–nucleus data of Fig. 4 and for various theoretical descriptions of possible "direct" photon production.

[5] M. Bourquin and J. Gaillard, Nucl. Phys. B114, 334 (1976).

[6] T. C. Awes et al. (WA80 Collaboration), preprint, January, 1994.

[7] S. Pratt, private communication, 1992.

Hadron Production in S+nucleus Collisions at 200 GeV/nucleon

S. MARGETIS for the NA35 Collaboration
Lawrence Berkeley Laboratory
Berkeley, CA 94720, USA

ABSTRACT

We present recent results on strange particle production and baryon rapidity distributions from the experiment NA35 at CERN.

1. Introduction

Central collisions between nuclei at relativistic energies form a hot and dense hadronic system over a large volume. Phenomenological models, as well as QCD calculations on the lattice, predict a phase transition in nuclear matter leading to deconfinement, a state of matter in which quarks and gluons are free to move inside the entire volume of the deconfined region. This new state was given the name of Quark Gluon Plasma (QGP). Based on the conjecture that collisions of heavy nuclei at high energies can create the necessary condition of a high energy density thermalized system, a series of experiments were built to search for possible signals of QGP creation.

The NA35 collaboration uses a wide acceptance apparatus at the CERN SPS which detects the majority of charged hadrons ($h^{\pm}$), and neutral strange particles produced in reactions of p, ^{16}O and ^{32}S projectiles at 60 and 200 GeV/nucleon lab momentum on different targets[1,2]. It consists of two major tracking devices which provide the momentum measurement of the charged particles: a 2 m long streamer chamber (SC) which is placed inside a 1.5 T vertex magnet, viewed by three cameras, and a 2.5x1.5x1.0 m^3 Time Projection Chamber (TPC)[3]. The SC covers the backward and the TPC the forward rapidities[4]. A set of calorimeters was used as the basic trigger device of the experiment, to select central (small impact parameter) collisions[2]. In the data presented here, a calorimeter placed in the beam path selects near head-on collisions, i.e. events where only a small amount of energy (mostly spectator nucleon energy) was detected in an angular acceptance of less than 0.3 degrees around the beam axis. The data sample consists of three systems: S+S, S+Ag and S+Au at 200 GeV/c with trigger cross section of 3, 3.2 and 6% of the total inelastic cross section, respectively.

All the TPC data were computer analysed. In the SC either a computer or an operator performed the pattern recognition task, i.e. to measure the trajectories of charged particles. No particle identification information was obtained on a track-by-track basis. Charged kaons (K^+, K^-) were identified by their characteristic one and three prong decays in the SC^5 and the TPC^6, as well as V_0 decays (K_s^0, Λ and $\overline{\Lambda}$ particles) by their characteristic decay topology in the SC.

2. Net Baryon Production

The distribution of the initial baryons in the final state can provide valuable information about the reaction dynamics of the colliding system, such as the fraction of the initial energy deposited in the interaction volume as well as the nature of the environment (baryon rich or baryon free). Large degrees of inelasticity, or stopping, are indicated by a large shift in rapidity of the participant nucleons towards midrapidity. The 'lost' or deposited energy is mostly used in particle production, which is discussed below.

In a system like S+S, which has zero net isospin, information about the net distribution of protons ($p-\overline{p}$) can be extracted using the charge excess technique[7]:

$$\frac{dN_{+,-}}{dy}(AA \rightarrow p) = \frac{dN}{dy}(AA \rightarrow h^+) - \frac{dN}{dy}(AA \rightarrow h^-)$$

This is a good approximation over all rapidities for collisions between isoscalar nuclei. It is also valid for asymmetric systems when the projectile is an isoscalar, if the measurement is performed in the projectile fragmentation region, i.e. forward of midrapidity[(1)]. The raw signal is corrected for geometrical acceptance asymmetries, effective rapidity shifts in weak decays (e.g. $\Lambda \rightarrow p\pi^-$), the excess of protons produced in secondary hadron–nucleus interactions, and the observed excess of K^+ over K^- in this experiment.

Figure 1 shows the preliminary net proton and Λ rapidity distributions for S, Ag and Au targets. The net baryon number around midrapidity is large and it increases with the target mass. The total net baryon number (participating baryons) is calculated using the following equation,

$$< B - \overline{B} >= 2(p - \overline{p}) + 1.6(\Lambda - \overline{\Lambda})$$

with the factor of 2 accounting for the neutrons and the empirical factor 1.6 for the unobserved charged Σ decays[8]. The total net baryon number results are summarized in the first two columns of Table 1.

The extrapolation to 4π in the S+S system was made by reflecting the measured distributions around midrapidity for this symmetric system. In S+Ag the average of the S+S and S+Au distributions was taken, since the mass of Ag is the average of S and Au masses. The quoted errors in the table reflect the systematic uncertainties of this method. The numbers in parenthesis in the first column are the total number of

---

[1]In this article, the term midrapidity denotes half the beam rapidity and not the *c.m.* rapidity of the system which is slightly different for asymmetric systems.

	$<B-\bar{B}>_{4\pi}$	$d(B-\bar{B})/dy \; y=y_{NN}$	$\frac{(\Lambda-\bar{\Lambda})}{(p-\bar{p})} \; 4\pi$	$<h^->$	$<h^->/<B-\bar{B}>$
S+S	52±5 (56)	8.6±1 (0.15)	0.24±0.02	95±5	1.8±0.2
S+Ag	90±10 (93)	15.0±1 (0.16)	0.22±0.04	160±8	1.8±0.2
S+Au	— (110)	22.0±1 (0.20)	—	—	—

Table 1: Net baryon, negative hadron multiplicities and rapidity densities in S+S, Ag, Au collisions (see text).

participants predicted by a clean–cut geometry model. The numbers in parenthesis in the second column are the net baryon density around midrapidity per participant nucleon. We observe that it increases with increasing target mass, indicating an increasing stopping for the heavier systems. This can be seen, in a qualitative manner in fig. 1a where for the heavier targets a change of the slope of the distribution forward of midrapidity brings more protons in the region around midrapidity depleting at the same time the region close to beam rapidity. In S+Ag, the large number of target protons results in the high peak at small rapidities.

In order to be more quantitative the mean rapidity shift ($<\Delta y>$) of the participating nucleons in the reaction has been calculated. It was found to be $<\Delta y> = 1.78 \pm 0.09$ for the system S+Au[9] and $<\Delta y> = 1.58 \pm 0.15$ in S+S[7]. This can be compared to $<\Delta y> \approx 1$ in nucleon–nucleon collisions. A relatively large amount of stopping is observed especially for the S+Au system.

The two last columns of Table 1 give information about the negative hadron (mostly pions) production in these systems. Notice the same number of produced negative hadrons per net baryon for the S+S and S+Ag systems.

3. Strange particle production

New results on strangeness production in the S+Ag system exhibit an enhancement relative to nucleon–nucleon collisions, which was previously observed in ref. [10] for central S+S collisions. In order to make comparisons, the measured yields are extrapolated over the full phase space. In the case of S+Ag the measurements cover the rapidity region up to midrapidity with a low p_T cut of 0.5 GeV/c. In order to be able to measure S+Au interactions forward of midrapidity, an additional magnet was placed upstream of the SC. The extrapolation in p_T was done by first fitting the transverse mass distributions m_T with an exponential function of the form:

$$\frac{1}{m_T}\frac{dN}{dm_T} = A \cdot e^{-m_T/T}$$

and then calculating the yield in 4π by integrating the fitted function. The effective temperature parameter T was found to be almost the same for every strange particle species we measured, around 200 MeV.

The extrapolation in rapidity was based upon the observation that the strange particle production in the forward hemisphere in systems with the same projectile nucleus is roughly independent of the target. This can be seen in fig. 2a,b,c where preliminary rapidity distributions for Λ, K_s^0 and $\overline{\Lambda}$ are shown. We observe that above rapidity 3 the yield of Λ, $\overline{\Lambda}$ and K_s^0 is similar in the S+S and S+Au systems, therefore, for the S+Ag system, the average of the above systems was taken. The curves in the Λ and K_s^0 distributions show p+S data multiplied by a factor of 29. It was found that multiplying the negative hadron distributions in p+S by this factor one could reproduce those in S+Ag reactions. This comparison is made model independent because models do not satisfactorily describe strangeness production at the fundamental nucleon–nucleon level[12].

Figures 3a,b show the preliminary K^+ and K^- rapidity distributions for the S+S and S+Ag system. The p+S data are also shown scaled up by the same factor 29 as before. We observe that the S+Ag data are systematically higher than the scaled p+S data and that the difference reaches its maximum for the Λ and K^+ particles. This is qualitatively expected since the quark content of a baryon rich environment, like the target fragmentation region of the heavy Ag target, favors interactions of an 'associated production' type, i.e. $p+X \rightarrow K^+\Lambda+X$. This is further

	$<h^->$	$<\Lambda+\Sigma^0>$	$<\overline{\Lambda}+\overline{\Sigma}^0>$	$<K_s^0>$	$<K^+>$	$<K^->$
S+Ag	160±8	13.0±0.7	2.4±0.4	13.2±1.1	17.4±1.0	9.6±1.0
50*NN	—	4.8±0.5	0.7±0.2	10.0±1.0	12.0±3.0	8.5±2.5
p+S	5.7±0.2	0.28±0.02	0.043±0.003	0.38±0.04	—	—
S+S	95±5	8.2±0.9	1.5±0.4	10.6±2.0	12.5±0.4	6.9±0.4
30*NN	—	2.9±0.3	0.4±0.1	6.0±0.6	7.2±1.9	5.1±1.6

Table 2: Mean total particle multiplicities in p+S, S+Ag and S+S collisions.

supported by fig. 1b which shows that these net Λs are concentrated around the target fragmentation region of the S+Ag system. Also, the third column of Table 1 shows that the net number of produced Λs per participant nucleon is the same in S+S and S+Ag. This scaling might indicate a relationship between these two particles. It might also explain why the extra particles are appearing in the same rapidity region where the participating nucleons are finally recorded.

How close does the excess of Λ and K^+ particles, as expressed by the distribution of the difference of $\Lambda-\overline{\Lambda}$ and K^+-K^-, follow the distribution of net protons $(p-\overline{p})$ in rapidity? Figure 3c shows the resulting distribution for the K^+-K^- for the systems S+S and S+Ag. This should be compared to fig. 1a and 1b. We observe that not only do the distributions of net Λ and K^+ closely follow the distribution of the net protons in the final state, but also the abundance of these extra Λ and K^+s are similar. This is a further indication that these extra particles might be produced in pairs. The above considerations support the idea that most of the enhancement of the Λ and K^+ particles can be attributed to processes similar to associated production. Of course, the exact mechanisms involved cannot be deduced by these experimental observations.

The degree to which flavor symmetry is achieved at the quark level in the final state is usually characterized by the *strangeness suppression* factor λ_s,

$$\lambda_s = \frac{<s+\bar{s}>}{0.5 \cdot (<u+\bar{u}> + <d+\bar{d}>)}.$$

The quantity λ_s can be calculated according to ref. [14] from the numbers in Table 2. We found that λ_s is half as strong in the S+S and S+Ag systems as compared to nucleon–nucleon[13] and nucleon–nucleus[14] systems.

Phenomenological, as well as microscopic string models, have great difficulties in consistently reproducing the observed enhancement in the two systems we have studied thus far. New mechanisms at the parton level are needed, for example, color rope (RQMD[15]) or double string formation (VENUS[16]). For a more detailed comparison with models, see ref. [11].

4. Summary

The following remarks summarize the above discussion:

- NA35 data on baryon distributions in S+Au collisions show high degrees of stopping, manifested in the large mean rapidity shift of the participant protons.
- We observe an increasing mean rapidity shift of the participant protons with increasing target mass.
- The region around midrapidity appears to be baryon rich for all systems.
- New data on strange particle production in S+Ag show a strangeness enhancement relative to nucleon–nucleon and nucleon–nucleus collisions.
- It appears that most of the excess of Λ over $\bar{\Lambda}$, and K^+ over K^-, can be explained by an 'associate production'–like process.

5. References

[1] A. Sandoval *et al.*; NA35 collaboration, Nucl.Phys. A461 (1987) 465c
[2] S. Margetis; Ph.D thesis, GSI–91-04
[3] P. Jacobs *et al.*; NA35 collaboration, LBL–33810
[4] D. Röhrich *et al.*, NA35 collaboration, IKF–HENPG/93-8 and
    Proc. of Quark Matter 93, Nucl.Phys. A566 (1994) 35c
[5] M. Kowalski *et al.*; NA35 collaboration, Nucl. Phys. A544 (1992) 609c
[6] S. Margetis; work in progress
[7] J. Bächler *et al.*; NA35 collaboration, to appear Phys. Rev. Lett.,
    February (1994)

[8] A.K. Wróblewski; Acta Phys. Pol. B16 (1985) 379

[9] J.T. Mitchell *et al.*; NA35 collaboration,
     Proc. of Quark Matter 93, Nucl.Phys. A566 (1994) 415c and LBL–34727

[10] A. Bamberger *et al.*; NA35 collaboration, Z. Phys. C43 (1989) 25,
      J. Bartke *et al.*; NA35 collaboration, Z. Phys. C48 (1990) 191

[11] M. Gaździcki *et al.*; NA35 collaboration, *Proc. of Quark Matter 93*,
      Nucl.Phys. A566 (1994) 503c Also paper in preparation.

[12] I. Sakrejda, proceedings of this workshop.

[13] M. Gaździcki and O. Hansen; Nucl. Phys. A528 (1991) 754

[14] H. Bialkowska *et al.*; Z. Phys. C55 (1992) 491

[15] H. Sorge *et al.*; Phys. Lett. B289 (1992) 6

[16] K. Werner and J. Aichelin, Phys. Lett. B308 (1993) 372

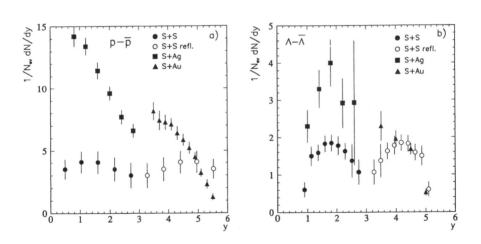

Fig. 1: Rapidity distributions of, a), net protons and, b), net Λ, in S+nucleus collisions.

134

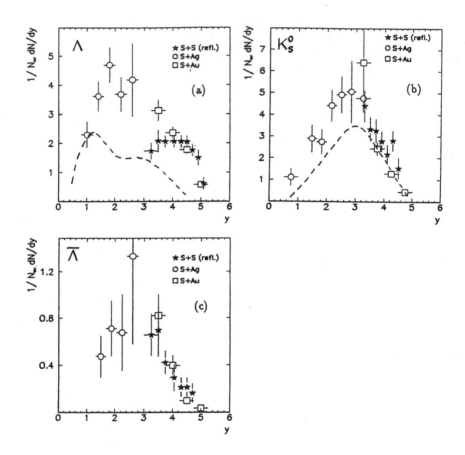

Fig. 2: Rapidity distributions of Λ, K_s^0 and $\overline{\Lambda}$ particles in S+nucleus collisions. The dashed curves represent p+S results scaled up by a factor of 29 (see text).

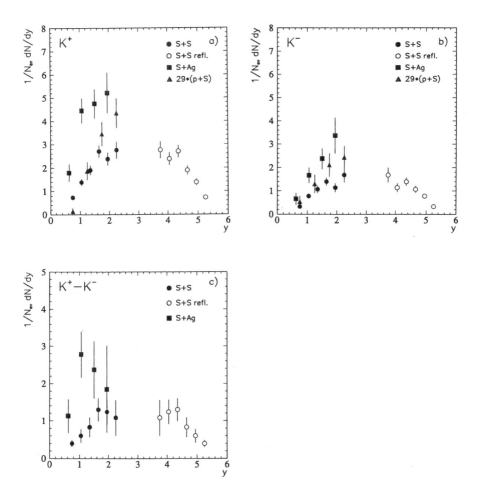

Fig. 3: Rapidity distributions of, a), K^+, b), K^- and, c), net charged kaons in S+nucleus collisions. The p+S results scaled up by a factor of 29 are also shown for comparison.

Probing the Baryon Density in Au-Au Collisions at 11.6 GeV/A: Pion Multiplicity Distributions

R. SETO

Univ. of CA, Riverside, Riverside, CA 92521, USA

for the *E802* Collaboration

ABSTRACT

In the past year, as Au beams have become available at the AGS, data have become available that indicate that there is a large degree of "stopping" in Au-Au collisions at 11.6 GeV/A. This would imply that baryon densities of the order of 5-10 ρ_0 are reached. E866 plans a multiplicity array which will allow the experiment to fix the geometry of the collision using the zero-degree calorimeter and study central events which have unusually high multiplicities. High multiplicities would be associated with the production of entropy which may be an indication of a transition to a QGP and/or unusually high baryon densities. A successful prototype of such a detector was installed for the E866 run in 1993. The charged particle multiplicity distribution agrees nicely with spectrometer measurements. A preliminary distribution of the photons from π^0's was also measured. The full device is now under construction and will be ready for the next Au run at the AGS in the fall of 1994.

1. Introduction

Recently, high energy gold beams have become available at the Brookhaven National Laboratories Alternating Gradient Synchrotron (AGS). This has provided a unique opportunity for physicists to study large systems of highly compressed nuclear matter. Simple arguments, such as the one proposed by Goldhaber in 1978 predict that the highest Baryon densities available in the laboratory are being produced by this machine.[1] In this paper he points out that the collision of two relativistic nuclei will produce a system with $2\gamma_{cm}$ times normal nuclear density due to the Lorentz factor. For Au-Au collisions at the AGS (11.6 GeV/A), this leads to a baryon density of about $6\rho_0$. There are, of course, various assumptions in this rather simple argument that must be tested. Ultimately, models which simulate the individual particle collisions must make a more realistic estimate of the baryon density. Predictions made by these models can then be compared to experimental data.

2. Stopping

One of the important ingredients to these arguments is that of stopping. At very high energies such as that being envisioned for RHIC, the nucleons are expected to be

Proton Distributions

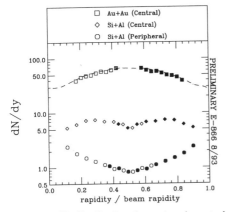

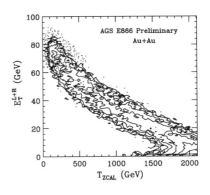

Fig. 1: dN/dy distributions for protons in central Au-Au, Si-Al, and peripheral Si-Au collisions.

Fig. 2: Transverse energy distributions in the lead glass vs. zero degree calorimeter energy. A special run was done to obtain this data in which the spectrometers were swung out of the way and the lead glass arrays put into place.

transparent, and the colliding nuclei simply pass though each other leaving a debris of high temperature, baryon free matter. At AGS energies, however, it is hoped that the nuclei stop each other almost completely. E866 has been able to make a test of this hypothesis using data from the first Gold beam run in 1992. The detector[2] is a magnetic spectrometer, capable of particle identification up to 4.7 GeV/c. A zero-degree calorimeter is used for centrality measurements. The proton rapidity distribution for the most central 4% of Au-Au events is shown in figure 1. The distribution has been folded around mid-rapidity. A new spectrometer commissioned in the 1993 run will fill in the missing data points at mid-rapidity. Also shown are proton distributions for central and peripheral Si+Al events. The dotted line shows a comparison to the ARC model,[3] which will be mentioned later. Assuming that the trend of the data continues, one concludes that there is no dip at mid-rapidity in the cross section for protons in central Au-Au collisions, which would be characteristic of a more transparent reaction as in the case of Si+Al.

In order make a quantitative measure of the "stopping", one would like to identify whether each nuceleon came from the target or the projectile and find the average $< \Delta y >$ of all nucleons. Since this is not possible, we use the data to calculate $< \Delta y >$ assuming that the protons with y greater than the mid-rapidity value to be from the projectile, and that the remainder are from the target. We then compare this rapidity shift to an isotropic fireball model. The Au on Au data gives a value of $< \Delta y > = 1.1$, as compared to 1.6 for an isotropic fireball. In contrast, collisions of Si+Al, give a value of 0.85. This means that there is more stopping in Au-Au collisions than in Si-Au and that this stopping is rather large, although not complete. The system appears to be expanding in the longitudinal direction with about 300 MeV more momentum than in the transverse direction. Various other methods of analysis yield similar conclusions, although the quantitative values are somewhat different.[4]

Several monte-carlo simulations have now been written which have been able to model heavy ion collisions with varying degrees of success. One of the best at these AGS energies is ARC.[3] Using primarily two body processes with vacuum values of the cross sections, they have been able to reproduce kinematic distributions and yields of various particles with surprising accuracy. It is reasonable then, to use this model to give us an estimate of the baryon density as an improvement over the simple arguments mentioned in the introduction. The model indicates a region of volume 3 fm^3 remains at greater than $6\rho_0$ for about 5 fm. Though the model probably cannot be trusted to take us through the high baryon density phase, it does tell us that conditions are such that high baryon density matter is formed at the AGS.

At these extremes, various physical phenomena are expected to occur. Particle masses are expected to shift since the inter-quark forces holding the baryons and mesons together will be affected by the external color field - a sort of color Stark effect.[5] Of course the infamous quark-gluon plasma may also be formed. Another intriguing possibility is that there may be events with an unusual fluctuation in the number of neutral pions, from coherent effects.[6]

One of the interesting effects seen in the Au-Au data is shown in figure 2. The relationship between the zero-degree calorimeter and the lead glass array is expected to be linear. This is true for the Si-Al data. The Fritiof model also gives a linear relationship even for Au-Au collisions. The data show a rise for the lowest values of the zero-degree calorimeter energy. That is, for the most central events there is a unusually large range of energies at central rapidity. This is what one might expect if some new physics were entering into the picture giving a large fluctuation in particle multiplicity. Interestingly, the ARC model also shows this effect.

It now becomes important for experiments to isolate events of unusually high baryon density. A subgroup of E866 from UC Riverside, MIT, and BNL have undertaken to build a new multiplicity array (NMA) for this purpose. The idea is to use the zero-degree calorimeter to fix the geometry of the event - i.e. to select central events. Then the multiplicity counter would be used to separate the central events into groups of varying multiplicity. High multiplicities would be associated with the production of entropy. In the case of the transition to a QGP even for a limited region of the fireball, the entropy is expected to increase since the number of degrees of freedom available to 3 quarks is more that is available to nucleons. Even without this phase transition large multiplicities would be associated with events which

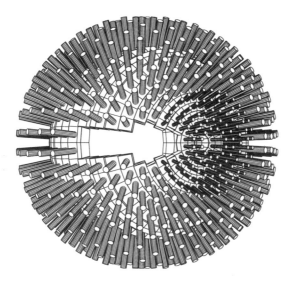

Fig. 3: The E866 New Multiplicity Array. The target will be in the center of the detector. Note the slot for the spectrometer.

had an unusually high number of collisions. Such a situation would arise if the the nuclei were more highly compressed resulting in an abnormally high baryon density. We would then be able to examine these data samples for various signatures such as strangeness production, particle slope parameters and other quantities, looking for variations as a function of multiplicity. These variations should be not be due to geometry, but to the physics of the fireball.

The detector is shown in figure 3. It is composed of about 500 modules of lucite Cerenkov counters surrounding the target area. The detector is azimuthally symmetric except for holes left for the spectrometer and the phoswich array, making the device useful for reaction plane studies. The η coverage is between -0.8 and 2.4; the forward coverage is limited by background considerations for the spectrometers. The modules vary in size between $3 \times 3 \times 6 cm^3$ and $8 \times 8 \times 6 cm^3$. Each module is read out by a phototube which is connected to an ADC. The occupancy in the most forward modules is less than 2. The selection of Cerenkov light as the basic detection principle allows us to reject slow particles (mostly spectator protons). A second consideration in the design of the module was to convert photons so that π^0's would be included in the multiplicity distributions. This is done by putting 0.3 radiation lengths (2 mm) of lead converter in front of the module. Each photon will have a 25% chance of conversion. Since every second π^0 will contribute an e^+e^- pair, on the average each

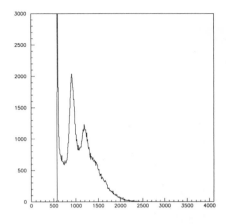

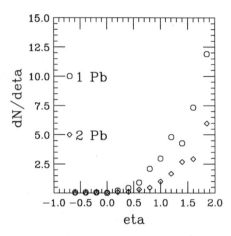

Fig. 4: The ADC spectrum from one of the prototype NMA modules at η=1.6. The data is from a central Au-Au sample.

Fig. 5: dN/dη distributions in the prototype NMA for triggers with no interaction in the target, i.e. δ electrons. Shown is the distribution with one and two layers of lead in front of each module.

π^0 will contribute one minimum ionizing electron to the signal. The lead has the additional advantage of stopping δ rays. In combination with a requirement that an electron penetrate at least 2 cm of the lucite to have a signal that is mistaken for a minimum ionizing particle, δ electrons with energy below 10-20 MeV will be rejected. This means that for a 1% interaction length gold target, and a central collision with about 700 particles in the final state, the error in the measured multiplicity introduced by δ electrons is less than 3 percent. Our present plans include some running with a checkerboard of 1 and 2 layers of lead radiator so that an event by event π^0 distribution can be measured. The detector is designed such that the resolution is dominated by the poisson fluctuations in the fraction of the total number of particles in the geometric acceptance of the device. For a central Au-Au event, about 350 particles will be in the geometric acceptance of the detector. The resolution on the total multiplicity will be about 5%.

For the run in September 1993, a prototype of about 30 modules was installed. Figure 4 shows the ADC distribution from a module at η=1.6, clearly showing the 1 and 2 particle peaks with a tail at higher values. The distribution is well modeled by GEANT simulations. δ electron distributions (fig. 5) were measured using events in which there there was no interaction in the target as measured by the beam counters. As expected the δ ray background with one layer of lead will give us less than a 2% error in the multiplicity measurement for central Au-Au collisions. Runs were taken using several thickness of lead radiator. By comparing these measurements, we were able to produce η distributions for charged particles and for π^0's. Figure 6

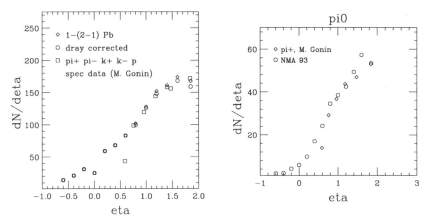

Fig. 6: $dN/d\eta$ distributions for central Au-Au events as measured by the NMA prototype. On the left is the distribution for charged particles as compared to the spectrometer measurement. On the right is the distribution of photons from π^0's as compared to the π^+ distribution.

shows $dN/d\eta$ for charged particles as compared to that measurement obtained by the spectrometer. The agreement is excellent. Also shown in figure 6 is the distribution of photons from π^0's together with the π^+ distribution. Because of slight differences in the normalization of the zero degree calorimeter in the two data sets the comparison is only qualitative.

3. Conclusions

In the past year, as Au beams have become available at the AGS, data have become available that indicate that there is a large degree of "stopping" in Au-Au collisions at 11.6 GeV/A. This would imply that baryon densities of the order of 5-10 ρ_0 are reached. Models such as ARC also support this conclusion. E866 plans a multiplicity array which will allow the experiment to fix the geometry of the collision using the zero-degree calorimeter and study central events which have unusually high multiplicities. High multiplicities would be associated with the production of entropy. In the case of the transition to a QGP even for a limited region of the fireball, the entropy is expected to increase since the number of degrees of freedom available to 3 quarks is more that is available to nucleons. Even without this phase transition large multiplicities would be associated with events which had an unusually high number of collisions. Such a situation would arise if the the nuclei were more highly compressed resulting in an abnormally high baryon density.

A successful prototype of such a detector was installed for the E866 run in 1993. The δ ray background, and module resolution were measured and found to be consistent with our expectations indicating that the resolution of the full device will be dominated by the poisson distribution of particles entering into the geometric acceptance of the detector. The charged particle multiplicity distribution agrees nicely with spectrometer measurements. A preliminary distribution of the photons from π^0's was also measured.

The full device is now under construction and will be ready for the next Au run at the AGS in the fall of 1994.

4. References

[1] A. S. Goldhaber, *Nature* **275**, 114 (1978).

[2] T. Abbott *et al.*, E802 collaboration, *Nucl. Inst. Meth.* **A290**, 41 (1990).

[3] S. Kahana, *Proceedings of Heavy Ion Physics at the AGS, HIPAGS '93*, **MITLNS-2158**, 263 (1993).

[4] the argument is from C. Parsons, in *Proceedings of Heavy Ion Physics at the AGS, HIPAGS '93*, **MITLNS-2158**, 72 (1993). See also F. Videbaek in the same volume.

[5] for a review see T. Hatsuda, *Nucl. Phys.* A **544**, 27c (1992).

[6] S. Pratt, in Proceedings of Quark Matter 1993, Borlange, Sweden. see also K. Rajagopal and F. Wilczek, *Nucl. Phys.* **B404**, 577(1993), although this is for the case of a baryon free system.

Fission Dynamics Studied by Neutron and GDR Gamma Emission

B. B. Back
Argonne National Laboratory
Argonne, IL 60439

D. J. Hofman and P. Paul
Department of Physics
State University of New York at Stony Brook
Stony Brook, New York 11794

ABSTRACT

We analyze recent data on the multiplicity of pre-scission γ-rays and neutrons to study the time-scale for fission of hot nuclei formed in heavy ion reactions. At the highest excitation energies measured, the data are well accounted for by the full one-body dissipation mechanism, but at lower excitation energies the dissipation appears to be absent. A disparity in the excitation energy dependence of the dissipation seen in neutron and γ-ray experiments is observed.

1. Introduction

The dynamics of large-scale re-arrangement of nuclear matter at low excitation energies has been the subject of intense studies for many years. Such processes occur in the fusion and quasi-fission processes between heavy ions as well as in the fission process. Much information has been gained from recent studies of the extra-push energy required for the fusion process to occur with heavy projectiles[1] and from time-scale measurements of the mass equilibration in quasifission reactions[2,3]. Theoretical descriptions of such processes require three ingredients, namely the potential energy surface, the mass parameter and the mechanism for energy dissipation associated with the motion. The potential energy and the mass parameter are well described by the Liquid Drop Model, or later refinements thereof, and the Werner-Wheeler approximation is believed to give a good description of the inertial parameter. In contrast, many different models have been proposed to describe the dissipation mechanism[4,5,6]. One of these, namely the simple one-body dissipation model proposed by Błocki, Randrup and Swiatecki[4] has been surprisingly successful in the description of heavy-ion fusion and quasi-fission reactions, despite serious theoretical problems of its basic assumptions[7]. In this work we explore the information in the dissipation that can be obtained from the study of pre-scission γ- and neutron emission in fission of hot nuclei.

144

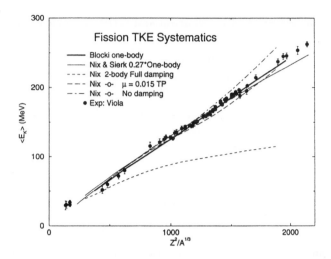

Fig. 1: Experimental data on total kinetic energies in fission (solid points) [8] are compared to the predictions based on full [4] (thick solid curve) and reduced [6] (thin solid curve) one-body dissipation. Also calculations [5] for no (dashed-dot) dissipation and two-body dissipation with a viscosity of $\mu = 0.015$ TP (long dashed) and full dissipation (short dashed) are shown.

2. Total kinetic energy in fission - time scales

Recently it has been realized that the time-scales of fission processes provides an observable, which is much more sensitive to the dissipation mechanism than the more traditional measurements of total kinetic release and mass distributions. This point is illustrated in Fig. 1, where the experimental data (solid points), as compiled by Viola et al.[8], are compared with different dissipation models[4,5,6]. It is evident that several models are able to describe the total kinetic energies in fission quite well. However, they predict very different time scales for the motion from the saddle to scission at shown in Table 1. Measurements of this observable is therefore expected to differentiate better between these mechanisms.

Several experimental methods have been employed to measure the time-scale of the fission process. The emission of pre-scission neutrons and charged particles (p,α) has been studied by several authors [10,11,12] using kinematics to separate pre-scission particles from those emitted from the fission fragments after separation. Pre-scission γ-rays can also be used because the γ-emission rate is well understood and the classical sum-rule is obeyed. Furthermore, since there is a good correlation between the nuclear symmetry axis prior to scission and the emission direction of the fission fragments one obtains a distinct anisotropy of the pre-scission γ-emission relative to this axis. In principle, the magnitude of this anisotropy contains information about the

Table 1: Predicted Saddle-to-scission times for ^{238}U. The viscosity, μ, is measured in Terapoise (TP), and, k, denotes the scaling factor for the one-body dissipation.

Dissipation	Saddle-to-Scission time (10^{-21}) s	Reference
None	2.5	Davies *et al* [5]
Two-body, $\mu = 0.015$ TP	3.5	Davies *et al* [5]
One-body, k = 0.27	8.5	Nix & Sierk [6]
One-body, k = 1.0	30 - 35	Carjan *et al* [9]

deformation of the system[13].

3. Pre-scission γ-ray emission

Recently, the saddle-to-scission time in ^{240}Cf initially formed at an excitation energy of $E_{exc} = 93$ MeV has been determined from an analysis of the pre-scission γ-rays in the giant dipole resonance (GDR) region[14]. It was found that the best fit to the observed γ-spectra were obtained for saddle-to-scission times of $t_{ss} = 26\text{-}30 \times 10^{-21}$ sec depending on the assumed dissipation strength inside the barrier as shown in Fig. 2.

Comparing this result with the various theoretical predictions listed in Table. 1, we see that it strongly favors the full one-body dissipation mechamism. The shorter saddle-to-scission times predicted by the other dissipation mechanisms, which reproduce the TKE distributions quite well, are not long enough to explain the observed spectrum. However, as illustrated in Fig. 3, we see that the standard statistical model is perfectly able to account for the observed spectrum at a 30 MeV lower bombarding energy, namely $E_{beam} = 200$ MeV, which corresponds to an excitation energy of 67 MeV. This sudden increase in dissipation with excitation energy is very surprising, although there is other empirical evidence to suggest that the dissipation is much weaker at low excitation energies. Thus Gavron *et al*[15] found that in order to reproduce the fission probabilities of actinide nuclei in the excitation energy range $E_{exc} = 6\text{-}12$ MeV, it was necessary to increase the fission width by assuming an increased density of transition states at the saddle point arising from non-symmetric shapes. According to the diffusion model for the fission process, the dissipation would reduce the fission width below the normal Bohr-Wheeler expression used in the statistical model calculations[14]. Such a reduction would lower the predicted fission probability in this excitation energy region. This placed an upper limit on the dissipation strength in this excitation energy regime of about $\gamma \leq 1$, i.e. less than 20% of the value for one-body dissipation.

146

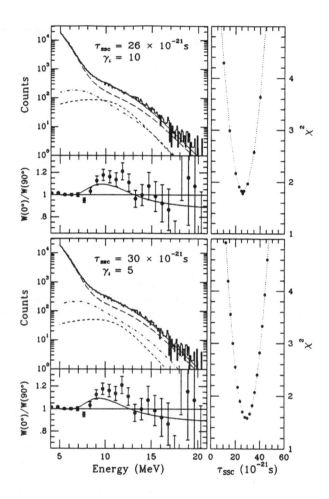

Fig. 2: Experimental γ-spectra (solid histograms) obtained in coincidence with fission fragments for the reaction $^{32}S+^{208}Pb$ at $E_{beam} = 230$ MeV are compared to statistical model calculations [14] including the effects of dissipation of different strengths inside the barrier, namely $\gamma = 10$ (upper panel) and $\gamma = 5$ (lower panel). Correspondng best fits are obtained by minimizing the χ^2 functions, (right panels).

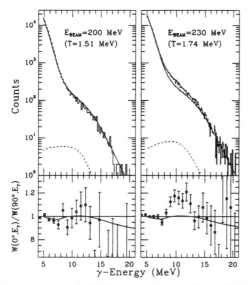

Fig. 3: Comparison of γ-spectra (solid histograms) in concidence with fission fragments obtained in the $^{32}S+^{208}Pb$ reaction at $E_{beam} = 200$ MeV (left panels) and $E_{beam} = 230$ MeV (right panels). The solid curves represent standard statistical model calculations without dissipation [14].

In addition, by analyzing the systematics of the width of iso-scalar quadrupole and octupole giant resonances, Nix & Sierk[6] have found that a reduction (of about a factor of four) of the full one-body dissipation strength is required to reproduce the data. In summary, there is good evidence that at low excitation energies the dissipation is substantially lower than predicted by the one-body mechanism. The rapid increase from a minimal strength to the full one-body value over the excitation energy range from 67 to 93 MeV does seem surprising, and suggests that further experimental study of this phenomenon is required.

4. Pre-scission neutron emission

Much data on the emission of pre-scission neutrons have been obtained in recent years, and, in general, a large excess of pre-scission neutrons has been observed indicating the role of dissipation in the fission process[11]. Recently, Thoennessen and Bertsch[16] have analyzed much of the available data and found that for most systems there exists a threshold energy E_{thresh}, above which the multiplicity of pre-scission neutrons exceed the standard statistical model predictions. They find, that this threshold decreases with increasing fissility of the system from a value of about $E_{thresh} = 90$ MeV for rare earth nuclei to about $E_{thresh} = 20$ MeV for ^{251}Es (see Table 2). This decrease occurs over the range where the fission barrier drops from being much larger than the neutron binding energy to values lower than the neutron

148

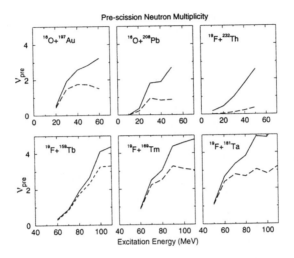

Fig. 4: Statistical model calculations of pre-scission neutron multiplicities illustrating the effect of dissipation. See text for details.

binding energy, allowing the fission decay to compete more efficiently with neutron emission. We expect that this change may affect the threshold energy at which the dissipation in the fission process would produce a noticeable effect in the pre-scission neutron multiplicity. We have explored this hypothesis by calculating the pre-scission multiplicity for different systems in the region of interest in a simple statistical model with no dissipation inside the saddle point, and a saddle-to-scission time of $t_{ss} = 2 \times 10^{-21}$ sec. The results are shown as dashed curves in Fig. 4.

The solid curves represent calculations, where the effects of dissipation are included assuming a reduced dissipation coefficient of $\gamma = 5$ and a saddle-to-scission time of $t_{ss} = 20 \times 10^{-21}$ sec. These parameters are in approximate agreement with full one-body dissipation. We observe that the effects of dissipation is generally much stronger for the heavier systems (large fissilities). For the lighter systems, the additional multiplicity of pre-scission neutrons is discernable only at high excitation energy, whereas it becomes apparent already at low excitation energy for the heaviest system. Adopting the criteria that an additional pre-scission neutron multiplicity of 0.5 unit is required for the excess to be experimentally significant, we find that the threshold energies, E_{obs}, for the observation of the effect of dissipation in the pre-scission neutrons are as listed in Table 2.

We notice a very strong similarity between the expected observation threshold and the observed excitation energy where the multiplicity of pre-scission neutrons deviate from the prediction of the pure sta tistical model. However, it should be

Table 2: Excitation energy threshold for the observation of dissipation effects

Reaction	Fissility x	Observation threshold E_{obs} (MeV)	E_{thresh} [16] (MeV)
$^{19}F + ^{159}Tb \rightarrow ^{178}W$	0.64	92	80 ± 10
$^{19}F + ^{169}Tm \rightarrow ^{188}Pt$	0.67	78	80 ± 5
$^{19}F + ^{181}Ta \rightarrow ^{200}Pb$	0.70	67	65 ± 5
$^{16}O + ^{197}Au \rightarrow ^{213}Fr$	0.74	32	45 ± 5
$^{16}O + ^{208}Pb \rightarrow ^{224}Th$	0.76	25	30 ± 5
$^{19}F + ^{232}Th \rightarrow ^{251}Es$	0.83	22	20 ± 10

pointed out that for some of the data points the extracted threshold is based on a large extrapolation from measured values. Thus the lowest point for the $^{19}F + ^{232}Th$ reaction was taken at $E_{exc} = 52.8$ MeV, whereas the extracted threshold is given[16] as $E_{thresh} = 20 \pm 10$ MeV. However, the fact remains that the energy spectra of γ-rays are compatible with the pure statistical model prediction up to an excitation energy of $E_{exc} = 70$ MeV for the $^{32}S + ^{208}Pb$ reaction leading to a composite system of almost identical fissility. At present this discrepancy represents a serious puzzle, and further experiments in which both the pre-scission neutronms and γ-rays are measured in the same system.

5. Conclusion

From the analysis of the multiplicity of GDR γ-rays emitted prior to scission, we are able to obtain a sensitive measure of the time-scale of fission and therefore the dissipation associated with this process. It is found that the one-body dissipation mechanism gives a good account of the observations at high excitation energies - an observation, which may be surprising in view of the theoretical shortcomings of the basic assumptions of this model [7,17].

The dependence on excitation energy has been studied in a few cases, and it appears that there is a rapid increase in dissipation strength in the range $E_{exc} = 60$-80 MeV. This is in contrast to the observation from measurements of pre-scission neutrons, where the full one-body dissipation strength appear to persist down to near the observational threshold, i.e. $E_{obs} \approx 20$ MeV for the heaviest systems studied. Evidence from fission probability measurements and the analysis of the widths of iso-scalar giant quadrupole and octupole resonances indicate, however, that the dissipation at low excitation energies $E_{exc} = 6$-12 MeV is insignificant. At the present time this situation is not clear. Further experiments are clearly warrented to try to settle this important aspect of nuclear dynamics.

One of the authors (BBB) gratefully acknowledges the hospitality extended to him during his stay at SUNY at Stony Brook. This work was supported in part by the U.S. Department of Energy, Nuclear Physics Division, under contract No. W-31-109-ENG-38 and in part by the U. S. National Science Foundation.

6. References

[1] R. Bock, Y. T. Chu, M. Dakowski, A. Gobbi, E. Grosse, A. Olmi, H. Sann, D. Schwalm, U. Lynen, W. F. J. Müller, S. Bjørnholm, H. Esbensen, W. Wölfli,and E. Morenzoni, Nucl. Phys. **A388**, 334 (1982)

[2] J. Tōke, R. Bock, G. X. Dai, A. Gobbi, S. Gralla, K. D. Hildenbrand, J. Kuzminski, W. F. J. Müller, A. Olmi, H. Stelzer, B. B. Back and S. Bjørnholm, Nucl. Phys. **A440**, 327 (1985)

[3] W. Q. Shen, J. Albinski, A. Gobbi, S. Gralla, K. D. Hildenbrand, N. Herrmann, J. Kuzminski, W. F. J. Müller, H. Stelzer, J. Tōke, B. B. Back, S. Bjørnholm, and S. P. Sørensen, Phys. Rev. **C36**, 115 (1987)

[4] J. Błocki, Y. Boneh, J. R. Nix, J. Randrup, M. Robel, A. J. Sierk, and W, J, Swiatecki, Ann. Phys. **113**, 330 (1978)

[5] K. T. R. Davie, A. J. Sierk, and J. R. Nix, Phys. Rev. **C13** 2385 (1976); A. J. Sierk and J. R. Nix, Phys. Rev.**C21** 982 (1980)

[6] J. R. Nix and A. J. Sierk, in proceedings "Winter Workshop on Nuclear Dynamics IV", Copper Mountain, Colorado, 1986

[7] J. J. Griffin and M. Dworzecka, Phys. Lett. **156B**, 139 (1985)

[8] V. E. Viola, K. Kwiatkowski, and M. Walker, Phys. Rev. **C31**, 1550 (1985)

[9] N. Carjan, A. J. Sierk, and J. R. Nix, Nucl. Phys. **A452**, 381 (1986)

[10] A. Gavron et al., Phys. Rev. **C35**, 579 (1987)

[11] D. J. Hinde et al., Phys. Rev **C45**, 1229 (1992)

[12] J. P. Lestone, Phys. Rev. Lett. **70**, 2245 (1993)

[13] R. Butsch, M. Thoennessen, D. R. Chakrabarty, M. G. Herman, and P. Paul, Phys. Rev. **C41**, 1530 (1990)

[14] D. J. Hofman, B. B. Back, I. Diószgi, C. P. Montoya, S. Schadmand, R. Varma, and P. Paul, Phys. Rev. Lett. **72**, 470 (1994)

[15] A. Gavron, H. C. Britt, E. Konecny, J. Weber, and J. Wilhelmy, Phys. Rev. **C13**, 2374 (1976)

[16] M. Thoennessen and G. F. Bertsch, Phys. Rev. Lett. **71**, 4303 (1993)

[17] W. Bauer, D. McGrew, V. Zelevinsky, and P. Schuck, these proceedings

CHARACTERIZATION OF PROJECTILE-LIKE PRODUCTS OF PERIPHERAL Xe-INDUCED REACTIONS

H. Madani, A.C. Mignerey, D.E. Russ, and J.Y. Shea
University of Maryland, College Park
G. Westfall, W.J. Llope, D. Craig, E. Gualtieri, S. Hannuschke,
T. Li, A. Vandermolen, and J. Yee
National Superconducting Cyclotron Laboratory
and Department of Physics and Astronomy
Michigan State University, East Lansing
E. Norbeck, R. Pedroni
Department of Physics
University of Iowa

ABSTRACT

Projectile-like fragments (PLF) produced by the 50-MeV/u ^{129}Xe on ^{27}Al, $^{nat.}$Cu, ^{139}La, and ^{165}Ho reactions were detected using the Maryland Forward Array (MFA) inside the MSU 4π array to determine the types of mechanisms occurring in peripheral heavy-ion collisions at intermediate energies. The experimental results are compared to model predictions based on the nucleon exchange process that is generally used to describe peripheral heavy-ion reactions at low bombarding energies.

1. Introduction

The study of intermediate energy heavy-ion reactions is a good means of gaining insight into the transition from attractive mean-field interactions, prevalent at low bombarding energies, to nucleon-nucleon collisions, dominant at higher bombarding energies. Extensive studies of peripheral heavy-ion reactions at low bombarding energies have succeeded in characterizing the deep-inelastic mechanism occurring in this energy regime in terms of a dissipative binary process [1-3]. The features of this process are fairly well described by nucleon exchange models [3,4]. Experimental results from reactions at higher bombarding energies, around the Fermi energy, have shown that the characteristics of a damped binary process are still present at intermediate energies [5-7]. The goal of the present study of Xe-induced peripheral reactions at E/A = 50 MeV is to explore the presence of deep-inelastic processes in the Fermi energy regime by comparing experimental charge and velocity distributions to model predictions.

2. Experimental Set-up

Table II.B-1 Reaction Parameters for 50-MeV/u ^{129}Xe on ^{27}Al, $^{nat.}$Cu, ^{139}La, and ^{165}Ho systems.

^{129}Xe + X	^{27}Al	$^{nat.}$Cu	^{139}La	^{165}Ho
E_{cm} - V_c (MeV)	1026	1938	3036	3266
θ_{gr} (Lab) (°)	0.83	1.66	5.56	5.88
E^*_{PLF}/A_{PLF} (MeV/u) at TKEL: 350-450 MeV	1.67	1.51	NA	1.56
E^*_{PLF}/A_{PLF} (MeV/u) at TKEL: 650-750 MeV	3.26	2.77	2.78	2.67
E^*_{PLF}/A_{PLF} (MeV/u) at TKEL: 1450-1550 MeV	NA	NA	5.67	5.63

An experiment to characterize the products of the 50-MeV/u ^{129}Xe on ^{27}Al, $^{nat.}$Cu, ^{139}La, and ^{165}Ho reactions was performed at the Michigan State University Superconducting Cyclotron Laboratory (MSUNSCL). The Maryland Forward Array (MFA)[8] was used to measure projectile-like fragments in coincidence with target-like fragments and light-charge particles that end up in the MSU 4π array [9].

The MFA consists of a 300-μ annular Si detector divided into 16 segments on one face and 16 concentric rings on the other face for position information. This Si detector is backed by a matching 16-element plastic phoswich, consisting of fast and slow plastic scintillators of thicknesses 1 mm and 10 cm, respectively.

The MFA was positioned inside the 4π array at 94.5 cm behind the target, with its symmetry axis along the beam axis. In this position, laboratory angles from 1.5° to 2.9° could be covered and, therefore, identification of projectile-like fragments in the grazing angle range was possible. Good Z resolution was achieved[8]. The reaction parameters: laboratory grazing angle θ_{gr}, available kinetic energy at entrance channel, E_{cm}-V_c (center-of-mass energy - Coulomb barrier), and representative PLF excitation energies at a given total kinetic energy loss (TKEL) are summarized in Table II.B-1 for the four systems employed in the present study. More experimental details are given in Section II.A.

3. Results and Discussion

3.1. PLF Charge Variances

In a deep-inelastic process, the evolution of the system with interaction time can be described by the correlations between the fragment's size (A or N and Z) and the total kinetic energy loss (TKEL). The mass (neutron number) and charge widths of the nuclide distributions are a function of the number of nucleon exchanges between the two interacting nuclei. At low bombarding energies, experimental evidence shows that the charge and mass distributions of projectile-like fragments (PLF's) broaden with increasing TKEL. This is attributed to an increase in nucleon exchanges between the two ions with interaction time. In the present analysis, the evolution of the experimental charge distributions of the PLF's obtained in the 50-MeV/u ^{129}Xe on

^{27}Al, $^{nat.}$Cu, ^{139}La, and ^{165}Ho systems, as a function of TKEL, is examined and compared to the predictions of Tassan-Got's nucleon exchange model.

A density plot of the charge as a function of total kinetic energy loss (TKEL) is shown in Figure 1 for the Xe on Cu system. It is evident that the charge distribution is narrow at all values of TKEL. The same feature was observed for the other three systems. The centroids $<Z>$ and variances σ_Z^2 of these distributions were determined as a function of TKEL and are shown in Figure 2 for all four systems. A maximum TKEL range as defined by the entrance channel Coulomb barrier, is indicated by the arrows. The experimental charge centroids (circles) decrease smoothly with increasing TKEL. The experimental charge variances, σ_Z^2, are narrow and lower than 2 for TKEL values lower than the maximum TKEL range for all four systems. Above this limit, σ_Z^2 increases for the Xe + Al and Xe + Cu systems.

The reactions were modeled using the nucleon exchange model of Tassan-Got, which simulates the stochastic transfer between two interacting heavy ions by a Monte Carlo method[3]. Using the statistical decay code GEMINI, The primary model predictions (solid lines) have been corrected for light particle evaporation (dotted lines), they are compared to experimental results. The model predictions do not reproduce the experimental $<Z>$, even after correcting for evaporation, especially for the heavier targets. A difference between the experimental values of σ_Z^2 and those predicted by Tassan-Got's nucleon exchange model is observed for the Xe + Al and Xe + Cu systems. For these two systems, the values of σ_Z^2 before (solid lines) and after (dotted lines) evaporation corrections are not significantly different. On the other hand, the secondary σ_Z^2 obtained for the Xe + La system decreases with increasing TKEL, until it reaches experimental values. A comparison of model calculated secondary σ_Z^2 among all four systems is not presently possible, since the predictions are yet incomplete. However, the present results show that it is possible that model predictions of secondary σ_Z^2 would exhibit a similar behavior for all systems.

A priori, it seems that the experimental results reported in the present study cannot be described in terms of a nucleon exchange mechanism. However, it is essential to realize that the determination of TKEL contains some caveats, the most important of which is probably neglecting preequilibrium emission, which is expected to have a significant effect at the bombarding energies used in the present study. To investigate the possibility of a transition between different mechanisms with a change of system size, and hence of excitation energy, a comparison among the PLF charge centroids and variances obtained with the four different systems is in order. However, because of the different TKEL ranges available in each case, it is appropriate to define a variable that would bring all the systems in the same range for comparison purposes. Since the number of nucleons exchanged between the two interacting ions is proportional to the system's relative velocity, the variable T_{red} defined as the ratio of the system relative velocity, $\sqrt{E_{cm} - Vc - TKEL}$, to the the entrance channel relative velocity, $\sqrt{E_{cm} - Vc}$. Thus, T_{red} is expressed as

$$T_{red} = 1 - \sqrt{\frac{E_{cm} - V_c - TKEL}{E_{cm} - V_c}},$$

and has values from 0 to 1.

A plot of σ_Z^2 as a function of T_{red} is displayed in Figure 3. A gradual increase of σ_Z^2 with increasing T_{red} is observed for all four systems. However, σ_Z^2 seems to saturate for the La and Ho targets, while it continues to increase for the lighter targets, Al and Cu. This difference of the behavior of σ_Z^2 with T_{red} could indicate that a contribution from deep-inelastic events decreases with increasing system size.

3.2. PLF Velocities

Another extreme mechanism to be considered is cold fragmentation of the projectile-like fragment. In this scenario, a piece of the projectile is sheared off without affecting its momentum. Therefore, the PLF's obtained are expected to continue moving with beam velocities. A three-dimensional plot of counts versus the ratio of PLF velocity to beam velocity, V_{PLF}/V_{beam}, as a function of the multiplicity of charged particles detected in the 4π array is displayed in Figure 4 for the Xe + Al system. The presence of beam velocity components at high multiplicities is observed; however, the V_{PLF}/V_{beam} centroids decrease with decreasing PLF charge, as shown in Figure 5 for all four systems. It is interesting to note that while the results for the three heavier targets fall on a common line, the $<V_{PLF}/V_{beam}>$ values for the Al target are consistently higher. Tassan-Got's model predictions of $<V_{PLF}/V_{beam}>$, after evaporation correction, are indicated by the solid line in Figure 5 for the PLF's emitted in the Xe + Cu reaction. Even though the model slightly underpredicts the experimental velocities, a good qualitative agreement between data (crosses) and model predictions is observed.

4. Summary

The experimental results reported in the present study of 50-MeV/u ^{129}Xe on ^{27}Al, $^{nat.}$Cu, ^{139}La, and ^{165}Ho reactions have been compared to the features of deep-inelastic mechanisms that describe peripheral heavy-ion reactions at low bombarding energies. Tassan-Got's model predictions, corrected for evaporation, underpredict the experimental $<Z>$ values for all systems, however, the difference is minimal for the Al target. Some of the differences between the experimental charge distributions and Tassan-Got's model predictions could be due to the fact that preequilibrium emission, which becomes significant at high bombarding energies, is not taken into account in the model calculations. However, this effect cannot explain the discrepancy in the charge variances.

The presence of near beam-velocity events at high charged particle multiplicities, is observed. This is indicative of the persistence of a binary mechanism even at bombarding energies much higher than the Coulomb barrier. This is in agreement with previous studies in which the features of dissipative binary processes have been observed for reactions at bombarding energies in the Fermi energy regime [5-7].

5. References

[1] H. Freiesleben *et al.*, *Phys. Rep.* **106**, 1 (1984).
[2] W.U. Schröder, and J.R. Huizenga, *Treatise on Heavy-Ion Science*, edited by D.A. Bromley (Plenum: New York, 1984), Vol. 2, p. 115.
[3] L. Tassan-Got and C. Stephan, Internal Report INPO-DRE-89- 46, 1989. *Nucl. Phys.* **A524**, 121 (1991).
[4] H. Madani *et al.*, submitted to *Phys. Rev. C* (1994), Ph. D. Thesis. # ER-4021-11, University of Maryland (1993)
[5] B. Lott *et al.*, *Phys. Rev. Lett.* **68**, 3141(1992).
[6] S. Baldwin *et al.*, Proceedings to the 9th Winter Workshop in Nuclear Dynamics, Key West, Florida, USA (1993)
[7] B. Quednau *et al.*, *Phys. Lett. B* **309**, 10 (1993).
[8] A.C. Mignerey *et al.*, Annual Progress Report # ER/40321-9, University of Maryland (1992), Annual Progress Report # ER/40321-12, University of Maryland (1993).
[9] G. Westfall *et al.*, *Nucl. Inst. and Meth. A* **238** , 347(1985).

Xe + Cu at 50 MeV/u
(Multiplicity > 4)

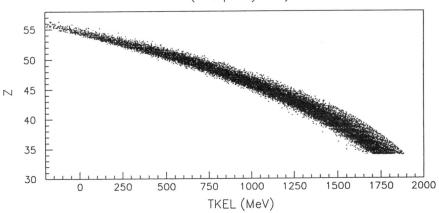

Fig. 1: Density plot of Z versus TKEL for the 50-MeV/u ^{129}Xe on ^{nat}Cu system.

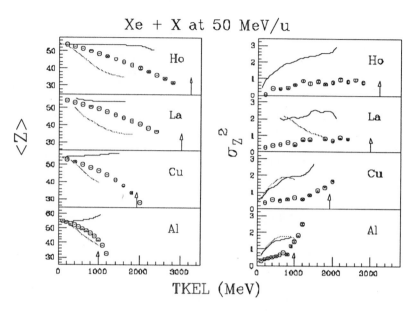

Fig. 2: Experimental charge centroids and variances (circles) as a function of TKEL. Tassan-Got's model predictions before and after evaporation corrections are indicated by the solid and dotted lines, respectively.

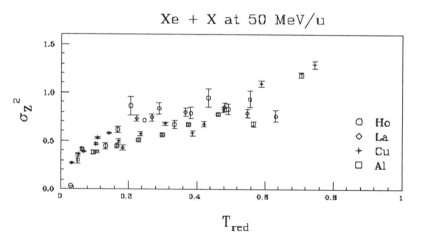

Fig. 3: Experimental charge variances as a function of T_{red} for 50-Mev/u ^{129}Xe on ^{29}Al (squares), $^{nat.}$Cu (crosses), ^{139}La (diamonds), and ^{165}Ho (circles).

Xe + Al at 50 MeV/u

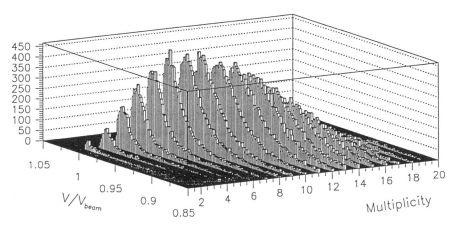

Fig. 4: A three-dimensional plot of V_{PLF}/V_{beam} as a function of the multiplicity of charged particles detected in the 4π ball for the 50-MeV/u ^{129}Xe on ^{29}Al system.

Xe + X at 50 MeV/u

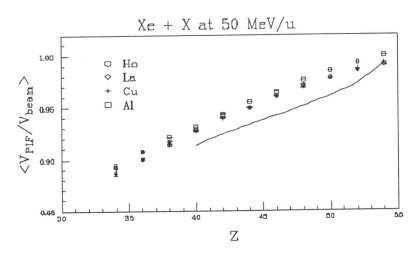

Fig. 5: The experimental V_{PLF}/V_{beam} centroids as a function of PLF charge for the 50-MeV/u ^{129}Xe on ^{27}Al, $^{nat.}$Cu, ^{139}La, and ^{165}Ho systems. Model predictions of $<V_{PLF}/V_{beam}>$ are indicated by the solid line for the ^{129}Xe on $^{nat.}$Cu system.

CHARACTERIZING FRAGMENT EMISSION IN INTERMEDIATE ENERGY HEAVY-ION REACTIONS

R. T. DE SOUZA and D. FOX

Department of Chemistry
and Indiana University Cyclotron Facility
Indiana University, Bloomington, IN 47405, USA

ABSTRACT

Experimental fragment-fragment correlations are used to probe the spatial-temporal extent the emitting source for central ^{36}Ar+^{197}Au collisions at E/A=35, 50, 80, and 110 MeV. The experimental correlations are compared with the Koonin-Pratt two-body formalism as well as with a 3-body Coulomb trajectory calculation. With increasing incident energy the spatial-temporal extent of the emitting system decreases. Comparison with a generalized N-body Coulomb trajectory model reveals that effects of interactions with other emitted particles are negligible.

1. Introduction

Highly excited nuclear matter can decay by the emission of multiple intermediate mass fragments (IMFs: $3 \leq Z \leq 20$) [1,2,3,4,5]. The experimental observation of multiple fragment emission in intermediate energy heavy-ion collisions has re-stimulated considerable speculation as to the mechanism underlying this phenomenon. Two radically different scenarios have been proposed to explain this phenomenon. A multifragment final state might arise either from a nearly simultaneous disintegration of the highly excited system [6,7,8,9,10], or from a series of sequential binary decays [11,12,13,14,15]. In both scenarios the formation and emission of multiple fragments is substantially enhanced by a reduction in the density of the nuclear system [6,7,8,9,15,16,17,18]. The density reduction might be driven by either a rapid dynamical compression decompression cycle [6,7,8,9,10] or a thermal pressure burst[19,20,21].

To constrain both statistical and dynamical models determination of the fragment emission time scale is important. To better characterize the evolution of the spatial and temporal extent of the decaying system with increasing incident energy, we have measured an excitation function for the system ^{36}Ar+^{197}Au at E/A=50, 80, and 110 MeV. Previously reported values for the mean emission time were based on an assumed value of the source radius [25]. In the present paper we will more closely examine the dependence of the deduced spatial-temporal extent of the decaying system on various assumptions.

2. Experimental Setup

The experiment was performed using the K1200 cyclotron at the National Superconducting Cyclotron Laboratory at Michigan State University. ^{36}Ar beams at E/A=50, 80, and 110 MeV, with an intensity of approximately 1x10^8 particles per second impinged on a ^{197}Au target with an areal density of 1 mg/cm^2. Charged particles emitted into the angular range $9° \leq \theta_{lab} \leq 160°$ were detected using the MSU Miniball array [29]. Particles were identified by atomic number up to Z=18 and by mass for Z=1 and 2. The approximate energy thresholds were $E_{th}/A \approx 2$ MeV for Z=3, $E_{th}/A \approx 3$ MeV for Z=10, and $E_{th}/A \approx 4$ MeV for Z=18. For detectors in rings 1-4, $9° \leq \theta_{lab} \leq 40°$, these calibrations are estimated to be accurate to within 5%. Data was taken with a trigger condition which required that at least two detectors in the Miniball were triggered in order to record an event.

3. General Properties

^{36}Ar+^{197}Au collisions at E/A=50, 80, and 110 MeV may lead to the emission of a large number of charged fragments, N_C, [25,30]. At all beam energies the N_C distribution is roughly characterized by a plateau followed by a steeply falling tail. The end of the plateau increases from $N_C \approx$20 for E/A = 50 MeV to $N_C \approx$32 for E/A = 110 MeV. The mean IMF multiplicity, $< N_{IMF} >$, increases with both increasing N_C and beam energy [3]. For the most central collisions, high N_C, $< N_{IMF} >$ reaches 2, 3, and 4 for E/A=50, 80, and 110 MeV respectively. The measured charged particle multiplicity can be used as a measure of the centrality of the collision. Small impact parameters, b, correspond to composite systems with high excitation and consequently large charged particle multiplicity. Using a simple geometrical model [31],the total charged particle multiplicity measured in the Miniball can be related to the impacat parameter of the collision. In this prescription, for each value of N_C the impact parameter, b, is found in terms of the maximum impact parameter, b_{max}, for which at least two charged particles are detected in the Miniball. In an attempt to isolate a single source of fragments as well as focus on systems where the energy deposition is presumably largest, the fragment correlations have been constructed for central collisions $b/b_{max} \leq 0.2$.

4. Spatial-Temporal Extent of the Emitting System

The strength of the final-state interaction between two fragments is measured using the two IMF correlation function $R(v_{red})$ which is defined in terms of the ratio of the coincidence yield, Y_{12}, to the product of the single particle yields, Y_1 and Y_2.

$$\sum_{\mathbf{p_1,p_2}} Y_{12}(\mathbf{p_1 p_2}) = C[1 + R(v_{red})] \sum_{\mathbf{p_1,p_2}} Y_1(\mathbf{p_1})Y_2(\mathbf{p_2}), \tag{1}$$

where $\mathbf{p_1}$ and $\mathbf{p_2}$ are the laboratory momenta of IMFs 1 and 2, the relative velocity, $v_{rel} = (\mathbf{p_1}/m_1 - \mathbf{p_2}/m_2)$, and the reduced velocity $v_{red} = v_{rel}/\sqrt{Z_1 + Z_2}$. C is a normalization constant determined by requiring $< R(v_{red}) >=0$ at large relative momenta where the final state interaction is small. The masses of the IMFs, m_1 and m_2, are assumed to be twice their atomic number times the nucleon rest mass. All correlation functions presented in this paper have been normalized to one in the region $0.026c \leq v_{red} \leq 0.035c$. Construction of the correlation function as a function of v_{red} allows summation over all combinations of IMF pairs with $4 \leq Z_1, Z_2 \leq 9$ [23,24].

The strength of the IMF-IMF final state interaction can be related to the initial spatial-temporal separation of the two fragments. Fragments emitted from a source with a relatively short mean IMF emission time, τ, will experience a stronger final state interaction than fragments emitted from a source with a longer mean IMF emission time. Similarly, fragments which are emitted from a source with a small radius, R_S, will have a stronger final state interaction than fragments emitted from a source with a larger radius. In order to extract information about the mean emission time comparisons between the experimental correlation functions and theoretical calculations are needed.

4.1. Koonin-Pratt Model

The simplest theoretical model for two particle correlations is the two particle Koonin-Pratt formalism [32] in which only the final-state interactions between the two emitted IMFs are considered. Effects due to the presence of other charged bodies on the two fragments are assumed to be negligible in the Koonin-Pratt formalism.

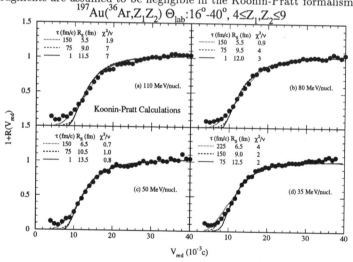

Fig. 1: Comparisons of Koonin-Pratt calculations with the best χ^2/ν values for central ^{36}Ar+^{197}Au collisions at E/A=35, 50, 80, and 110 MeV to the experimental correlation functions.

Previous comparisons between the experimental correlation functions and calculations based on the Koonin-Pratt formalism have been based on a limited number of combinations of the source size, R_S and the mean emission time, τ [22,23,24,25]. To more fully probe the spatial-temporal extent of the emitting system, we have carried out a series of calculations using the Koonin-Pratt formalism. In these calculations, the source size was varied in 0.5 fm steps over the range $5.5 \leq R_S \leq 14$ fm, and the mean emission time was varied in 25 fm/c steps over the range $1 \leq \tau \leq 250$ fm/c. For each calculation, the χ^2 per degree of freedom, χ^2/ν, between the Koonin-Pratt calculation and the experimental correlation function was calculated. At each energy there is a clear valley in χ^2/ν running from very short values of τ and large values of R_S, to larger values of τ and smaller values of R_S. An example of the agreement reached between the Koonin-Pratt formalism and the experimental final state interaction, a slice along the valley for each beam energy is shown in Figure 1. While at $E/A = 80$ and 110 MeV the calculations with $\tau = 150$ fm/c do a somewhat better job of reproducing the data for $v_{red} \geq 0.010c$, all the calculations shown in Fig. 1 do a reasonable job of reproducing the data for $v_{red} \geq 0.010c$.

The calculations indicate a clear preference for reducing R_S with increasing beam energy while leaving τ unchanged. However, the size of the simplest geometrical configuration (two touching IMFs) $R_{IMF} = 2.4\text{-}3.1$ fm however places certain minimum constraints on the size of the source. In addition, both dynamical multifragmentation models and statistical decay models require a significant reduction in the density of the system in order to produce multiple IMFs [6,7,8,9,15,16,17,18].

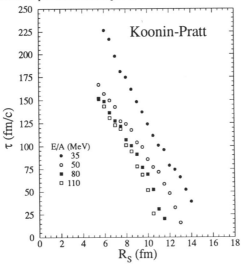

Fig. 2: Mean emission time τ as a function of R_S for central $^{36}\text{Ar}+^{197}\text{Au}$ collisions at $E/A=35$, 50, 80, and 110 MeV.

The dependence of τ on R_S at all four beam energies is shown in Fig. 2. The value of τ shown in Fig. 2 is the weighted average of the three calculations with the best χ^2/ν for each value of R_S. From Fig. 2 it can be seen that if one assumes a given source radius, as has been done previously [22,23,24,25], then τ decreases with increasing beam energy. In order to uniquely determine the mean emission time, an independent measurement of the source radius is required. For any reasonable value of R_S, $R_S \geq_{IMF1} +R_{IMF2}$, $\tau \leq 225$ fm/c at $E/A=35$ MeV and ≤ 145 fm/c at $E/A=110$ MeV. Furthermore if we assume a reduction in the density of the emitting source, a condition required be theoretical models to account for the observed high fragment yield [6,7,8,9,15,16,17,18], then assuming emission from a Au nucleus with $\rho/\rho_0 \approx 0.3$, we find that $\tau \approx 110$ and 25 fm/c for $E/A=35$ and 110 MeV respectively.

4.2. 3-Body Coulomb Trajectory Model

The Koonin-Pratt calculations presented in the previous section involve only the final-state interaction between two IMFs emitted from a decaying system. To examine the effect of the presence of the emitting system on the IMF-IMF correlation functions a 3-body Coulomb trajectory calculation is necessary. In the 3-body Coulomb trajectory calculation, the decaying system, with initial charge Z_S, initial mass number A_S, and radius R_S, emits the first IMF and then recoils. The trajectories of the emitted IMF and the source residue are calculated up to time t when the second IMF is emitted. The emission time t of the second IMF is assumed to be exponential with a mean time τ. The charge and energy of the two IMFs sample the measured Z-distribution and energy spectra. The mass numbers of the IMFs are taken to be twice their charge. Previously published comparisons between 3-body trajectory calculations and the data were restricted to only a few combinations of R_S and τ [25]. Analagously to the Koonin-Pratt calculations the three body model was run for a grid in R_S and τ and χ^2/ν was calculated. At each beam energy the calculations were carried out for two different emitting source sizes, $Z_S=40$ and $A_S=96$, and $Z_S=79$ and $A_S=197$. As with the Koonin-Pratt calculations, a well defined minimum is observed in the χ^2/ν as a function of R_S and τ. For a fixed source size, Z_S and A_S, the spatial-temporal extent of the emitting system decreases with increasing beam energy.

The behavior of τ as a function of R_S for the 3-body trajectory calculations is shown in Fig. 3(a). The solid points represent the 3-body trajectory calculations carried out with $Z_S=40$ and $A_S=96$, the open points are for $Z_S=79$ and $A_S=197$, and the lines show the corresponding results from the Koonin-Pratt formalism. The 3-body trajectory calculations indicate a decrease in the spatial-temporal extent of the emitting system with increasing beam energy. This result is in agreement with the Koonin-Pratt calculations.

For most cases the 3-body trajectory calculations indicate longer mean emission times than those inferred from the Koonin-Pratt calculations. The 3-body trajectory calculations for the larger initial source, $Z_S=79$ and $A_S=197$, show a slightly weaker dependence on R_S than the calculations for the smaller source. Examining the effect of the two different initial choices of Z_S and Z_S we see that for $R_S \leq 8$ fm using the larger source results in the extraction of a shorter mean emission time, while for

R_S >10 fm the larger source indicates a longer mean emission time. In Fig. 3(b) τ has been replotted in terms of the ratio of the density of the decaying system, ρ, to normal nuclear density, ρ_0. For all four cases shown in Fig. 3(b) the mean emission time initially rises steeply with increasing density until $\rho/\rho_0 \approx 0.3$. For $\rho/\rho_0 > 0.3$ the mean emission time is almost independent of the density of the emitting system. Further, for a given density the calculation for the smaller initial source leads to a longer extracted mean emission time. It is interesting to note that many theoretical calculations require freeze out densities of $\approx 0.3\rho_0$. For $\rho/\rho_0 \approx 0.3$, the mean emission time for E/A=50 MeV is 115-150 fm/c. The mean emission time decreases with increasing beam energy where we have $\tau \approx$75-100 fm/c at E/A=110 MeV.

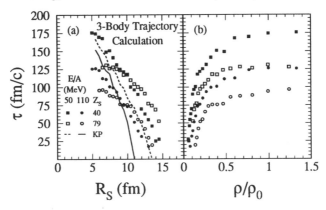

Fig. 3: Mean emission times as a function of (a) R_S and (b) ρ/ρ_0 extracted from the data using a 3-body trajectory model. Extracted mean emission times for central ^{36}Ar+^{197}Au collisions at E/A=50 (squares) and 110 MeV (circles) are shown. The open symbols are for a source with initial Z_S=79 and A_S=197, the solid symbols are for a source with initial Z_S=40 and A_S=96. The lines are for the Koonin-Pratt calculations from Fig. 2.

5. Discussion

The comparisons to both the Koonin-Pratt and 3-body Coulomb trajectory models made in the previous section show a clear decrease in the mean IMF emission time with increasing beam energy. The mean emission times based on the Kooin-Pratt model are comparable to the transit time of an Ar projectile past a Au nucleus, $\tau = 45 - 80$ fm/c. A comparison of the mean emission times from the Koonin-Pratt and 3-body Coulomb trajectory models indicates that the effect of including a target residue on the extracted mean emission time is an $\approx$50 fm/c increase in the mean emission time if the residue is assumed to be a Au nucleus and the source has expanded to $\rho/\rho_0 \approx 0.3$. Both model comparisons show that for $E/A = 50-110$ MeV the mean emission times are consistent with emission times expected from dynamical models [8,9,10] or an expanding-evaporating source model [15,33]. The decrease in the deduced emission time with increasing beam energy is in agreement with a transition from sequential emission to simultaneous multifragmentation.

The results presented so far have been obtained using models which consider only two or three of the particles involved in the reactions being studied. The effect of the presence of other charged bodies on the fragment correlation function has been neglected. While these effects might be expected to be small at low energies where most of the charge is contained in the two observed fragments and a large residue, at higher energies, $E/A=110$ MeV, an average of four IMFs and at least 28 light charged particles are observed in the Miniball acceptance for central collisions [3].

5.1. N-Body Trajectory Calculations

In order to study the possible effects of the interaction between all charged fragments on the correlations functions we have written a generalized N-body Coulomb trajectory code. In this calculation we start with a heavy residue of charge Z_{res} located at the center of a sphere of radius R. The rest of the charge of the initial system, $Z_{tot}=97$, is then distributed to additional particles based on the measured Z-distribution. The particles are assigned random, non-overlapping positions within the sphere and are emitted simultaneously with an isotropic angular distribution. The masses of all particles are taken from the minimum of the β stable valley. The initial energy of each fragment is chosen to sample a Maxwell-Boltzman distribution with temperature T. The trajectories of all charged bodies are then integrated and the correlation function is constructed based upon the asymptotic momenta of the fragments after filtering for the detector acceptance.

It is important to note that there are some major differences between the 3-body and N-body Coulomb trajectory models which make direct comparisons between the models difficult. In the 3-body model the IMFs are emitted from the surface of a sphere with an initial momentum away from the residue. In the N-body calculation all particles are emitted from within the volume of a sphere with no correlation between the position and momentum of the particle. It is possible in the N-body model for a particle to start out with an initial momentum vector pointing towards the residue at the center of the sphere. The effects of these differences between the models is, to first order, to reduce the effective source radius.

The results of four different N-body calculations are compared in Fig, 4. The four calculations are:

(a) A full N-body calculations in which the residue, $Z_{res} = 30$, and all charged bodies are tracked, dashed line.

(b) All particles are generated, but only the residue, $Z_{res} = 30$, and IMFs with $4 \leq Z \leq 9$ are tracked, solid line.

(c) Only the residue, $Z_{res} = 30$, and two IMFs with $4 \leq Z \leq 9$ are generated and tracked, dash-dotted line.

(d) Only the residue, $Z_{res} = 44$, and four IMFs with $4 \leq Z \leq 9$ are generated and tracked, dotted line.

For all four cases T=20 MeV, R=11.26 fm and only IMFs with $4 \leq Z \leq 9$ are used to construct the correlation functions. A value of Z_{res}=30 in the N-body Coulomb calculation is comparable to Z_S=40 for the 3-body Coulomb trajectory calculation because Z_S in the 3-body calculation is the charge of the emitting system prior to emission of the IMFs.

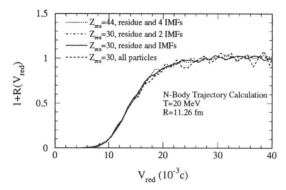

Fig. 4: Results of N-body Coulomb trajectory calculations with T=20 MeV, R=11.26 fm. The dashed line is for the full calculation with Z_{res}=30 in which all charged particles are included. The other line types represent limited calculations in which only the residue and IMFs with $4 \leq Z \leq 9$ are included in the calculation. For the solid line fragments were formed using the measured Z distribution for all charges, but only those IMFs with $4 \leq Z \leq 9$ were then tracked along with a Z_{res}=30 residue. The dash-dotted line represents a calculation in which only two IMFs with $4 \leq Z \leq 9$ and the residue, with Z_{res}=30 were considered. The dotted line is for a calculation with Z_{res}=44 and four IMFs with $4 \leq Z \leq 9$.

In Fig. 4 we see very little difference between the four cases. The large number of charged particles, include in case (a) and neglected in the other cases have no effect on the correlation function. In addition only tracking two IMFs, case (c), instead of all the IMFs also has no effect on the correlation functions. It can, therefore, be concluded that only the residue and two IMFs need be considered, all other particles have a negligible effect on the correlation functions.

6. Conclusions

The spatial-temporal extent of the emitting source for central ^{36}Ar+^{197}Au interactions has been shown to decrease with increasing beam energy. Comparisons with a 3-body Coulomb trajectory model show that the mean emission time rises steeply as a function of the density of the emitting system until $\rho/\rho_0 \approx 0.3$. For higher densities the mean emission time is essentially flat as a function of ρ. While extraction of a unique mean emission time requires an independent measure of the source size, or density, we have shown that for any reasonable value of R_S, $R_S \geq R_{IMF1} + R_{IMF2}$, comparing the measured correlation functions to calculations of the Koonin-Pratt model yields $\tau \leq 225$ fm/c at $E/A=35$ and $\tau \leq 145$ fm/c at $E/A=110$ MeV. Using a 3-body Coulomb trajectory model and assuming $\rho/\rho_0 \approx 0.3$ leads to mean emission times of 115-150 and 75-100 fm/c at $E/A=50$ and 110 MeV respectively. Calculations with an N-body trajectory code indicates that light charged particles and additional IMFs have a negligible effect on the measured

7. Acknowledgements

We would like to thank Dr. T. Glasmacher for the use of his N-body trajectory calculation code. This work was supported by the U.S. Department of Energy under DE-FG02-92ER40714 and the National Science Foundation under Grant No. PHY-89-13815 and PHY-90-15957.

8. References

[1] J. W. Harris et al., Nucl. Phys. **A471**, 241c (1987).

[2] K. G. R. Doss et al., Phys. Rev. Lett. **59**, 2720 (1987).

[3] R.T. de Souza et al., Phys. Lett. **B268**, 6 (1991).

[4] C. A. Ogilvie et al., Phys. Rev. Lett. **67** 1214 (1991).

[5] D. R. Bowman et al., Phys. Rev. Lett. **67**, 1527 (1991).

[6] W. Bauer et al., Phys. Rev. Lett. **58**, 863 (1987).

[7] K. Sneppen and L. Vinet, Nucl. Phys. **A480** 342 (1988).

[8] J. Aichelin et al., Phys. Rev. **C37**, 2451 (1988).

[9] G. Peilert et al., Phys. Rev. **C39**, 1402 (1989).

[10] D. H. Boal and J. N. Glosli, Phys. Rev. **C37**, 91 (1988).

[11] L. G. Moretto, Nucl. Phys. **A247**, 211 (1975).

[12] W. A. Friedman and W. G. Lynch, Phys. Rev. **C28**, 16 (1983).

[13] W. A. Friedman and W. G. Lynch, Phys. Rev. **C28**, 950 (1983).

[14] W. A. Friedman, Phys. Rev. **C40**, 2055 (1989).

[15] W. A. Friedman, Phys. Rev. **C42**, 667 (1990).

[16] A. Vicentini et al.,Phys. Rev. **C31**, 1783 (1985).

[17] R. J. Lenk and V. R. Pandharipande, Phys. Rev. **C34**, 177 (1986).

[18] T. J. Schlagel and V. R. Pandharipande, Phys. Rev. **C36**, 162 (1987).

[19] G. Bertsch and P. J. Siemens, Phys. Lett. **B126**, 9 (1983).

[20] H. Sagawa and G. F. Bertsch, Phys. Lett. **B155**, 11 (1985).

[21] H. Schulz, B. Kämpfer, H. W. Barz, G. Röpke, and J. Bondorf, Phys. Lett. **B147**, 17 (1984).

[22] Y. D. Kim et al., Phys. Rev. Lett.**67**, 14 (1991).

[23] Y. D. Kim et al., Phys. Rev. **C45**, 387 (1992).

[24] Y. D. Kim et al, Phys. Rev. **C45**, 338 (1992).

[25] D. Fox et al., Phys. Rev. **C47**, R421 (1993).

[26] T. C. Sangster et al., Phys. Rev. **C47**, R2457 (1993).

[27] D. R. Bowman et al., Phys. Rev. Lett. **70**, 3534 (1993).

[28] The mean emission time, τ, and the "half-life" for intermediate mass fragment (IMF) emission, $\tau_{IMF-IMF}$ are related by $\tau = 1.44\tau_{IMF-IMF}$

[29] R. T. de Souza et al., Nucl. Instr. and Meth. **A295**, 109 (1990).

[30] R. T. de Souza et al., Phys. Lett. **B300**, 29 (1993).

[31] C. Cavata et al., Phys. Rev. **C42**, 1760 (1990).

[32] W. G. Gong, et al., Phys. Rev. **C43**, 781 (1991).

[33] W. A. Friedman, *Proceedings of the International Symposium Towards a Unified Picture of Nuclear Dynamics, Nikko, Japan, June 1991*, edited by Y. Abe, S. M. Lee, and F. Sakata, AIP Conf. Proc. No. 250 (AIP, New York, 1992) p. 422.

FRAGMENTATION FROM NECK-LIKE STRUCTURES

C.P. MONTOYA, W.G. LYNCH, D.R. BOWMAN, P. DANIELEWICZ,
G.F. PEASLEE,N. CARLIN, R.T. DE SOUZA, C.K. GELBKE, W.G. GONG,
Y.D. KIM, M.A. LISA, L. PHAIR, M.B. TSANG, C. WILLIAMS
National Superconducting Cyclotron Laboratory
and Department of Physics and Astronomy
Michigan State University, East Lansing, MI 48824-1321, USA

N. COLONNA, K. HANOLD, M.A. MCMAHAN, G.J. WOZNIAK, L.G.
MORETTO
Nuclear Science Division
and Accelerator and Fusion Research Division,
Lawrence Berkeley Laboratory, Berkeley, CA 94720, USA

ABSTRACT

A quantitative analysis of the fragment production mechanisms in mid-peripheral collisions for the ^{129}Xe+^{nat}Cu at E/A=50 MeV reaction has been carried out. The experimental data collected show an enhancement of intermediate mass fragment (IMF, $3 \geq Z \leq 20$) emission between the projectile (PLF) and target residues. Efficiency corrected IMF yields are extracted for PLF evaporative decay and emission focused towards the target. The focused distribution is consistent with a neck fragmentation picture. The data clearly demonstrate enhanced fragment production from a non-compact breakup geometry. The "neck" width can be inferred from a comparison to evaporation.

1. Introduction

Large numbers of IMFs are observed in intermediate energy heavy ion reactions and the mechanisms driving their production are not yet well understood. Recently, it has been shown that in central collisions of symmetric nuclei BUU codes predict the formation of rather exotic shapes such as bubbles and donuts[2]. In addition, percolation studies suggest that non-compact geometries are capable of producing more fragments than compact ones at the same given temperature[3]. This naturally raises the question of just how much of a role does breakup geometry play in the fragmentation of real nuclei. The resolution of this question can be sought in an experimental examination of the fragmentation of such complex shapes, should they prove to exist. Experiments will attempt to confirm the existence of such exotic shapes, from their fragmentation patterns[1]; however, an alternative is the analysis of fragmentation in mid-peripheral collisions. In such collisions it is believed that instabilities in the neck formed between the target and projectile could result in emission of fragments from the neck region[4]. Neck formation is a rather general process observed in finite systems which behave hydrodynamically. In fact it is easily observed, visually, in macroscopic colliding liquid drop experiments.[5] Nuclear matter, also, displays hydrodynamic behavior and neck formation is an integral part of theoretical descriptions of nuclear fission and mass equilibration in deep-inelastic collisions. In most cases the neck is absorbed by the final fragments after scission takes place. There is, however, the possibility for fragment emission from the neck region as it ruptures. Emission of alpha particles[6] and IMFs[7] from the neck region has been observed with small probability in ternary fission studies. In mid-peripheral nuclear collisions, where the neck is forced to stretch faster with increasing beam energy, one can expect more fragment emission from the neck region due to Rayleigh instabilities. From BUU calculations it is believed that this will occur for reactions similar to ^{129}Xe+^{nat}Cu at E/A=50 MeV[4]. The analysis of this reaction for information regarding the fragmentation of its neck geometry and the results are presented below.

2. Experimental Setup

The experimental data were collected at the National Superconducting Cyclotron Laboratory at Michigan State University. The K1200 cyclotron provided a 49.8 MeV per nucleon ^{129}Xe beam with an intensity of $\sim 10^7$ particles per second. The beam was used to bombard a 2.3 mg/cm^2 ^{nat}Cu target in the 92 in. scattering chamber.

The resulting light charged particles and fragments were detected by the MSU Miniball[8] and LBL forward array[9]. The Miniball covered 87% of 4π between laboratory angles of 16° and 160° with 171 phoswich detector elements. The charge identification thresholds were about 2, 3, and 4 MeV/nucleon for Z=3, 10, and 18, respectively.

The LBL forward array was used to identify particles with charges, Z=1-54, between the laboratory angles of 2° and 16°. The geometrical efficiency of the array was 64%. The detection thresholds were approximately 6, 13, 21, and 27 MeV/nucleon for Z= 2, 8, 20, and 54, respectively. Taken together the two detector systems had a geometric acceptance greater than 88% and covered angles between 2° and 160°. Further details about the experimental setup can be found in Ref. 10.

3. Analysis

While the slow,heavy target residue is difficult to observe in the Miniball, the projectile residue is easily observed in the forward array. Fig. 1 shows the charge and energy per nucleon of the heaviest detected residue as a function of the measured charge particle multiplicity. The points with error bars, superimposed on the contour plots represent the means and standard deviations, respectively, of those quantities. The figure shows a monotonic decrease in residue charge and energy per nucleon with

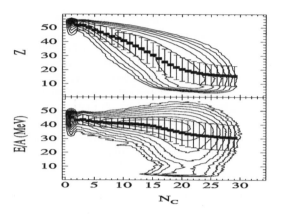

Fig. 1: Contour plots of the heaviest detected residue charge (top) and its energy per nucleon (bottom) plotted against measured charged particle multiplicity. Points and error bars represent means and variances for a given multiplicity, respectively.

increasing multiplicity. However, the velocity dissipation observed is much less than would be expected in a deep inelastic model.

The analysis was limited to events with one and only one detected IMF (ternary events), in order to specify the breakup geometry more completely. The measured multiplicity for such events is gaussian distributed with a mean of 13 and standard deviation of 4, independent of the IMF charge. If one assumes a monotonic dependence of the charge particle multiplicity upon impact parameter to construct an impact parameter filter, this mean multiplicity corresponds to $b \sim 6$ fm. For these mid-impact parameter events the projectile residue is still reasonably large ($< Z >= 37$) and the dissipation still moderate ($< E/A >= 41$MeV).

In order to correct for the detection efficiency; one must simulate events and filter them through the acceptance of the detector array. For these simulations we adopt the coordinate system depicted in Fig. 2. Here, $\vec{V}_t$ is the velocity of the target-like residue relative to the center of momentum frame for the 3-body system consisting of the target-like residue, projectile-like residue, and the IMF. Relative to this total CM system, $\vec{V}_p$ denotes the velocity of the center of momentum for the two-body system of the PLF and the IMF. A reaction plane is defined by the beam axis and $\vec{V}_p$, with $\hat{n}$ denoting the corresponding reaction normal. The IMF and PLF velocities in the two-body frame are represented by $\vec{V}_{IMF}$ and $\vec{V}_h$; θ' is the polar angle between $\vec{V}_{IMF}$ and $\hat{n}$; ϕ' is the azimuthal angle for the projection of $\vec{V}_{IMF}$ on the reaction plane. The azimuthal angle is zero when $\vec{V}_{IMF}$ and $\vec{V}_t$ point in the same direction and a positive rotation is in the beam axis direction.

This coordinate system was defined to facilitate the simulation of IMF evaporation from the particle-unstable PLF. Fig. 3 shows measured velocity plots for different

171

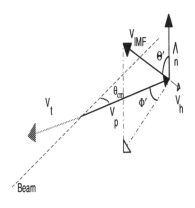

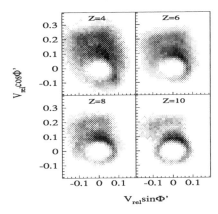

Fig. 2: The coordinate system used to define the breakup geometry of particle unstable projectile residue. The various variables are explained below.

Fig. 3: Grey scale velocity plots of breakup in coordinate system defined by Fig. 2.

IMF emissions in this new coordinate system. For negative values of $V_{rel} \cos \phi'$ one observes an evaporation component, albeit modified by the detection efficiency, as well as a strongly enhanced probability for emission directed at the target ($\phi' = 0$). The separation of these two components requires detailed simulations of the breakup mechanisms.

An analytic model for the evaporation from a hot rotating nucleus[11,12] was used to simulate the evaporative component of Fig. 3. The differential multiplicity used for the monte-carlo simulation of the evaporation was given by:

$$\frac{d^2 M}{dEd\Omega}(E, \theta', \phi') = N_0 (E - V_c) \exp[-(E - V_c)/T] \frac{J_1(iK)}{iK} \qquad (1)$$

where:

$$K = \frac{R\omega}{T} [2m(E - V_c)]^{1/2} \sin \theta' \qquad (2)$$

Here, E is the energy of relative motion between the PLF and IMF; the variables T and $R\omega$, are fit parameters representing temperature and the tangential velocity at the surface of the rotating nucleus; V_c was varied to improve the description of the energy spectra near the coulomb barrier. The two fragments are then transformed into the lab frame using the velocity and angle of the composite system. The latter two quantities are simply assumed to be gaussian distributed with means and variances determined from the experimental data. The parameters of this evaporation simulation were adjusted to optimize the comparison with the experimental data where the evaporative component is dominant ($100° < \phi' < 280°$). The velocity and angle of the particle unstable projectile-like residue are represented by the solid points in the

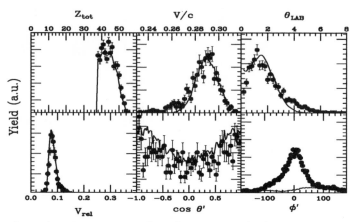

Fig. 4: Comparison of measured data (points) to simulation (line) for evaporative decay ($100° < \phi' < 280°$). Details concerning the various distributions are given in the text.

upper middle and right panels of Fig. 4 for $Z_{IMF} = 8$. The corresponding simulations are denoted by the solid lines. A comparison of the measured and assumed distributions for $Z_{tot}=Z_{IMF}+Z_{PLF}$ is shown in the upper right panel: both distributions were subject to a threshold at $Z_{tot} = 39$. Measured and simulated $V_{rel} = |\vec{V}_{IMF} - \vec{V}_h|$ distributions for the evaporative component are shown in the lower left panel of Fig 4. The lower central panel illustrates the $\cos \theta'$ distribution. It should be noted that the strong dip at $\cos \theta' = 0$ is an artifact of the detection geometry, which favors out of plane emission. The unfiltered distribution actually has a slight maximum near $\cos \theta' = 0$, since rotation enhances in-plane emission. The measured and simulated distributions for ϕ' over the full angular range are shown in the lower right panel. Clearly there is a strong, nearly gaussian distributed contribution near $\phi' = 0$, which is not described by evaporative decay.

Relative velocity distributions over the full ϕ' angular range are shown for Z=4, 6, 8, 10, and 12 in Fig. 5. The solid line depicts the total distribution, the dotted line the evaporative component and the dashed line the excess contribution near $\phi' = 0$. This additional component, broadly distributed over V_{rel} for the smaller values of Z_{IMF}, is more localized near the PLF for larger values of Z_{IMF}. Since some of the emission at $V_{rel} > 0.15$ may originate from evaporative decay of the particle unstable target-like residue, an upper threshold at $V_{rel} = 0.15$ was imposed on the subsequent analysis to allow a clean comparison of the evaporative and non-evaporative projectile decay. The non-evaporative component was simulated by multiplying Eqn. 1 by a gaussian of width σ_ϕ in ϕ'.

The middle panel of Fig. 6 shows the efficiency corrected yields for the evaporative (solid circles) and non-evaporative (open squares) decay components. As expected

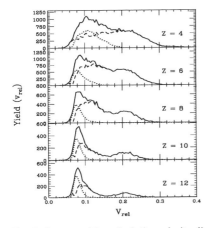

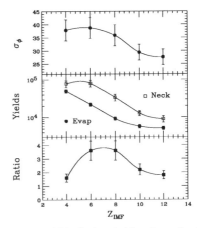

Fig. 5: Decomposition of relative velocity distributions (solid) into evaporative decay (dotted) and non-evaporative decay (dashed) for a series of different Z_{IMF}.

Fig. 6: Width of azimuthal focusing as function of Z_{IMF} (top). Efficiency corrected yields (middle) for evaporation (solid circles) and neck-like decays (open squares). Ratio of neck-like emission to evaporation (bottom).

for systems with larger surface to volume ratios the neck-like emission shows a weaker dependence of the IMF yield on Z_{IMF}. The actual yields for neck-like emission are also considerably larger as well; the bottom panel shows the explicit ratio of the neck-like to evaporative emission. The ratio reaches a maximum for $6 \leq Z_{IMF} \leq 8$ and declines thereafter[14]. A decline is also observed in the width of the azimuthal angle distribution (upper panel). For the heaviest IMF's the angular widths for both θ' and ϕ' are comparable; the fragments are clearly focused at velocities intermediate between those of the projectile and target-like residues strongly suggesting their origin as remnants of a neck temporarily formed between the two residues.

4. Conclusions

The non-equilibrated emission noted in Ref. 13 for peripheral collisions has been attributed to emission from a "neck region" temporarily formed between target and projectile. The distribution of masses emitted from the neck region is extracted and compared to the distribution of masses coming from evaporative processes. The distributions are consistent with a neck fragmentation picture. The enhanced probability for "neck" emission indicates that non-compact geometries play a role in enhancing fragment production in mid-peripheral collisions between nuclei. An examination of the dynamical effects leading to such non-compact geometries may lead to information regarding the incompressibility and viscosity of nuclear matter.

174

5. Acknowledgements

This work was supported by the National Science Foundation under Grant PHY-89-13815 and the U.S. Department of Energy under Contract DE-AC03-76SF00098. W.G.L acknowledges support from the U.S. Presidential Young Investigator Program and N.C. acknowledges partial support by the FAPESP, Brazil.

6. References

[1] T. Glasmacher, C.K. Gelbke, S. Pratt Phys. Lett. B **314**, 265 (1993).

[2] W. Bauer, G.F. Bertsch, H. Schulz Phys. Lett. B **69**, 1888 (1992).

[3] L. Phair, W. Bauer, C.K. Gelbke, Phys. Lett. B **314**, 271 (1993).

[4] M. Colonna, M. Di Toro, V. Latora, A. Smerzi, Prog. Part. Nucl. Phys. **30**, 17 (1992).

[5] A. Menchaca-Rocha et al., Phys. Rev. E **47**, 1433 (1993).

[6] Fluss et al., Phys. Rev. C **7** 353 (1973)

[7] D.E. Fields, K.Kwiatkowski, K.B. Morley, E. Renshaw, J.L. Wile S.J. Yennello, V.E. Viola, R.G. Korteling, Phys. Rev. Lett. **69**, 3713 (1992).

[8] R.T. de Souza, N. Carlin, Y.D. Kim, J. Ottarson, L. Phair, D.R. Bowman, C.K. Gelbke, W.G. Gong, W.G. Lynch, R. A. Pelak, T. Peterson, G. Poggi, M.B. Tsang, H.M. Xu, Nucl. Instr. Meth A **295**, 109 (1990).

[9] W.L. Kehoe, A.C. Mignerey, A. Moroni, I. Iori, G.F. Peaslee, N. Colonna, K. Hanold, D.R. Bowman, L.G. Moretto, M.A. McMahan, J.T. Walton, G.J. Wozniak, Nucl. Instr. Meth A **311**, 258 (1992).

[10] D.R. Bowman, C.M. Mader, G.F. Peaslee, W. Bauer, N. Carlin, R.T. de Souza, C.K. Gelbke, W.G. Gong, Y.D. Kim, M.A. Lisa, W.G. Lynch, L. Phair, M.B. Tsang, C. Williams, N. Colonna, K. Hanold, M. A. McMahan, G.J. Wozniak, L.G. Moretto, W.A. Friedman, Phys. Rev. C **46**, 1834 (1992).

[11] M.B. Tsang, C.B. Chitwood, D.J. Fields, C.K. Gelbke, D.R. Klesch, W.G. Lynch, K. Kwiatkowski, V.E. Viola,Jr, Phys. Rev. Lett. **52**, 1967 (1984).

[12] C.B. Chitwood, D.J. Fields, C.K. Gelbke, D.R. Klesch, W.G. Lynch, M.B. Tsang, T.C. Awes, R.L. Ferguson, F.E. Obershain, F. Plasil, R.L. Robinson, G.R. Young, Phys. Rev. C **34** 858 (1986)

[13] D.R. Bowman, G.F. Peaslee, N. Carlin, R.T. de Souza, C.K. Gelbke, W.G. Gong, Y.D. Kim, M.A. Lisa, W.G. Lynch, L. Phair, M.B. Tsang, C. Williams, N. Colonna, K. Hanold, M. A. McMahan, G.J. Wozniak, L.G. Moretto, W.A. Friedman, Phys. Rev. Lett. **70**, 3534 (1993).

[14] A decreasing ratio for more symmetric PLF breakup was also noted in: G. Cassini, P.G. Bizzeti, P.R. Maurenzig, A. Olmi, A.A. Stefanini, J.P. Wessels, R.J. Charity, R. Freifelder, A. Gobbi, N. Herrmann, K.D. Hildebrand, H. Stelzer, Phys. Rev. Lett. **71**, 2567 (1993).

ISOMER PRODUCTION IN FRAGMENTATION REACTIONS

W. BENENSON[a], D. BAZIN[†], J.H. KELLEY[a], D.J. MORRISSEY[b],
N.A.ORR[‡], R. RONNINGEN, B.M. SHERRILL[a], M. STEINER[a],
M. THOENNESSEN[a], J.A. WINGER, and S.J. YENNELLO[††], B.M. YOUNG[a]
National Superconducting Cyclotron Laboratory,
(a) Department of Physics and Astronomy
and (b) Department of Chemistry
Michigan State University, East Lansing, MI 48824-1321, USA
I. TANIHATA, X.X. BAI*, N. INABE, T. KUBO, C.-B. MOON,
S. SHIMOURA**, T. SUZUKI
The Institute of Physical and Chemical Research, RIKEN
Wako Saitama 351-01, Japan
R. N. BOYD
Department of Physics, Ohio State University, Columbus, Ohio 43210
K. SUBOTIC
Boris Kidrich Inst., POB 522, 11001 Beograd, Yugoslavia

ABSTRACT

The population of the 0^+ 0.228 MeV isomer of ^{26}Al has been measured in projectile fragmentation reactions at E/A = 65, 80, and 100 MeV. For beams of ^{27}Al and ^{28}Si, the yield of the isomer was found to be approximately equal to that of the ground state, whereas for ^{40}Ar beams it was much weaker. The significance and utility of the surprisingly high isomer yields is discussed.

1. Introduction

Projectile fragmentation is one of the main methods for the production of secondary radioactive beams, and therefore the understanding of the reaction mechanism of projectile fragmentation is important for future work in this field. Studies of the yield and the angular and momentum distributions of projectile fragments have given a coherent picture of the process, which has been reviewed recently[1]. Production of nuclei in excited states is normally not considered either experimentally or theoretically, but in the case of nuclei with long lived isomeric states, there is the possibility that the radioactive beams produced by fragmentation could be in excited states when they arrive at the secondary target. In the present paper measurements of the production of beams in excited isomeric states in projectile fragmentation is described for the first time. Isomeric state production has been studied for transfer reactions by Brown et al.[2], and isomerism in the target residue has been investigated to study angular momentum effects in sub-barrier fusion[3] and high energy fragmentation reactions[4]. Excited state production in general has been used as a revealing probe of nuclear temperature and dynamics in heavy ion reactions at intermediate energies[5].

A secondary beam with a high isomeric state content could be used for the investigation of reactions with one of the reaction partners in an excited state. For example, the $^{26}Al^m(p,\gamma)^{27}Si$ reaction is of considerable importance in understanding the high-temperature burning cycle of ^{26}Al[6] which is of particular interest to γ-ray astronomy[7]. The measurement of cross sections involving nuclei in excited states would be made feasible by such a beam. Since isomers exist with high and low spins compared to the ground state, production of isomer beams might provide an insight into the importance of angular momentum in the fragmentation process. Unique information on the reaction mechanism of the projectile fragmentation process should also be available from isomeric state production. Since all other states decay away immediately in the target by particle or γ-decay, isomeric states are one of the only means available for the determination of the high temperature hydrogen burning cycle.

In addition, the production ratio of $^{26}Al^m$ to the ^{26}Al ground state could have interest for nuclear structure if the mechanism were proven to be a simple one nucleon removal from the ^{27}Al beam. The ground state is a stretched (5^+) configuration whereas in the isomer the valence nucleons have antiparallel spins. In the case of single step one neutron removal, differences (exclusive of the 2J+1 factors) between the ground and isomeric state could be due to the effect of the alignment of the neutrons with respect to the single unpaired valence proton. However, as will be discussed below, it is very unlikely that the process is a single one nucleon removal at the intermediate energies employed in the experiments described in this paper.

The nucleus ^{26}Al is the lightest known with a long-lived ($\tau \geq 1$ sec) isomeric state. Its low lying level and decay scheme[8] are well known. The ground state is long-lived $(7.2 \times 10^5 y)$, whereas the 0^+ first excited state at 228 keV decays with a half-life of 6.3 s by a single β^+ transition to the ground state of ^{26}Mg with an end point energy of 2.98 MeV. This decay energy and lifetime are very close to those of the ground states of neighboring nuclei, for example ^{25}Al and ^{27}Si.

2. EXPERIMENT

Three separate experiments were performed to measure the population of the isomeric state of ^{26}Al in projectile fragmentation. At the National Superconducting Cyclotron Laboratory (NSCL) beams of ^{28}Si and ^{40}Ar were used in conjunction with the A1200 separator. A computer controlled wheel was used to collect the activity and move it to a point between to NaI γ-ray detectors. At the Institute for Physical and Chemical Research (RIKEN), a beam of ^{27}Al was used in conjunction with the RIPS separator. A telescope of plastic scintillators was used both to stop the ^{26}Al

Table 1: Isomer fraction for the three systems studied

System	E/A (MeV)	F (%)	δF (%)
^{40}Ar+^{12}C	65	24	1
^{28}Si+^{9}Be	80	56	3
^{27}Al+^{9}Be	100	49	2

beam and to record the β decay of the implanted ions. The goal of both experiments was to determine F, the fraction of the ^{26}Al beam in the isomeric state. Both of these experiments are described in a recent letter[9]. The quantity F is given by

$$F = \frac{N_{isomer}}{N_{total}}$$

where N_{isomer} and N_{total} are the number of ions in the beam in the isomeric state and the total number of ^{26}Al in the beam, respectively.

3. RESULTS

The results given in Table I are quite limited in scope because these are the first measurements of this type to have been carried out. The dependence of the isomer fraction on, for example, projectile, bombarding energy, and ejectile momentum and angle is as yet unmeasured. If the decay were statistical in nature, then the fraction, ignoring γ-ray feeding, would be given by the (2J+1) factors to be 1/12. However, when γ-ray feeding is taken into account, the fraction saturates at a value close to 0.32 for temperatures greater than a few MeV. The fact that the ^{27}Al and ^{28}Si beam measurements exceed this value means that a more detailed examination of the process is required for understanding. However, it is clear that there is a distinct difference between the ^{27}Al and ^{28}Si beam points and the one obtained with ^{40}Ar. For the intermediate energy heavy ions employed here, it is well known that nucleon transfers from the target occur. Souliotis et al.[10] found a significantly lower peak velocity for nuclei produced by transfers from the target to the projectile as compared to pure fragmentation products. The yield of transfer products, the production of which was forbidden in a purely projectile fragmentation process, is appreciable. Reactions which produce fragments close to the projectile (^{27}Al and ^{28}Si) may well include such non-fragmentation processes, and thus the qualitative difference between these data points and that with the ^{40}Ar beam may be due to such effects.

It is very difficult to understand the very large fraction of isomers produced with the ^{27}Al and ^{28}Si beams. The energy levels of ^{26}Al are quite well known, and in particular the gamma ray branching has been measured up to the 6.82 MeV level (^{26}Al becomes unstable to proton emission at 6.13 MeV.) The decay of virtually every gamma emitting level can be followed to either the ground state or the isomeric state. Although there are individual levels that decay preferentially to the isomeric state, there is no region of excitation energy above 3 MeV in ^{26}Al which would end up decaying equally to ground and first excited state. In general, if a Gaussian distribution of strengths is assigned to an excitation region with a width of 100 keV or more, the calculated fraction ends up with values near 0.3, a value similar to what one would get under the assumption of a single nuclear temperature above 2 MeV.

Calculations were carried out to compare to the measured yields of isomeric states. The standard model used to describe high energy reactions is that of a fast internuclear cascade followed by a slower evaporative process. The ISABEL internuclear cascade code by Yariv and Fraenkel[11] was used to calculate a set of highly excited projectile fragments from the fast stage of the reaction. This distribution of primary fragments has a broad range of mass, charge, excitation energy, and angular momentum. These fragments will deexcite very rapidly into the isomeric and ground state nuclei observed in the present measurements. Simple statistical evaporation calculations were performed[12] with different statistical deexcitation models that explicitly conserve angular momentum, the CASCADE code by Pühlhoffer[13] and the PACE code by Gavron[14]. Both of these calculations determine the distribution of (cold) residues by repeatedly deexciting the starting nucleus. The output of the fast stage was coupled to the slow stage by performing five deexcitations per primary fragment. The number of final residues in the isomeric and ground states were summed for direct comparison to the measured ratio.

Only preliminary results from the calculations described above can be given here; more detailed results will be presented in a subsequent study. The results show distinct differences between the isomer yield from beams far from ^{26}Al and those nearby in mass. This comes from the fact that for the case of the beams far from ^{26}Al in mass there are many particle decay channels available through nuclei with mass between 28 and that of the beam. These decays go preferentially to the ground state of ^{26}Al due to phase space reasons and therefore attenuate the isomer fraction significantly. In the case of production by beams just above ^{26}Al in mass, the main path is through highly excited states of ^{26}Al. Thus the difference between the light and heavy beam data is well understood, but the absolute value of the fraction is not predicted and remains a mystery. However, in the calculations described above, transfer products, which are known to contribute, are neglected. If these reactions favor the production of low spin states in ^{26}Al, then the resulting cascades would favor the isomeric state over the ground state.

4. CONCLUSIONS

The first studies of isomer production in heavy ion fragmentation have been presented. They show that for beams close to the fragment in mass a very high yield of isomers can be obtained. Using this effect to carry out experiments will be a difficult experimental challenge, but one that promises unique information. A reaction model which is capable of predicting the magnitude of the isomer fraction could add greatly to the knowledge of the detailed reaction mechanism for fragmentation at intermediate energies.

5. ACKNOWLEDGEMENTS

The work was supported by National Science Foundation under Grant No. PHY-89-13815 and PHY89-20606

† on leave from Centre d'Etudes Nucléaires de Bordeaux-Gradignan, present address: GANIL, B.P. 5027, 14021 Caen CEDEX, France
‡ present address LPC-ISMRA, Boulevard du Marechal Juin, 14050 Caen Cedex, France
†† present address: Cyclotron Institute, Texas A&M University, College Station, TX 77843
* present address: Institute of Atomic Energy, Beijing, China
** present address: Rikkyo University, Oshima-ku, Tokyo 171, Japan

6. References

[1] D.J. Morrissey et al., Phys. Rev. C 39(1989)460
[2] J.A. Brown et al., Proceedings of the Second International Conference on Radioactive Nuclear Beams, Th Delbar, ed., Adam Hilger, pub., Bristol
[3] D.E. DiGregorio et al. Phys.Rev. C 42(1990)2108
[4] K. Sümmerer, W. Brüchle, D.J. Morrissey, M. Schädel, B. Szweryn, and Y. Weifan, Phys. Rev. C 42(1990)2546
[5] D.J. Morrissey W. Benenson, and W. Friedman, Ann. Rev. Nucl. Sci., to be published and D.J. Morrissey et al., Phys. Rev. C 34(1968)761
[6] R.A. Ward and W.A. Fowler, Astrophys. J. 238(1980)266 W.A. Fowler, Rev. Mod. Phys. 56(1984)149
[7] W.A. Mahoney, J.D. Ling, W.A. Wheaton, and A.S. Jacobson, Astrophys. J. 268(1984)578
[8] P.M. Endt. Nuc. Phys. A 521(1990)1
[9] B.M Young et al., Phys. Lett. B 311(1993)22
[10] G.A.Souliotis, D.J.Morrissey, N.A.Orr, B.M.Sherrill, J.A.Winger, Phys.Rev. C 46(1992)1383
[11] Y.Yariv and Z. Fraenkel, Phys. Rev. C 20(1979)2227, and ibid. 24(1981)488
[12] D.J.Morrissey et al., Phys. Rev C 43(1979)1139
[13] F. Pühlhofer, Nucl. Phys. A 280(1977)267
[14] A.Gavron, Phys. Rev. C 21(1980)230

PHENIX

Richard Seto
University of California, Riverside
Riverside, CA 92521, USA

for the PHENIX collaboration

Abstract

The PHENIX detector is one of two large detectors planned for the Relativistic Heavy Ion Collider. Its goal is to detect the quark-gluon plasma by the measurement of many signatures as a function of dE_T/dy to look for a simultaneous anomaly due to the formation of the QGP. It will be capable of measuring electrons, photons and hadrons in the midrapidity region at high rates with good momentum resolution.

1 Introduction

The PHENIX experiment is one of two large detectors being planned for the Relativistic Heavy Ion collider expected to begin operation in 1999. The center of mass energy for gold on gold collisions is 200 GeV/nucleon with higher energies possible for lower mass ions depending on the charge to mass ratio. The primary goal of the experiment is to detect the quark-gluon plasma (QGP) and to measure its properties. It is expected that this new phase of matter will manifest itself in many ways. However, there is no theoretical guidance indicating which experimental signature is most likely to yield a conclusive result. Phase transitions have traditionally shown themselves as changes in the character of some quantity such as the entropy or the temperature, as a function of the energy density. PHENIX has chosen the route of fixing the geometry of events from the total multiplicity and measuring many different signatures as a function of the the transverse energy which can be related to the energy density via the Bjorken formula.[1] The simultaneous change of several signatures would be a clear indication of a phase change.

PHENIX has chosen to limit its solid angle coverage and measure both hadronic and leptonic signatures. Leptons and photons are expected to provide information on the early stages of the collision and thermalization, while hadrons provide complementary information about the later hadronization of the QGP. In order to have uniform transverse momentum acceptance for dielectrons, an axial field was chosen. The detector, shown in figure 1, is a double arm spectrometer. The two arms each subtend $90°$ in ϕ and ± 0.35 units in pseudorapidity, and are separated by $45°$ in azimuth. They are instrumented with drift chambers (DC), pad chambers (PC), a ring imaging Cerenkov counter (RICH), a multi-layer time expansion chamber/transition radiation detector (TEC/TRD) and an electromagnetic calorimeter (EMcal). Within

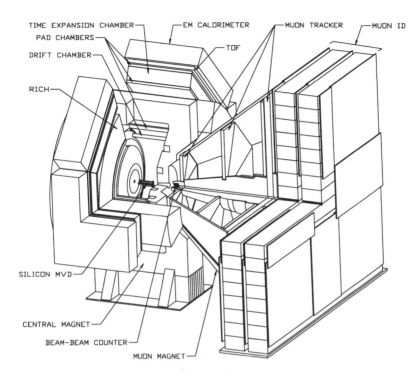

TIME EXPANSION CHAMBER
PAD CHAMBERS
DRIFT CHAMBER
RICH
EM CALORIMETER
TOF
MUON TRACKER
MUON ID
SILICON MVD
CENTRAL MAGNET
BEAM-BEAM COUNTER
MUON MAGNET

Figure 1: A three-dimensional cutaway view of the PHENIX detector, showing the location of the various detector subsystems.

the aperture of the spectrometer are limited acceptance detector subsystems for the detection of photons using lead glass (PbGl) and for hadron identification using a a high resolution time of flight wall. A forward spectrometer which covers from 10° to 30° around the beam axis is used to track and identify muons. Finally there is a silicon multiplicity/vertex detector around the beam pipe with large pseudorapidity coverage for the measurement of the event by event multiplicity and for the location of the z position of the vertex. Beam counters are placed at forward angles to give a start time for the electronics and to give a crude z position for a first level trigger.

2   Physics

The technique used by the PHENIX collaboration will be to initially fix the geometry of the collision, namely the centrality, by measuring the total multiplicity in the silicon detectors over a wide range of rapidity to eliminate possible fluctuations in rapidity. Then, $dE_T/d\eta$ will be measured via an electromagnetic calorimeter over a narrow range of rapidity to find the energy density. As previously stated, we will look at a variety of signatures in the leptonic, photonic and hadronic channels for systematic changes as a function of the energy density. We have developed a sophisticated monte-carlo simulation of the detector based on GEANT called PISA (PHENIX Integrated Simulation Application) and a companion analysis package called PISORP (PISA Output Readback Program). In what follows, I will describe a sample of these signatures and the sensitivity we will have as simulated using PISA.

2.1 Di-leptons

The dilepton spectrum is of interest both for the more easily measurable vector mesons and the di-lepton continuum. Decays of the ω, ϕ, and the J/ψ family will be accessible in the mid-rapidity region via their decay to electrons. The ρ/ω, the ψ family and the Υ family will be accessible in the higher pseudorapidity region from 1.2 to 2.4 via their decay to muons.

Vector mesons provide several avenues of interest. At high temperatures it is expected that chiral symmetry is restored and that the masses and widths of all particles will be shifted. This may be a very dramatic signature of new physics. Although this phase change is not identical to the phase change to the QGP, it is expected that the two phenomena occur at the same or nearly the same temperature. These shifts could be observed directly in the measurement of the mass and widths of the ω and ϕ. The mass resolution of the dielectron measurement (about 4 MeV at the ϕ mass) makes the experiment sensitive to relatively small changes. In addition, the ratio $\sigma(\phi \to e^+e^-)/\sigma(\phi \to K^+K^-)$ would be very sensitive to changes in the mass of the kaons. Of course this requires that we measure the decay of the ϕ in the K^+K^- channel as well. The overall production rate of the ϕ may itself be interesting since it may increase significantly when the temperature becomes comparable to the mass of the strange quark. This effect has already been seen at CERN.[2] The $\rho \to$ dilepton rate is expected to be sensitive to the lifetime of the hadronic phase of the fireball since several generations of thermal ρ mesons would contribute to the signal. The ρ/ω will be accessible in the muon channel where the loss in mass resolution does not hurt the signal/background because of the large ρ width.

The QCD Debye screening effect will also be studied using the rate of production of both the ψ and Υ families. A uniform acceptance in p_T is required since these effects are expected to disappear at higher values of p_T where the effect can be normalized. The charm content of the suppressed vector mesons should show up as additional D mesons. PHENIX will have the ability to measure the total rate of charm production via e-μ coincidence in the relevant mass range between $1.5 < m_{e-\mu} < 7GeV$.

The measurement of the continuum due to quark-antiquark annihilation in the thermal plasma will be possible in the initial stage of the experiment only if the rate is above the Dalitz background. Present upgrade plans include an Dalitz rejection device near the vertex which would allow the detection of the thermal background at levels presently calculated by theorists.[3]

The measurement of the di-electron spectrum presents an experimental challenge because of the large backgrounds due to dalitz decays of π^0's. Since the level of contamination of dalitz pairs to the electron spectrum is about 1% of the π^0 yield with a spectrum softer than the pions, and since invariant mass techniques can reduce the dalitz background by another factor of between 3 and 10 (with a Dalitz rejection device planned for an upgrade, this can be a factor of 100), we require that the rate of mis-identification of charged pions as electrons be less than 10^{-4}. Electron identification is done via three main detector systems: the RICH, the TEC/TRD, and the electromagnetic calorimeter. The RICH gives us a factor of 10^{-3} up to about 3 GeV/c. The additional factor of 10 is obtained from the the energy loss in the TEC at low energies and from the E-M calorimeter at higher energies. At the cross-over points where the energy loss curves in the TEC intersect other particles, time of flight is used from the E-M calorimeter, which has a timing resolution of better than 500 ps. Above 3 GeV where the RICH cuts off, the Transition Radiation Detector (TEC/TRD) will extend the range of electron identification to high energies. Mass resolutions for the vector mesons range between 0.4% and 0.8% depending on the transverse momentum. Dimuons are measured in the forward arm where the rapidity boost eases the identification of muons from hadrons. The forward arm consists of a magnetic spectrometer with a hadron absorber for muon identification. The mass resolution of the muon spectrometer is approximately 6%$\sqrt{m}$. Figure 2 shows the expected invariant mass spectra. The running time assumed for the electrons is about one month at the designed luminosity, while the assumption for muons is about one year. A like-sign subtraction for the muons makes the vector mesons clearly visible.

2.2   *Hadrons*

It is crucial to measure hadrons, not only to look for QGP signatures, but also to study the basic collision dynamics of the system via p_T spectra and HBT. To this end a high resolution time of flight system is installed in one of the electron arms with low occupancy and timing resolution of better than 100 ps. This would allow pion and kaon identification to a p_T of about 2.5 GeV/c and p and $\bar{p}$ identification to about 4 GeV/c. The solid angle covered by the high resolution TOF is large enough to measure the ϕ decay to K^+K^- important to the ratio $\sigma(\phi \to e^+e^-)/\sigma(\phi \to K^+K^-)$ mentioned previously.

Because of the large multiplicities of bosons in one event, sufficient pair statistics are available to construct the correlation functions necessary to construct precise information of both the spacial and time distributions using the Hanbury-Brown-Twiss effect. The resolution is sufficient to measure the expected source size of 10 fm.

184

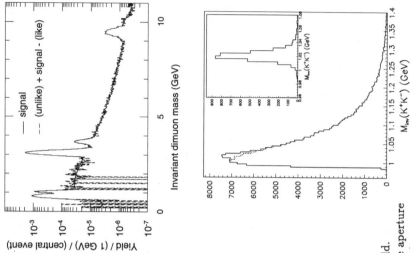

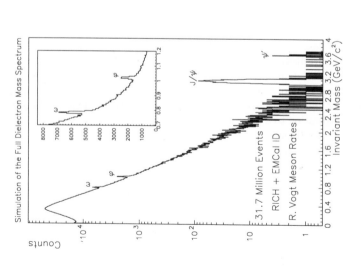

Figure 2: Left: The full dielectron pair mass spectrum.
Right-Top: Results from a like-sign subtraction in the dimuon yield.
Right-Bottom: The invariant mass spectrum of K^+K^- pairs in the aperture of the high resolution TOF. The dashed line shows the combinatoric background. The inset shows the ϕ peak after a like-sign subtraction.

3 Detector Components

Because of the limited space available for this article I will comment only on a subset of the detector components.

3.1 *Tracking*

The tracking system consists of several devices with the task of pattern recognition and tracking divided between devices. The majority of the magnetic field is within a radius of 2.5 meters, before the front face of the RICH. Two layers of drift chambers are located at 2.05 m and 2.35 m. These drift chambers have a spacial resolution of 150 μm in $r\phi$ and a two-track separation of better than 1.5 mm. Small angle stereo provides z resolution of about 2 mm. The momentum measurement of charged particles is done primarily with the drift chambers, assuming that the track originates from the vertex. A system of pad chambers gives non-projective 3 dimensional space points. They are particularly useful for pattern recognition and track matching with the RICH and EM Cal at the trigger level. The spacial resolution is between 2 and 4 mm (depending on the plane) and the two track separation is between 8 and 16 mm. The occupancy is about 6%. The time expansion chamber is placed between the RICH and the EM Cal and is essentially a TPC whose drift is radial rather than axial. From the energy deposited in the gas an e/π separation of about 10^{-2} is achieved. Three dimensional space points are also available from the TEC allowing one to reconstruct track direction-vector information in each TEC plane. Radiators are placed between the latter planes of the forming the basis for the TRD. The transition radiation produced in the radiator by ultra-relativistic particles will allow e/π separation of better than 2×10^{-3} up to 50 GeV/c. Prototype detectors for the drift chambers are currently under construction at PNPI in Russia. Pad chambers similar to the ones proposed for PHENIX have been constructed and have been operating at BNL for several years.[6] The TEC/TRD has undergone extensive testing with a prototype at BNL.

3.2 *The Ring Imaging Cerenkov Counter*

At the heart of the electron identification scheme is the ring imaging Cerenkov counter filled with either ethane or methane at atmospheric pressure. It is essentially blind to low momentum hadrons and will identify electrons up a momentum of about 3.5 GeV/c. The Cerenkov photons are reflected by two spherical mirrors onto two arrays of 3200 phototubes. These PMT arrays are placed behind the central magnet and are thus shielded from hadrons by the pole tips. A prototype has been tested at KEK with an array of 25 PMT's. Figure 3 shows examples of electron rings from this prototype. These initial tests showed an e/π rejection of better than 10^{-4} with almost no loss of electrons.

186

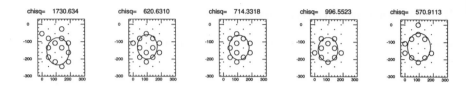

Figure 3: A typical PMT hit pattern observed in the prototype RICH detector with a 1.2-m ethane radiator, and from events with 5 consecutive electron triggers.

4 Conclusion

The PHENIX collaboration now numbers over 300 members from 43 institutions and 10 different countries. The leadership or Project Management (PM) of the collaboration is shared between the spokesman, Shoji Nagamiya of Columbia University, the Project Director, Sam Aronson of Brookhaven National Lab, the Deputy Project Director, Glenn Young of Oakridge, and the Project Engineer, Leo Paffrath, also from BNL. An executive council, led by the spokesman, monitors the performance of the PM and gives advice on collaboration matters. It is made up of the PM and selected representatives from the collaboration. A second group, the Detector Council is made up of the leaders of the various subsystems and is led by the Project Director. Its primary responsibility is the design and construction of the detector.

The PHENIX collaboration has just completed its conceptual design report which is available from BNL.[4]Significant progress has been made toward the design and prototyping of the various parts of the detector. We look forward to many years of exciting, new physics opportunities.

References

[1] J. D. Bjorken,*Phys. Rev.* **D27**, 140 (1983).

[2] P. Koch, U.Heinz, *Nucl. Phys.* **A525**, 293c (1991).

[3] K. Kajantie et. al., *Phys. Rev.* **D34**, 2746 (1986); E. V. Shuryak, *Phys. Lett* **B78**, 15 (1978); R. Hwa and K. Kajantie, *Phys. Rev.* **D32**, 1109 (1985);

P. V. Ruuskanen, in "Quark Gluon Plasma",*Adv. Series on Directions in High Energy Physics*, Vol. 6, R. C. Hwa, ed., World Scientific, Singapore, 1990, p. 519.

[4] References for these physics processes can be found in the *PHENIX Conceptual Design Report*, BNL (1993).

The STAR Experiment at RHIC

JAY N. MARX
for the STAR Collaboration
Lawrence Berkeley Laboratory
University of California, Berkeley, CA 94720, USA

1. Introduction

STAR (Solenoidal Tracker at RHIC) will be one of two large, sophisticated experiments ready to take data when the Relativistic Heavy Ion Collider (RHIC) comes on-line in 1999. The design of STAR, its construction and commissioning and the physics program using the detector are the responsibility of a collaboration of over 250 members from 30 institutions, world-wide.

The overall approach of the STAR Collaboration to the physics challenge of studying collisions of highly relativistic nuclei is to focus on measurements of the properties of the many hadrons produced in the collisions. The STAR detector is optimized to detect and identify hadrons over a large solid angle so that individual events can be characterized, in detail, based on their hadronic content. The broad capabilities of the STAR detector will permit an examination of a wide variety of proposed signatures for the Quark Gluon Plasma (QGP), using the sample of events which, on an event-by-event basis, appear to come from collisions resulting in a large energy density over a nuclear volume.

In order to achieve this goal, the STAR experiment is based on a solenoid geometry with tracking detectors using the time projection chamber[1] approach and covering a large range of pseudo-rapidity so that individual tracks can be seen within the very high track density expected in central collisions at RHIC. STAR also uses particle identification by the dE/dx technique and by time-of-flight. Electromagnetic energy is detected in a large, solid-angle calorimeter. The construction of STAR, which will be located in the Wide Angle Hall at the 6 o'clock position at RHIC, formally began in early 1993.

2. Physics Program of STAR

The physics goals of the STAR Collaboration which includes searching for the Quark Gluon Plasma, studying the dynamics of relativistic nuclear collisions and studying the parton physics of extended hadronic matter is described in detail elsewhere[2,3,4,5]. The analysis approach for this experiment will be the correlation of many observables on an event-by-event basis. In order to understand the underlying physics, and to identify the QGP, the STAR physics program will require measurements using proton-proton, proton-nucleus, and nucleus-nucleus collisions under a wide variety of conditions so that the dependence on impact parameter, incident energy and nuclear species can be understood. The large acceptance, good momentum and two-track resolution and particle identification characteristics of the STAR experiment will permit event-by-event measurements of many observables related to QGP signatures and collision dynamics such as particle spectra, flavor composition, source size, and density fluctuations in energy, entropy and multiplicity in azimuth and pseudo-rapidity (η). In addition, measurements of the remnants of hard-scattered partons will provide important new information on the nucleon structure functions and parton shadowing in nuclei in a regime where there is a high enough energy for perturbative QCD to give a useful description of the hard parton scattering[6].

3. The STAR Detector

The configuration of the STAR detector is illustrated in figure 1. The RHIC colliding beams collide in the center of the detector. Moving radially outwards from the beampipe, particles first encounter a vertex chamber based on silicon drift technology. This device, which is called the Silicon Vertex Tracker (SVT), consists of three layers of ladders of silicon drift devices to provide three independent space points with a resolution of less than 50 microns and three samples of ionization for each track that traverses the device. In essence, the SVT is a silicon time projection chamber with over 100,000 channels of TPC-like electronics. The full SVT covers $|\eta| \leq 1$ with one layer covering $|\eta| \leq 2$.

Beginning at 50 cm radially from the beam collision point is the large, central TPC. The central TPC, which has a diameter of 4 m and a length of 4.2 m, will provide intrinsically three-dimensional space points and ionization samples for 45 radial points along tracks between $|\eta| \leq 1$ using 45 rows of cathode pads on the endcap sectors and fewer points and ionization samples for $1 \leq |\eta| \leq 2$. Each cathode pad is sampled 512 times during the 40 microsecond TPC drift time resulting in 71 million r-θ-ϕ pixels within the TPC. It is this high degree of segmentation that gives the TPC its unique power to detect individual tracks within the very high track multiplicity expected at RHIC.

The spatial resolution of the central TPC will be about 600 microns with a resolution in dE/dx of about 7%. Tracking efficiencies of approximately 90% should be achievable in the track densities that will be typical of RHIC events and particles identification should be effective for momenta up to about 600 MeV/c. The TPC is read out with over 140,000 channels of electronics (one channel per cathode pad) based in custom integrated circuits. Each channel consists of a low-noise, charge-sensitive preamplifier, a shaper amplifier, a 512 deep switched capacitor array for time sampling in the drift coordinate, and a digitizer. The TPC is immersed in very uniform 0.5 Tesla axial magnetic field provided by the conventional solenoid magnet.

Immediately outside of the central TPC will be an array of trigger scintillation counters and time-of-flight counters. The TOF system will include 10,000 counters with 100 picosecond time resolution covering the barrel for $|\eta| \leq 1$.

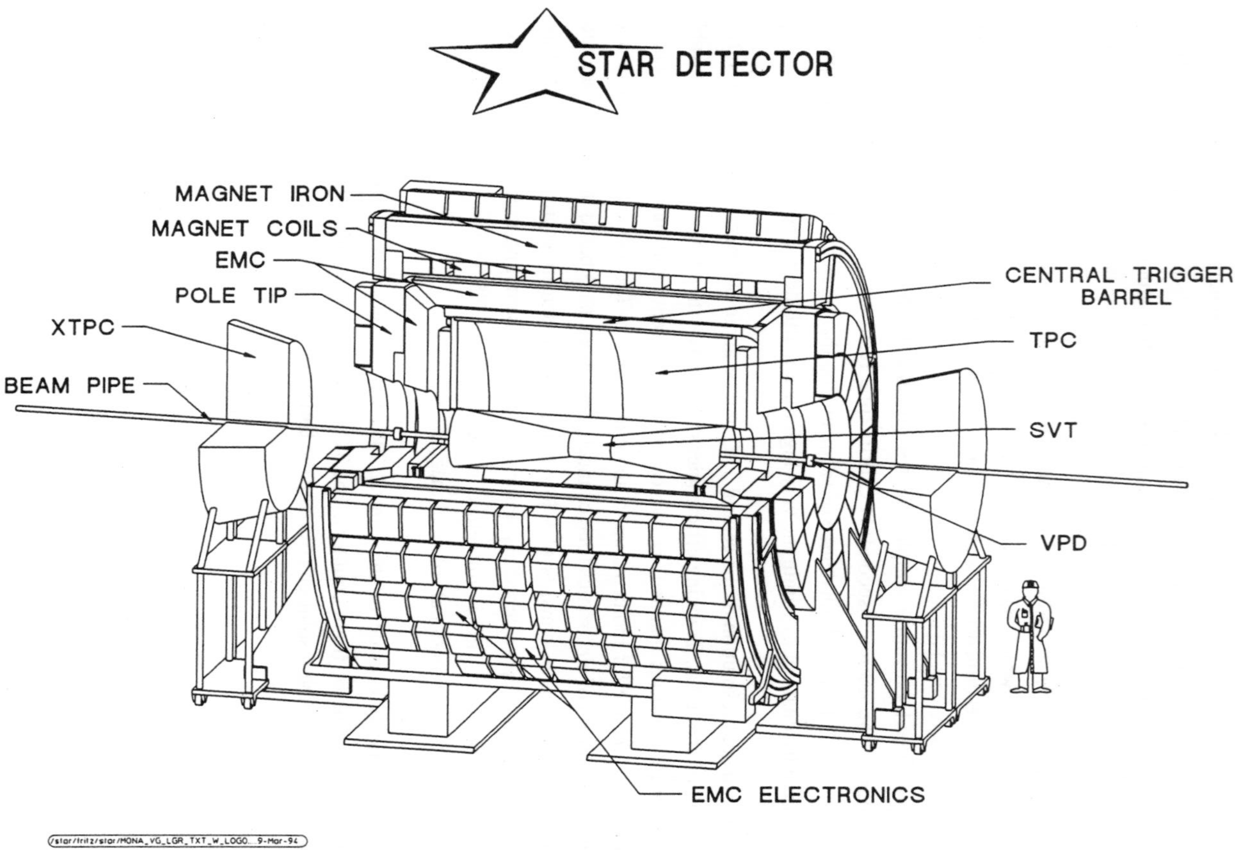

Figure 1 Overview of the STAR Detector

The TOF system will identify hadrons with p_t greater than 0.3 MeV/c with pions and kaons identified up to p_t=1.5GeV/c, and protons and antiprotons to p_t=2.3 GeV/c.

Outside of the TOF system, covering $|\eta| \leq 1$ is the barrel electromagnetic calorimeter (EMC). The barrel EMC is a multilayer lead/scintillator sampling calorimeter with 1200 towers arranged in η and ϕ. Signals from the barrel EMC are taken through the gaps between the magnet coil pancakes and through the iron yoke segments to phototubes on the outside of the detector.

The STAR detector also includes endcap electromagnetic calorimeters that cover the magnet poletips between $1 \leq |\eta| \leq 2$ and external time projection chambers (XTPC) to provide tracking information and charge identification between $2 \leq |\eta| \leq 4.5$. The XTPCs will use radial drift to allow separation of individual tracks, even in the very dense region at the highest pseudo-rapidity covered.

The STAR trigger is designed to operate at a number of levels, allowing more and more sophisticated decisions to be made as the information from various detectors becomes available for processing. The trigger approach includes the preservation of information from all trigger detectors in the data stream so that increasingly sophisticated and relevant algorithms can be developed as the appropriate physics observables are better understood. The basic trigger for STAR consists of a multiplicity trigger formed from 240 trigger elements around the TPC barrel and from fast signals from the anode wires on the TPC endcap. The overall detected multiplicity and the multiplicity distribution in the η and ϕ pixels defined by these elements forms the primary level 0 trigger. There is also a veto calorimeter at zero degrees and a vertex position detector that is used in the level 0 trigger to limit the interaction point within the interaction diamond. Fast signals from the calorimeter and TOF, and then slow signals from the TPC and SVT will be available for more sophisticated algorithms that operate at the higher trigger levels.

The STAR data acquisition system must have sufficient bandwidth to accommodate the tens of megabytes of data from each event at an input rate of up to 100 hertz, to operate on this data with sophisticated level 3 trigger algorithms in order to choose the most interesting events and then to write data into a storage medium at 1 hertz. A number of high-bandwidth communications networks are being evaluated as the backbone of the STAR data acquisition system.

Figure 2 illustrates the overall pseudo-rapidity coverage of the STAR tracking systems for a single FRITIOF Au-Au collision at a center of mass energy of 200 GeV per nucleon. The central TPC for tracks between $|\eta| \leq 2$ and the XTPC for $2 \leq |\eta| \leq 4.5$ covers a large fraction of the charged tracks in the event. Figure 3 illustrates the particle identification coverage as a function of the transverse momentum of both pions and kaons emitted in a single Au-Au collision. Most of the particles that are not identified either interact or decay in flight.

An overview of the STAR detector systems is given in Table 1.

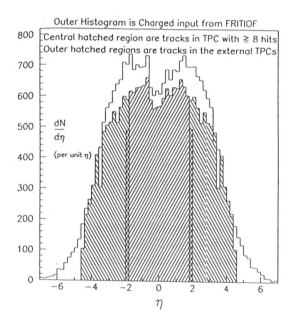

Fig. 2 The outer histogram displays the dN/dη distribution that is input from the FRITIOF event generator. The hatched histograms represent the dN/dη distributions reconstructed from tracking in the central and external TPCs.

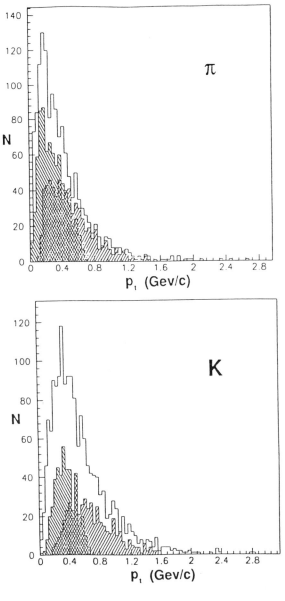

Fig. 3 Transverse momentum distribution, dn/dp$_t$, for charged pions (above) and kaons (below) from the primary interaction vertex generated by the Lund/FRITIOF nucleus-nucleus code and tracked through the experimental setup. The \\\\ region contains those particles identifiable with dE/dx measurements in the TPC. The //// region contains particles identified by TOF. The cross-hatched region has particles identified by both methods and the open region by neither. Most of the particles which are not identifiable and appear in the open region either interact or decay in flight.

Table 1 Overview of STAR detector systems

Solenoidal Magnet	
Coil Inner Radius	2.62 m
Length	6.9 m
Field Strength	0.25 - 0.5 T
Time Projection Chamber (TPC)	
Inner Radius	0.5 m
Outer Radius	2.0 m
Length	4.2 m
# of Channels	136,240
Acceptance	$\lvert \eta \rvert \leq 2.0$
Trigger Detectors	
Vertex Position Detectors	
Segmentation	20 elements
Acceptance	$3.3 < \lvert \eta \rvert < 3.8$
TPC Endcaps (MWPC)	
Segmentation	2,000
Acceptance	$1.0 \leq \lvert \eta \rvert \leq 2.0$
Veto Calorimeters	
Segmentation	to be determined
Acceptance	$\lvert \eta \rvert > 5$
Central Trigger Barrel (Scintillator)	
Segmentation	200
Acceptance	$\lvert \eta \rvert \leq 1$
Silicon Vertex Tracker (SVT) R&D Project	
Type	Si Drift Chambers
Radius 1st layer	5 cm
Radius 2nd layer	8 cm
Radius 3rd layer	11 cm
Length	42 cm
# of Channels	83,000
Acceptance - 1st, 2nd, 3rd layers	$\lvert \eta \rvert \leq 2.1, 1.7, 1.4$

EM Calorimeter Upgrade	
Inner Radius	2.20 m
Thickness	0.325 m
Segmentation	$\Delta\eta, \Delta\phi = 0.105, 0.105$
# of Channels	1200
Acceptance	$\lvert \eta \rvert \le 1.05$
Time-of-Flight Upgrade	
Type	shingle slats
Inner Radius	2.06 m
# of Channels	7776
Acceptance	$\lvert \eta \rvert \le 1.0$
External TPC Upgrade	
Number of Modules at Each End	4
Dimensions of Each Module(LxWxD)	$2.2 \times 2.0 \times 1.0 \ m^3$
Z Distance to Front Face	7.0 m
Total # of Channels	22,000
Acceptance	$2.0 \le \lvert \eta \rvert \le 4.5$

4. Status of STAR

STAR was approved for construction in early 1993 by the Brookhaven National Laboratory management based on advice from their Program Advisory Committee and from the RHIC Technical Advisory Committee. This approval followed many rounds of review of the STAR Collaboration's letters of intent, physics proposal and the conceptual design report and associated cost, schedule and management plans for STAR. The USDOE concurred with this approval in March 1993.

STAR will be funded and constructed in several phases. The so-called baseline is funded primarily from RHIC construction funds and from resources provided by the Collaboration. This baseline includes the solenoid magnet and associated support structures, the central TPC and its electronics, the data acquisition and trigger systems, and the computing and software development needed to begin the physics program. In addition, the engineering necessary to integrate the other detectors (SVT, EMC, TOF, XTPC) into the baseline in the future is being done to assure that all envisioned detector systems of STAR can be installed and operated without undo disruption of the physics program.

During the early years of STAR construction, R&D is proceeding on the SVT, TOF, EMC and XTPC so that the technologies of choice can be proven and a viable conceptual design for these systems can be developed. It is the goal of the STAR Collaboration to seek funding from U.S. and other sources so that these systems will be constructed and operating as close to the time of initial RHIC operations as possible.

At the present time, construction of the STAR baseline is moving ahead rapidly. Construction of the endcap sectors for the TPC should begin in a few months based on a successful prototype that is now undergoing detailed tests. A 90kV prototype high voltage insulator for the TPC is under test. The design and prototyping of the electronics for the TPC is almost complete and the first 15% of the TPC electronics will be procured from industry within the next nine months. The magnet design is essentially complete and the process of procuring the coil is taking place. In addition, the design of the additional assembly building, utilities and other infrastructure needed at the RHIC site for STAR is complete with construction on these items scheduled to begin in a few months.

We expect cosmic ray testing of the TPC to begin in early 1996, the magnet construction to be completed in mid-1997 followed by installation of the TPC, trigger detectors, electronics and data acquisition systems leading to cosmic rays tests of the baseline STAR detector in the second half of 1998. STAR will be ready to begin its physics program in March 1999 when beams first collide at RHIC and we expect that significant parts of the remaining systems (SVT, EMC, TOF and XTPC) will be on-line soon thereafter.

Acknowledgements

We thank Joan Tarzian for assistance with the manuscript, and the organizers of the 10th Winter Workshop on Nuclear Dynamics for having arranged a productive and interesting workshop in a very pleasant setting. This work is supported in part by the Director, Office of Energy Research, Division of Nuclear Physics of the Office of High Energy Physics of the U.S. Department of Energy under contract DE-AC03-76SF00098.

References

1. J. N. Marx and D. R.Nygren, Phys. Today 31(10), 46 (1978).

2. RHIC Letter of Intent for an Experiment on Particle and Jet Production at Midrapidity, The STAR Collaboration, Lawrence Berkeley Laboratory Report 29651.

3. Update to the RHIC Letter of Intent for an Experiment on Particle and Jet Production at Midrapidity, The STAR Collaboration, Lawrence Berkeley Laboratory Report 31040.

4. Conceptual Design Report for the Solenoid Tracker at RHIC, The STAR Collaboration, LBL-PUB5347 (1992).

5. J. W. Harris and the STAR Collaboration, Nucl. Phys. A566, 277c(1994).

6. K. Geiger and B. Mueller, Nucl. Phys. B369, 600(1992).

PHYSICS CAPABILITIES OF A SILICON VERTEX DETECTOR AT RHIC

R. BELLWIED
Wayne State University
Department of Physics and Astronomy
Detroit, MI 48201, USA

ABSTRACT

The physics capabilities of a Silicon vertex detector in the heavy ion environment of RHIC are discussed. As example the SVT (Silicon Vertex Tracker) in STAR is chosen. Physics issues related to strangeness production, charm production, and condensation effects are presented. Simulations show that a device like the SVT will contribute to the detection of a possible Quark Gluon Plasma.

1. Introduction

In the last decade Silicon based vertex devices have proven to be key detectors for the identification of charged particle trajectories in high energy particle physics. Information gathered from these detectors determines the primary and secondary vertices of emitted particles in two body collisions (pp, pA and AA collisions). In some instances they also provide momentum and de/dx information for particle identification.

Most detectors are based on Silicon strip technology (e.g. CDF, D0, ALEPH, and DELPHI, L3). A strip allows only the position determination in one dimension, so one is required to build several layers of detector in which each layer provides information in one dimension. In the low multiplicity environment of high energy pp-collisions this leads to satisfactory two dimensional position resolution, but for the heavy ion environment at RHIC the x-y ambiguities, caused by the layered information structure, are too big.

In recent years both SSC experiments (SDC and GEM) started research and development programs for Silicon pixel detectors. The pixel technology foresees small Silicon pads (on the order of 30*30 μm) which are bump bonded to readout electronics of the same size. The small pixel size is required to allow charge sharing between neighboring pixels, which leads to a two-dimensional position resolution smaller than the actual pixel size (around 10μm in each direction). The charge sharing effect is due to instant Coulomb repulsion in the induced electron cloud. The small pixel size leads to a high channel count and thus a high cost for a large area detector.

For RHIC therefore we chose alternately a Silicon drift detector. This technique was developed in the early 80's by Gatti and Rehak [1] and is based on the drift of the induced electron cloud in the bulk of the Silicon wafer. The drift of electrons in gas is a very common detection method, which is used in many applications (time projection chamber, wire drift chamber), but drifting electrons in solid state requires a special detector layout. The material used is high resistivity n-type Silicon (NTD material), in which a strong voltage gradient is supplied across the detector via cathode strips ($\Delta V = 3000$ V) to form a potential valley. The electron cloud spread during drift is not only due to Coulomb repulsion, but also diffusion. By applying the right voltage it can be optimized. The drift time is divided in time buckets, which correspond to a certain lateral pixel size. At the optimized voltage the drift takes about 1 μs per 1 cm. A detector has a drift length of 3 cm. Running the electronics at a certain frequency (40 MHz) leads to a pixel size of 250 micron by 165 micron. The 250 micron represent the anode pitch, whereas the 165 micron correspond to the time bucket length. Position resolution was measured on a prototype Silicon drift wafer at WSU. The accomplished resolution is comparable to a pixel detector, but at the same time the number of readout channels is reduced by a factor 6000! The proposed detector for STAR consists of 216 wafers (size = 6 by 6 cm) which are arranged on three honeycomb structured barrels around the interaction region (R $\approx$ 5, 10, 15 cm). The coverage is 2π in ϕ and 2 units in pseudorapidity η. It has 103680 readout channels and consists of 0.8 m^2 of Silicon. An obvious disadvantage of a drift detector is its drift time. Most Silicon devices operate in a high beam intensity environment in which the event rate is determined by the minimum readout rate of the slowest detector component. In STAR the Silicon Vertex Tracker (SVT) is teamed with a Time Projection Chamber (TPC) which has a drift time of around 50 μs, roughly a factor 10 longer than the SVT. So for the specific case of the STAR experiment the drift time of the electrons in the SVT does not affect the event rate.

2. Physics Motivation

The main motivation for a high precision tracking device close to the primary event vertex is the need for identification of decays of short lifetime particles. The TPC gas volume starts at a distance of 50 cm from the primary vertex. Particles with a $c\tau$ of less than 10 cm cannot be detected by the TPC in a direct way, because their decay occurs outside the detector volume. By simulating the STAR layout it was found that the TPC has good acceptance for K, π, p, d, t etc. Λ's can be retrieved from TPC information by identifying secondary protons as Λ decay products. In addition, all strange and multistrange baryons as well as shorter lived mesons will be detected by the SVT. Invariant mass reconstruction simulations for Λ, Ξ and Ω particles and their antiparticles proved to be successful by employing SVT and TPC tracking information. The impact parameter resolution from the SVT was simulated to be a few 100 μm, which allows to resolve secondary from primary vertices at close distances.

3. Physics goals

The physics goals of the STAR-SVT are described in detail in the Conceptual Design Report (CDR) of STAR [2]. The main emphasis is on flavor physics and low p_T-phenomena at RHIC. The SVT also supports the TPC physics by adding high resolution spacepoints and two track resolution to the TPC dataset. That is particularly interesting for HBT measurements and the momentum resolution of charged particles at high p_T. This contribution addresses just the physics that is accessible by the SVT itself. In the following chapters we discuss:

- strange and multistrange baryon/antibaryon ratios
- charm production
- low momentum effects in pion spectra

3.1. Strange and Multistrange Baryon/Antibaryon Ratios

The interest in the production of strange baryons was initiated by early theory papers [3] which predicted a quark gluon plasma (QGP) signature in the ratio of strange particles versus non strange particles. This was attributed to the equal production probability of u,d and s-quarks in a plasma state, whereas the s-quark production is highly suppressed in a hadronic phase. Subsequently experiment E802 at the AGS discovered an unusual high K/π ratio [4]. Realistic event generation codes [5] were able to describe this effect mainly on the basis of final state interactions of the primary produced Kaons and Pions (rescattering with spectator mass or co-movers). Theorists then proposed a QGP signature in the ratio of strange to antistrange baryons [6]. This theory which is still accepted as a hadronic measure of a phase transition, assumes that the QGP phase is characterized by a zero strange chemical potential, whereas hadronic phases have a non zero strange chemical potential. If $\mu_s=0$, then the formation of strange and antistrange baryons should have comparable probabilities and ratios like $\bar{\Lambda}/\Lambda$ or $\bar{\Xi}/\Xi$ would approach unity. Fig.1 shows a calculation by Peter Koch [7] in which he predicts particle ratios as a function of increasing strangeness content. Rafelski [6] calculated the formation of $s\bar{s}$ pairs in the plasma phase and concluded that the abundance of strangeness should be sufficient to allow a statistically significant measurement of strange particle ratios even at CERN energies. Preliminary results [8] hint at a unusually high ratio in one of the systems at CERN ($\bar{\Xi}/\bar{\Lambda}$ ratio in S+W collisions at 200 GeV/u). At RHIC the effect should be even more pronounced. Matter and antimatter will be reconstructed with the same accuracy in the SVT.

3.2. Charm Production

An even more exciting component of flavor physics is the charm production at relativistic incident energies in heavy ion systems. D-mesons constitute the only hadronic signal, whose final state interactions are negligible. If c-quarks are produced in the early stages of the interaction the slope parameter of their particle spectrum

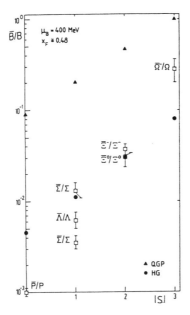

Fig. 1: Strange matter to antimatter ratio as a function of strangeness content in QGP and hadronic phase (from [7])

might provide a nearly unmodified signal of the incident collision temperature. Several theorists ([9],[10],[11]) pointed out that, based on parton cascade models, the early stages of the collision might be characterized by the thermal equilibration of the gluons after very short times (around 1 fm/c). No chemical equilibrium will be reached because of a strong suppression of quark production. If the initial temperature of the system exceeds 500 MeV, then the production of $c\bar{c}$ pairs is above threshold, which should yield a strong enhancement of D-mesons. The J/Ψ production is strongly suppressed due to screening effects [12]. First calculations by Shuryak [9], Mueller [10] and Geiger [11] predict that the hot glue stage has a very high initial temperature. This preequilibrium stage produces $c\,\bar{c}$ pairs solely in gluon-gluon interactions (gluon-gluon fusion, gluon-gluon excitation etc.). Depending on the initial configuration estimates for the D-meson enhancement factor per unit rapidity vary from 3-40. In all cases a sizeable increase in the D cross section is predicted. The slope parameter of the D spectrum conveys the initial phase temperature and is proposed to be between 500 and 1000 MeV. The enhanced production of D-mesons is not thermal production, though. Its origin is in secondary parton interactions (parton cascade) after the initial parton-parton collision (QCD production) which yields around 0.8 D's per interaction. Subsequent thermal equilibration of quarks might also contribute to the D abundance

Au+Au with QGP Au+Au w/o QGP

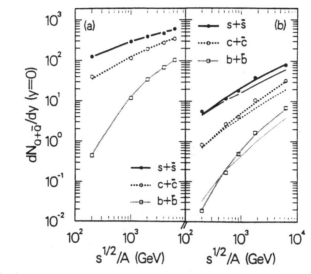

Fig. 2: D meson enhancement prediction in parton cascade models (from [11])

[13]. So overall the system passes through three stages of charm production:

1. the initial parton-parton interaction
2. the thermal gluon equilibration (preequilibrium, $\tau \approx 1$ fm/c)
3. the thermal quark equilibration (equilibrium, $\tau \approx 3$ fm/c)

The contribution from the third stage is modest, because the system expanded and cooled down to critical temperatures (around 250 MeV), which are well below the charm production threshold. Fig.2 shows a comparison of yield predictions for the quark pairs from the preequilibrium stage by Geiger.

The SVT should allow two different methods of D yield determination. First a full mass reconstruction via the two body decay of the D0 was attempted. The D0 has a very short lifetime ($c\tau$ =150 micron), but it decays with a strong branching ratio (5.4%) into K^- and π^+. Two major problems had to be addressed in the SVT simulations: a.) the primary vertex resolution has to be sufficient to distinguish the decay vertex from the primary vertex, b.) the random background of $K\pi$ pairs in the mass reconstruction has to be sufficiently low. Simulations of the impact parameter resolution show that by choosing only vertices beyond a certain cutoff parameter (e.g. 500 μm), primary and secondary vertices might be distinguishable. To reduce the random background independent phase-space cut parameters were chosen [14].

Based on these parameters, preliminary calculations reveal that a sufficient signal to noise ratio for the distinction between D mesons and random pairs requires an event rate of at least 10,000,000 events (assuming only D's from the initial parton-parton collision and therefore no enhancement). Although this rate can be recorded in STAR within one month of running a realistic simulation of that many central Au+Au collisions at RHIC is at the moment beyond the computing capabilities.

An alternate analysis for D yield determination is the background free reconstruction of K vertices which originate from D-decays. The comparable method of single particle vertex reconstruction was successfully applied by CDF to determine the B meson lifetime [15] from the J/ψ decay. Unlike the first method, this approach only measures the D cross section and not its four-momentum vector. Kinematic observables, like the slope parameter of the distribution are therefore not accessible. Still, a strong increase in the D yield hints at new physics. In the ideal case of a fully gaussian primary vertex distribution, the additional Kaons produced in D decays would result in an exponential tail to the primary vertex with a slope due to the D lifetime. No other decay, except the Ω decay, can obscure this signal. By choosing a secondary vertex window between 500 and 1000 micron from the primary vertex, simulations show that the Ω induced background is negligible. But in the specific case of the SVT, the radiation length induced by the Beryllium beam pipe as well as the detector layers themselves, introduces an exponential tail to the primary vertex distribution due to multiple Coulomb scattering, which is roughly comparable to the D meson distribution in yield and slope. Further studies to reduce the effect of multiple Coulomb scattering on the vertex resolution via software methods is underway. These methods are based on track information in addition to the SVT spacepoints (SVT dE/dx information, TPC spacepoint information).

3.3. Condensation effects

One of the main issues of the SVT as a standalone physics device is the reconstruction of low momentum primary particles. The low momentum part of the pion spectrum was found to be enhanced over simple Boltzmann like or even exponential particle emission spectra. The effect was first measured at CERN [16] and then corroborated at the AGS [17]. Many different theories were proposed to explain the effect in the pion spectrum [18]. At AGS energies it seems to be apparent that due to the large stopping power the baryon density is enhanced to a point were much of the baryon density converts into resonance matter, that means the pions and nucleons form Δ or N^* resonances which subsequently decay back into nucleons and pions. In this process the pion average momentum is reduced, which leads to an enhancement of low momentum pions [19]. For systems with higher incident energy various scenarios were postulated, one of them being Bose Einstein condensation (BEC) above a critical pion density.

BEC occurs because the density of bosons in excited states at given temperature

$T = \beta^{-1}$ is bounded above by

$$\rho_e^{\max}(T) = \rho_e(T, z=1) \quad , \tag{1}$$

where

$$\rho_e(T, z) = \int_0^\infty dE\, g(E)\, f_{T,z}(E) \quad , f_{T,z}(E) = \left[z^{-1} \exp(\beta E) - 1\right]^{-1} . \tag{2}$$

Here, the fugacity z (≤ 1) is related to the chemical potential by $z \equiv \exp\beta(\mu - E_0)$. The zero for the single particle energy E has been chosen at the value of the lowest energy level E_0, in the thermodynamic limit; $g(E)$ is the corresponding density of states. If the density ρ exceeds $\rho_e^{\max}(T)$, the ground state accepts the additional particles, and

$$\rho_0 = \rho - \rho_e = \frac{1}{V}\frac{z}{1-z} \tag{3}$$

becomes appreciable ($z = N_0/(N_0 + 1) \sim 1 - 1/N_0 \sim 1$). BEC is a momentum space effect.

Alternatively, for fixed ρ, BEC occurs for $T < T_c$, where T_c is the solution of $\rho = \rho_e^{\max}(T_c)$. The solid line in Fig. 3 shows T_c as a function of ρ, for ideal relativistic pions of one charge state in a large interaction volume. In this case condensation occurs when $\mu = m_\pi \equiv E_0$.

For RHIC–energies, estimates from the FRITIOF event generator (Au+Au) suggest that ~ 3000 pions per isospin state will be produced (assume isospin symmetry). Present calculations for RHIC indicate a 10–20 fm source radius. At $R = 10$ fm there is a sizeable probability for ρ to exceed $\rho_e^{\max}$ at freeze–out. If the freeze–out temperature is independent of the incident energy, the square in Fig. 3a represents the situation at RHIC: freeze–out occurs within the condensed phase. Thus, RHIC may provide an environment in which Bose-Einstein condensation is possible. In the extreme case, the mean pion momentum depends only on the freeze–out radius, and is proportional to h/R. All pions are emitted coherently, and the momentum distribution of the pions is strongly peaked at low momenta.

Of course, the (dynamical) situation is somewhat more complicated [20–23]. In the initial stages of the re-hadronized phase the total pion number is large, but they will reside in resonances. Any excess pion density will probably be removed by inelastic processes, since the average collisional c.m. energy $\sqrt{s}$ (or "temperature") is high ($\pi\pi \to \rho,\ \rho\rho \to \pi\pi,\ \pi\pi \to K\bar{K}$, etc.). Thus, the pion chemical potential will be close to zero. Later on, as $\sqrt{s}$ drops below the inelastic thresholds, the pion gas possibly has a chance to build up significant chemical potentials. In fact, for an adiabatic expansion, $\mu \to m_\pi$, but a more rapid evolution would lead to partial thermalization with $\mu \lesssim 130$ MeV.

The pion spectra will thus be affected. The dotted line in Fig. 3b shows a pion momentum distribution

$$\frac{dN}{dp} = A\, 4\pi\, p^2 \left[\exp\{\beta(\sqrt{p^2 - m_\pi^2} - m_\pi)\} - 1\right]^{-1} \tag{4}$$

for $\mu = 0$, and $T = 150$ MeV.[1] This curve describes a typical Au+Au RHIC event as generated by FRITIOF rather well. For large $\mu \geq 100$ MeV, the momentum spectrum is significantly shifted toward lower momenta: the dashed curve in Fig. 3b assumes the extreme case of single pion condensation, $\mu \to m_\pi$.

Near freeze–out, an additional effect, will influence pion distributions. As suggested by Pratt [24], if pion densities come close to, or exceed, ρ_c^{max} at any stage during the hadronic evolution, multiparticle interference plays a role in particle emission. The average particle separation is close to the thermal wavelength, and n–body symmetrization becomes important. The overall result of this "laser" effect on the spectrum is an increase in the low momentum component (dashed curve in Fig. 3b), though this factor depends on dynamical details.

In general most observables are affected by condensation effects:

- the multiplicity of pions increases dramatically;
- pion emission is coherent with a very low $<p_T>$ that depends strongly on the source size;
- the two pion correlation function approaches 1 at the intercept, characteristic of a coherent state;
- large isospin fluctuations may develop (e.g. unusual π^0/π^+ ratios);

Isospin fluctuations should occur in particular if the chiral condensate is disoriented, a theory suggested by Bjorken [25]. Of course, other mechanisms like in-medium dispersion relations [26], may well contribute to low momentum pions. It will require detailed modeling to disentangle the different contributions. Regardless, a sizeable low p enhancement over the Boltzmann distribution may be expected at RHIC.[2]

There are two ways of measuring low momentum pions at RHIC:

- (a) search for very low momentum π^0, by detecting the photons (with energies around 70 MeV) emitted from their decay;
- (b) search for very low momentum charged π's with a very high resolution tracking device.

Method (a) is suited for a high resolution electromagnetic calorimeter, as suggested by the PHENIX collaboration; (b) is an ideal measurement for the SVT detector in STAR. Its superior position resolution and its standalone tracking capability might allow to detect condensation effects on an event by event basis, by measuring the low momentum part of the charged pion spectrum in a central Au+Au collision at RHIC. Simulations show that 'condensation events' are characterized by the detection of

---

[1] In this non–degenerate limit ($z \ll 1$), f is practically a Boltzmann distribution.

[2] Charged pions will have an additional Coulomb repulsion effect, which might lead to a reversal of the condensation effect by broadening the momentum distribution, but this effect is rather small compared to the overall mean momentum reduction.

206

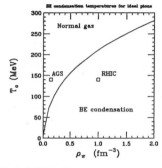
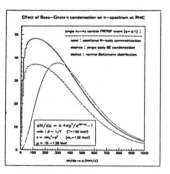

Fig. 3: Fig.3a.) Critical temperature for ideal pion BEC as a function of density
Fig.3b.) Effect of single body condensation and multibody symmetrization on the pion momentum
spectrum

an enhanced particle flux in the SVT and the simultaneous detection of low particle density in the TPC. The low momentum cutoff for identification in the TPC is around 150 MeV/c. If the bulk of the pion spectrum has lower momenta, then the above effect should occur. An event selection criterion, based on the multiplicity ratio between SVT and TPC, could even be integrated into the STAR trigger concept.

If the statistics in a single event is too small, an inclusive spectrum of the invariant pion cross section should exhibit an increased low $p_\perp$ enhancement due to Bose-Einstein condensation. Considering the diversity of explanations for the existing low $p_\perp$ effect, a measurement of an inclusive cross section spectrum might not be unambiguous, though.

4. Conclusions

A high resolution tracking device close to the primary vertex allows the measurement of secondary vertices with large efficiencies in the heavy ion environment at RHIC. In addition, the standalone tracking capabilities of the SVT convey the very low momentum part of the produced particle spectra. Strange and multistrange particle ratios will be measured with great accuracy. Novel physics that we plan to address with the SVT at RHIC includes the possible enhancement of the D-meson cross section and the search for condensation effects in the low momentum part of the pion spectrum.

5. Acknowledgments

I would like to thank the STAR collaboration, and in particular the SVT group for numerous discussions. Gerd Welke helped to shape up the low p_T physics part. Discussions with Tom Cormier, John Harris, Spiros Margetis, Tim Hallman and Claude Pruneau are gratefully acknowledged.

6. References

[1] P. Rehak, E. Gatti, *Nucl.Instr. and Meth.* **225** (1984) 608

[2] STAR Conceptual Design Report, LBL–PUB–5347 (1992)

[3] J. Rafelski, *Phys.Rep.* **88** (1982) 331

[4] T. Abbott *et al.*, E802 collaboration, *Phys. Rev. Lett.* **66**, 1567 (1991)

[5] H. Sorge et al., *Phys.Rev.Lett.* **68** (1992) 286

[6] J. Rafelski, *Nucl. Phys.* **A544** (1992) 279 c

[7] P. Koch, *Z. Phys.* **C38** (1988) 269

[8] J.B. Kinson, *Nucl.Phys.* **A544** (1992) 321c

[9] E. Shuryak, *Phys.Rev.Lett.* **68** (1992) 3270

[10] B. Mueller, X.N. Wang, *Phys.Rev.Lett.* **68** (1992) 2437

[11] K. Geiger, *Phys.Rev.* **D48** (1993) 4129

[12] T. Matsui, H. Satz, *Phys.Lett.* **178B** (1986) 416

[13] T. Altherr, D. Seibert, *Phys.Rev.* **D** to be published

[14] J.Grebieszkow, STAR note 121 (1993)

[15] F. Abe, *Phys.Rev.Lett.* **71** (1993) 3421

[16] B. Jacak, *Nucl.Phys.* **A525** (1991) 77c

[17] T.K. Hemmick for E814, *Proc. for HIPAGS, Cambridge* (1993)

[18] E. Shuryak, *Nucl.Phys.* **A525** (1991) 3c

[19] G. Brown, J. Stachel, and G. Welke, *Phys. Lett.* **B253** (1991) 19

[20] M. Kataja, and P. Ruuskanen, *Phys. Lett.* **B243** (1990) 181

[21] S. Gavin, and P. Ruuskanen, *Phys. Lett.* **B262** (1991) 326

[22] G. Welke, and G. Bertsch, *Phys. Rev.* **C45** (1992) 1403

[23] H.W. Barz, P. Danielewicz, H. Schulz, G. Welke, *Phys. Lett.* **B287** (1992) 40

[24] S. Pratt, *Phys. Lett.* **B301** (1993) 159

[25] J.D. Bjorken, *Acta Phys.Pol.* **B23** (1992) 637

[26] E. Shuryak, *Phys. Rev.* **D42** (1990) 1764

PARTON EQUILIBRATION IN ULTRARELATIVISTIC HEAVY ION COLLISIONS[1]

Xin-Nian Wang

Nuclear Science Division, MS 70A-3307
Lawrence Berkeley Laboratory, Berkeley, CA 94720, USA

ABSTRACT

Medium effects which include color screening and Landau-Pomeranchuk-Migdal suppression of induced radiation are discussed in connection with the equilibration of dense partonic system in ultrarelativistic heavy ion collisions. Taking into account of these medium effects, the equilibration rate for a gluonic gas are derived and the consequences are discussed.

1. Introduction

Strong interactions involved in hadronic collisions can be generally divided into two categories depending on the scale of momentum transfer Q^2 of the processes. When $Q^2 \sim \Lambda_{QCD}^2$, the collisions are nonperturbative in QCD and are considered soft. These interactions can be approximated by some effective theories in which partons inside a nucleon cannot be resolved and nucleons interact coherently by the exchange of mesons or soft Pomerons. These kind of coherent interactions will result in the collective excitations as observed in experiments at low and intermediate energies, $\sqrt{s} <$ a few GeV. On the other hand, if Q^2 is much larger than Λ_{QCD}^2, parton model becomes relevant and they interact approximately incoherently. Due to the high Q^2, parton interactions can be calculated via perturbative QCD (pQCD), given the initial parton distribution functions inside a nucleus. At high collider energies as RHIC and LHC, the hard processes of parton interactions become dominant and heavy ion collisions are generally very violent[1]. In these violent collisions, it is quite likely that colors are liberated during the collisions and we can start with a color deconfined system.

In the last few years, many efforts have been made to estimate the initial parton density by perturbative calculations[2,3,4,5]. Though the results depend on parton structure functions inside a nucleus and the modeling of the underlying soft interactions, a general consensus is that a dense partonic system will be produced during the early

---

[1]Collaborations with my colleagues are gratefully acknowledged. This work was supported by the Director, Office of Energy Research, Division of Nuclear Physics of the Office of High Energy and Nuclear Physics of the U.S. Department of Energy under Contract No. DE-AC03-76SF00098.

stage of ultrarelativistic heavy ion collisions. However, the system is not necessarily in both thermal and chemical equilibrium. The initial system is totally dominated by gluons and the quark and anti-quark densities are far below their chemical equilibrium values. Numerical simulations of parton cascade can be made with certain simplifications. However, the complexity of these calculations makes it difficult to obtain a clear understanding of the different time scales on various parameters and model assumptions. Most importantly, medium effects and interferences are difficult to incorporate. In this talk, I would like to describe two medium effects. Color screening is caused by the interaction of a color charge with the medium which gives a natural infrared cut-off to regularize the parton cross sections inside the dense system[6]. Another medium effect is the so-called Landau-Pomeranchuk-Migdal effect of the reduced radiation by multiple scattering through the medium[7]. These medium effects can be utilized to obtain a parameter-free set of equations, based on perturbative QCD in a dense partonic medium, that describes the evolution of quark and gluon distributions towards equilibrium[8].

2. Color Screening

Let us first look at the color screening in the initially produced parton system. Following the standard calculation(in Coulomb gauge) of screening in the time-like gluon propagator in a medium of gluonic excitation, we have the screening mass[6],

$$\mu_D^2 = -\frac{3\alpha_s}{\pi^2} \lim_{|\mathbf{q}|\to 0} \int d^3k \frac{|\mathbf{k}|}{\mathbf{q}\cdot\mathbf{k}} \mathbf{q}\cdot\nabla_{\mathbf{k}} f(\mathbf{k}). \tag{1}$$

Now, instead of using the Bose-Einstein distribution for the thermalized case, we simply relate the phase space density $f(\mathbf{k})$ to the initial gluon distribution calculated from pQCD. At high energies we find that the transverse and longitudinal screening lengths are very close. We assume that the gluon momentum distribution is similar to a thermal distribution

$$f_g(k) = \lambda_g e^{-u\cdot k/T}, \tag{2}$$

with λ_g and T characterizing the thermalization of the gluon gas. With this distribution, one obtain the effective color screening mass

$$\mu_D^2 = \lambda_g g^2 T^2. \tag{3}$$

We will use this effective color screening mass as an infrared cut-off for parton interaction cross sections inside a non-equilibrium gluonic gas.

3. Induced Radiation

The leading processes which contribute to the gluon equilibration are those with induced gluon radiation. Let us consider the simplest case of induced radiation from a two-quark scattering. The Born amplitude for two-quark scattering $(p_i, k_i) \to (p_f, k_f)$ through one gluon exchange is,

$$\mathcal{M}_{el} = ig^2 T^a_{AA'} T^a_{BB'} \frac{\bar{u}(p_f)\gamma_\mu u(p_i)\bar{u}(k_f)\gamma^\mu u(k_i)}{(k_i - k_f)^2}, \tag{4}$$

where A, A', B, and B' are the initial and final color indices of the beam and target partons. The corresponding elastic cross section is,

$$\frac{d\sigma_{el}}{dt} = C^{(1)}_{el} \frac{\pi\alpha_s^2}{s^2} 2 \frac{s^2 + u^2}{t^2}, \tag{5}$$

where $C^{(1)}_{el} = C_F/2N = 2/9$ is the color factor for a single elastic quark-quark scattering and s, u, and t are the Mandelstam variables.

Taking into account the dominant contribution, the radiation amplitude from a single scattering is,

$$\mathcal{M}_{rad} \equiv i\frac{\mathcal{M}_{el}}{T^a_{AA'} T^a_{BB'}}\mathcal{R}_1,$$

$$\mathcal{R}_1 \simeq 2ig\vec{\epsilon}_\perp \cdot \left[\frac{\mathbf{k}_\perp}{k_\perp^2} + \frac{\mathbf{q}_\perp - \mathbf{k}_\perp}{(\mathbf{q}_\perp - \mathbf{k}_\perp)^2}\right] T^a_{AA'}[T^a, T^b]_{BB'}, \tag{6}$$

where $\mathcal{M}_{el}$ is the elastic amplitude as given in Eq. (4), and $\mathcal{R}_1$ is defined as the radiation amplitude induced by a single scattering. For later convenience, all the color matrices are included in the definition of the radiation amplitude $\mathcal{R}_1$. With the above approximations, we then recover the differential cross section for induced gluon bremsstrahlung by a single collision as originally derived by Gunion and Bertsch [9],

$$\frac{d\sigma}{dt dy d^2 k_\perp} = \frac{d\sigma_{el}}{dt}\frac{dn^{(1)}}{dy d^2 k_\perp}, \tag{7}$$

where the spectrum for the radiated gluon is,

$$\frac{dn^{(1)}}{dy d^2 k_\perp} \equiv \frac{1}{2(2\pi)^3 C^{(1)}_{el}} \overline{|\mathcal{R}_1|^2} = \frac{C_A \alpha_s}{\pi^2} \frac{q_\perp^2}{k_\perp^2 (\mathbf{q}_\perp - \mathbf{k}_\perp)^2}. \tag{8}$$

In the square modulus of the radiation amplitude, an average and a sum over initial and final color indices and polarization are understood. The above formula is also approximately valid for induced radiation off a gluon line, except that the color factor $C^{(1)}_{el}$ in the elastic cross section has to be replaced by $1/2$ for gq and $9/8$ for gg scatterings.

One nonabelian feature in the induced gluon radiation amplitude, Eq. (6), is the singularity at $\mathbf{k}_\perp = \mathbf{q}_\perp$ due to induced radiation along the direction of the exchanged gluon. For $k_\perp \ll q_\perp$, we note that the induced radiation from a three gluon vertex can be neglected as compared to the leading contribution $1/k_\perp^2$. However, at large

$k_\perp \gg q_\perp$, this three gluon amplitude is important to change the gluon spectrum to a $1/k_\perp^4$ behavior, leading to a finite average transverse momentum. Therefore, $q_\perp$ may serve as a cut-off for $k_\perp$ when one neglects the amplitude with the three gluon vertices as we will do when we consider induced radiation by multiple scatterings in the next section.

4. EFFECTIVE FORMATION TIME

The radiation amplitude induced by multiple scatterings has been discussed in Ref. [7]. We here only briefly discuss the case associated with double scatterings. We consider two static potentials separated by a distance L which is much larger than the interaction length, $1/\mu$. For convenience of discussion we neglect the color indices in the case of an abelian interaction first. The radiation amplitude associated with double scatterings is,

$$\mathcal{R}_2^{\text{QED}} = ie\left[\left(\frac{\epsilon \cdot p_i}{k \cdot p_i} - \frac{\epsilon \cdot p}{k \cdot p}\right)e^{ik\cdot x_1} + \left(\frac{\epsilon \cdot p}{k \cdot p} - \frac{\epsilon \cdot p_f}{k \cdot p_f}\right)e^{ik\cdot x_2}\right], \tag{9}$$

where $p = (p_f^0, p_z, \mathbf{p}_\perp)$ is the four-momentum of the intermediate parton line which is put on mass shell by the pole in one of the parton propagators, $x_1 = (0, \mathbf{x}_1)$, and $x_2 = (t_2, \mathbf{x}_2)$ are the four-coordinates of the two potentials with $t_2 = (z_2 - z_1)/v = Lp^0/p_z$. We notice that the amplitude has two distinguished contributions from each scattering. Especially, the diagram with a gluon radiated from the intermediate line between the two scatterings contributes both as the final state radiation for the first scattering and the initial state radiation for the second scattering. The relative phase factor $k \cdot (x_2 - x_1) = \omega(1/v - \cos\theta)L$ then will determine the interference between radiations from the two scatterings. If we define the formation time as

$$\tau(k) = \frac{1}{\omega(1/v - \cos\theta)} \simeq \frac{2\omega}{k_\perp^2}, \tag{10}$$

then Bethe-Heitler limit is reached when $L \gg \tau(k)$. In this limit, the intensity of induced radiation is simply additive in the number of scatterings. However, when $L \ll \tau(k)$, the final state radiation from the first scattering completely cancels the initial state radiation from the second scattering. The radiation pattern looks as if the parton has only suffered a single scattering. This is often referred to as the Landau-Pomeranchuk-Migdal (LPM) effect[10]. The corresponding limit is usually called factorization limit.

The extrapolation to the general case of m number of scatterings gives us the radiation amplitude in QCD,

$$\mathcal{R}_m = i2g\frac{\vec{\epsilon} \cdot \mathbf{k}_\perp}{k_\perp^2}T_{A_1 A_1'}^{a_1} \cdots T_{A_m A_m'}^{a_m} \sum_{i=1}^{m} \left(T^{a_m} \cdots [T^{a_i}, T^b] \cdots T^{a_1}\right)_{BB'} e^{ik\cdot x_i}, \tag{11}$$

which contains m terms each having a common momentum dependence in the high energy limit, but with different color and phase factors. The above expression should also be valid for a gluon beam jet, with the corresponding color matrices replaced by those of an adjoint representation. The spectrum of soft bremsstrahlung associated with multiple scatterings in a color neutral ensemble is, similar to Eq. (8),

$$\frac{dn^{(m)}}{dy\,d^2k_\perp} = \frac{1}{2(2\pi)^2 C_{el}^{(m)}}\overline{|\mathcal{R}_m|^2} \equiv C_m(k)\frac{dn^{(1)}}{dy\,d^2k_\perp}, \tag{12}$$

where $C_{el}^{(m)} = (C_F/2N)^m$ is the color factor for the elastic scattering cross section without radiation. $C_m(k)$, defined as the "radiation formation factor" to characterize the interference pattern due to multiple scatterings, can be expressed as,

$$C_m(k) = \frac{1}{C_F^m C_A N}\sum_{i=1}^{m}\left[C_{ii} + 2Re\sum_{j=1}^{i-1}C_{ij}e^{ik\cdot(x_i-x_j)}\right], \tag{13}$$

where the color coefficients are defined as

$$C_{ij} = Tr\left(T^{a_m}\cdots[T^b,T^{a_i}]\cdots T^{a_1}T^{a_1}\cdots[T^{a_j},T^b]\cdots T^{a_m}\right). \tag{14}$$

If we average over the interaction points x_i according to a linear kinetic theory, we find that an effective formation time in QCD can be defined as,

$$\tau_{\text{QCD}}(k) = r_2\tau(k) = \frac{C_A}{2C_2}\frac{2\cosh y}{k_\perp}, \tag{15}$$

which depends on the color representation of the jet parton. The induced gluon radiation due to multiple scattering will be suppressed when the mean free path λ is much smaller than the effective formation time.

5. Equilibration Rates

The most important reactions for establishing gluon equilibrium are $gg \leftrightarrow ggg$. Elastic scattering processes, on the other hand, are crucial for maintaining local thermal equilibrium. Multi-gluon radiation is presumably suppressed by color screening, while radiative processes involving quarks have smaller cross sections due to QCD color factors.

The evolution of the gluon density n according to reactions mentioned above can be described by a rate equation. Adding the equation for energy conservation assuming only longitudinal expansion we end up with a closed set of equations determining the temperature $T(\tau)$ and the gluon "fugacity" $\lambda_g(\tau) \equiv n/n_{eq}(T)$ as a function of the proper time τ,

$$\frac{\dot{\lambda}_g}{\lambda_g} + 3\frac{\dot{T}}{T} + \frac{1}{\tau} = R_3(1 - \lambda_g), \tag{16}$$

$$\lambda_g^{3/4}T^3\tau = \text{const.} \tag{17}$$

This set of evolution equations is completely controlled by the gluon production rate, $R_3 = \frac{1}{2}\sigma_3 n$.
Taking into account of LPM effect, the cross section for $gg \to ggg$ processes can be written as,

$$\frac{d\sigma_3}{d^2q_\perp dy d^2k_\perp} = \frac{d\sigma_{el}}{d^2q_\perp}\frac{dn^{(1)}}{d^2k_\perp dy}k_\perp \cosh y \theta(\lambda - \tau_{\text{QCD}}(k))\theta(E - k_\perp \cosh y), \tag{18}$$

where $\tau_{\text{QCD}}(k)$ is given by Eq. (15), the second θ-function is for energy conservation. The gluon density distribution induced by a single scattering is given by Eq. (8) which must be regularized by the color screening mass μ_D in Eq. (3). Using the elastic cross section of gluon scattering regularized also by μ_D, we obtain a fugacity independent mean free path

$$\lambda^{-1} = \sigma^{2\to2}n = \frac{9}{8}a_1\alpha_s T, \tag{19}$$

Using these values we evaluate the chemical gluon equilibration rate $R_3 = \frac{1}{2}n\sigma_3$, as defined in Eq. (18), numerically. This rate scales with the temperature linearly but is a complicated function of the gluon fugacity. The result can approximated by an analytical fit,

$$R_3 = 2.1\alpha_s^2 T \left(2\lambda_g - \lambda_g^2\right)^{1/2}, \tag{20}$$

which will be used in solving the time dependent rate equations. The rate we thus obtained has a nonlinear dependence on the fugacity due to the inclusion of the LPM effect and the effective color screening mass. If LMP effect were not included, the rate whould be much larger, thus would lead to a fast equilibration.

6. References

[1] J. P. Blaizot and A. H. Mueller, Nucl. Phys. B **289**, 847 (1987).

[2] K. Kajantie, P. V. Landshoff, and Lindfors, Phys. Rev. Lett. **59**, 2527 (1987); K. J. Eskola, K. Kajantie, and J. Lindfors, Nucl. Phys. B **323**, 37 (1989).

[3] X.-N. Wang and M. Gyulassy, Phys. Rev. D **44**, 3501 (1991).

[4] K. Geiger and B Müller, Nucl. Phys. B **369**, 600 (1992).

[5] K. J. Eskola and X.-N. Wang, Report No. LBL-34156, 1993, Phys. Rev. D in press.

[6] T. S. Biró, B. Müller, and X.-N. Wang, Phys. Lett. B **283**, 171 (1992).

[7] M. Gyulassy and X.-N. Wang, Report No. LBL-32682, 1993, Nucl. Phys. B in press.

[8] T. S. Biró, E. van Doorn, B. Müller, M. H. Thoma, and X.-N. Wang, Phys. Rev. C **48**, 1275 (1993).

[9] J. F. Gunion and G. Bertsch, Phys. Rev. D **25**, 746 (1982).

[10] L. D. Landau and I. J. Pomeranchuk, Dolk. Akad. Nauk SSSR **92**, 92(1953); A. B. Migdal, Phys. Rev. **103**, 1811 (1956); E. L. Feinberg and I. J. Pomeranchuk, Suppl. Nuovo. Cimento **3**, 652 (1956); Phys. JETP **23**, 132 (1966).

QUANTUM SIZE EFFECTS IN CLASSICAL HADRODYNAMICS

J. Rayford NIX

Theoretical Division, Los Alamos National Laboratory
Los Alamos, New Mexico 87545, USA

ABSTRACT

We discuss future directions in the development of classical hadrodynamics for extended nucleons, corresponding to nucleons of finite size interacting with massive meson fields. This new theory provides a natural covariant microscopic approach to relativistic nucleus-nucleus collisions that includes automatically spacetime nonlocality and retardation, nonequilibrium phenomena, interactions among all nucleons, and particle production. The present version of our theory includes only the neutral scalar (σ) and neutral vector (ω) meson fields. In the future, additional isovector pseudoscalar (π^+, π^-, π^0), isovector vector (ρ^+, ρ^-, ρ^0), and neutral pseudoscalar (η) meson fields should be incorporated. Quantum size effects should be included in the equations of motion by use of the spreading function of Moniz and Sharp, which generates an effective nucleon mass density smeared out over a Compton wavelength. However, unlike the situation in electrodynamics, the Compton wavelength of the nucleon is small compared to its radius, so that effects due to the intrinsic size of the nucleon dominate.

1. Introduction

During three previous winter workshops we have discussed a new microscopic many-body approach to relativistic nucleus-nucleus collisions based on classical hadrodynamics for extended nucleons, corresponding to nucleons of finite size interacting with massive meson fields.[1-3] The underlying foundations of this new theory, as well as applications to soft nucleon-nucleon collisions, have been published recently.[4,5] In this contribution, we would like to discuss future directions in the systematic development of this theory.

2. Extended Nucleon

We all know that the nucleon is a composite particle made up of three valence quarks plus additional sea quarks and gluons. When nucleons collide at very high energies, a few rare events correspond to the head-on or hard collisions between the individual quarks and/or gluons. For describing these events, the underlying quark-gluon structure of the nucleon is of crucial importance. Yet such hard collisions are extremely rare, typically one in a billion.[6] The vast majority of events correspond to soft collisions not involving individual quarks or gluons. For the description of

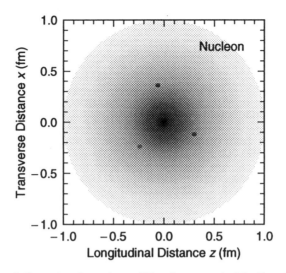

Fig. 1: Slice through the center of a nucleon. Although composed of the three indicated valence quarks plus sea quarks and gluons, for many purposes the nucleon can be regarded as a single extended object, with an exponentially decreasing mass density that is spherically symmetric in its instantaneous rest frame.

such events, an appealing idea is to regard the nucleon as a single extended object interacting with other nucleons through the conventional exchange of mesons (whose underlying quark-antiquark composition is ignored).

Experiments involving elastic electron scattering off protons have determined that the proton charge density is approximately exponential in shape,[7] with a root-mean-square radius of 0.862 ± 0.012 fm. Although many questions remain concerning the relationship between the proton charge density and the nucleon mass density,[8] it should be a fairly accurate approximation to regard them as equal. We therefore take the nucleon mass density to be

$$\rho(r) = \frac{\mu^3}{8\pi} \exp(-\mu r) \,, \tag{1}$$

with $\mu = \sqrt{12}/R_{rms}$ and $R_{rms} = 0.862$ fm. We show in fig. 1 a gray-scale plot of the mass density through the center of a nucleon calculated according to this exponential, with three valence quarks also indicated schematically.

3. Present Version

In the present version of our theory, we consider N extended, unexcited nucleons interacting with massive, neutral scalar (σ) and neutral vector (ω) meson fields. At bombarding energies of many GeV per nucleon, the de Broglie wavelength of projectile nucleons is extremely small compared to all other length scales in the problem. Furthermore, the Compton wavelength of the nucleon is small compared to its radius, so that effects due to the intrinsic size of the nucleon dominate those due to quantum uncertainty. The classical approximation for nucleon trajectories should therefore be valid, provided that the effect of the finite nucleon size on the equations of motion is taken into account. The resulting classical relativistic many-body equations of motion can be written as[4,5]

$$M_i^* a_i^\mu = f_{s,i}^\mu + f_{v,i}^\mu + f_{s,ext,i}^\mu + f_{v,ext,i}^\mu \ ,$$ (2)

where M_i^* is the effective mass, $f_{s,i}^\mu$ is the scalar self-force, $f_{v,i}^\mu$ is the vector self-force, $f_{s,ext,i}^\mu$ is the scalar external force, and $f_{v,ext,i}^\mu$ is the vector external force.

These classical relativistic many-body equations of motion can be solved numerically without further approximation. In particular, there is

- No mean-field approximation
- No perturbative expansion in coupling strength
- No superposition of two-body collisions

We have thus far solved these equations for soft nucleon-nucleon collisions at $p_{lab} = $ 14.6, 30, 60, 100, and 200 GeV/c to yield such physically observable quantities as scattering angle, transverse energy, radiated energy, and rapidity.[5] We found that the theory provides a physically reasonable description of gross features associated with the soft reactions that dominate nucleon-nucleon collisions. In addition, the present version of the theory permits a qualitative discussion of several important physical points:

- Effect of finite nucleon size on equations of motion
- Inherent spacetime nonlocality
- Particle production through massive bremsstrahlung

4. Additional Meson Fields

Nevertheless, from nucleon-nucleon scattering experiments we know that several additional meson fields are important and must be included for a quantitative description:[9]

- Isovector pseudoscalar (π^+, π^-, π^0)
- Isovector vector (ρ^+, ρ^-, ρ^0)
- Neutral pseudoscalar (η)

The next step in the systematic development of the theory should be the inclusion of these additional meson fields.

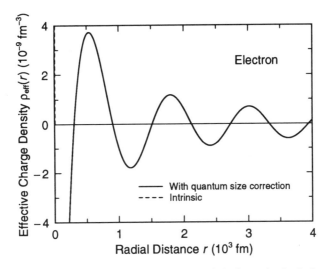

Fig. 2: Effective charge density for a point electron. The intrinsic charge density is the positive delta function indicated by the vertical dashed line at the origin.

5. Quantum Spreading Function

Although effects due to the intrinsic size of the nucleon dominate those due to quantum uncertainty, it is nevertheless important to provide an estimate of the effects of quantum uncertainty and to include them in the equations of motion if they are important. This should be possible by use of techniques analogous to those used by Moniz and Sharp[10] for nonrelativistic quantum electrodynamics.

5.1. Nonrelativistic Quantum Electrodynamics

There appear in the classical nonrelativistic equations of motion for an extended electron terms of the form $\int_\infty \int_\infty \rho(r)\, \mathcal{O} \rho(r')\, d^3r\, d^3r'$, where the operator $\mathcal{O}$ is a function of $\mathbf{r}$ and $\mathbf{r}'$. Moniz and Sharp[10] have shown in nonrelativistic quantum electrodynamics that the effect of quantum mechanics on the equations of motion is to replace such terms by terms of the form $\int_\infty \int_\infty \rho(r)\, \mathcal{O}_{\text{eff}}\, \rho_{\text{eff}}(r')\, d^3r\, d^3r'$ and derivatives with respect to λ of these terms. The effective operator $\mathcal{O}_{\text{eff}}$ is a function of $\mathbf{r}$, $\mathbf{r}'$, and $\lambda^2 \nabla_{\mathbf{r}'}{}^2$. The effective density ρ_{eff} is given by

$$\rho_{\text{eff}}(r) = \int_\infty S(|\mathbf{r} - \mathbf{r}'|)\, \rho(r')\, d^3r' \, , \tag{3}$$

where

$$S(|\mathbf{r} - \mathbf{r}'|) = -\frac{\cos(2|\mathbf{r} - \mathbf{r}'|/\lambda)}{\pi \lambda^2 \, |\mathbf{r} - \mathbf{r}'|} \tag{4}$$

218

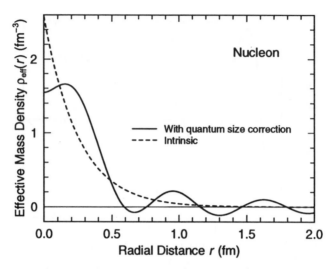

Fig. 3: Effective mass density for an extended nucleon. The intrinsic mass density is exponential with a root-mean-square radius of 0.862 fm.

and $\lambda = 1/(M_0)$ is the Compton wavelength associated with the particle's bare mass. [Please note that we have made some obvious corrections to eq. (3.27) of ref. 10.]

For a point electron, where the intrinsic charge density is $\rho_e(r) = \delta_+(r)/(4\pi r^2)$, eq. (3) reduces to

$$\rho_{\text{eff}}(r) = -\frac{\cos(2r/\lambda_e)}{\pi \lambda_e^{\,2} r} ,\qquad (5)$$

with a Compton wavelength $\lambda_e = 386.15933$ fm. As illustrated in fig. 2, the effective charge density for a point electron is a decreasing oscillatory function of radial distance from the origin.

5.2. Classical Hadrodynamics

For an extended nucleon, where the intrinsic mass density is given by the exponential (1), eq. (3) leads to

$$\rho_{\text{eff}}(r) = \frac{\mu^3}{8\pi\lambda r} \int_0^\infty r' \exp(-\mu r') \left\{ \sin(2|r - r'|/\lambda) - \sin[2(r + r')/\lambda] \right\} dr' .\qquad (6)$$

For a bare nucleon mass[4] $M_0 = 949.47$ MeV, the Compton wavelength $\lambda = 0.20783$ fm, which is much smaller than the root-mean-square radius. This has the consequence, as illustrated in fig. 3, that the effective mass density for an extended nucleon oscillates around the intrinsic mass density as a function of radial distance from the origin.

6. Conclusions

To conclude, we have shown that classical hadrodynamics

- Provides a manifestly Lorentz-covariant microscopic many-body approach to relativistic heavy-ion collisions
- Satisfies *a priori* the basic conditions that are present
- Requires minimal physical input
- Leads to equations of motion that can be solved numerically without further approximation
- Contains an inherent spacetime nonlocality that may be responsible for significant collective effects
- Provides in its present form a qualitative description of transverse momentum, radiated energy, and other gross features in nucleon-nucleon collisions
- Should be extended to include additional meson fields for the π, ρ, and η, as well as quantum size effects

7. Acknowledgments

We are grateful to R. J. Hughes and A. J. Sierk for their participation during the early stages of this investigation, to B. W. Bush for his pioneering contributions during later stages, and to D. H. Sharp for a stimulating discussion concerning his equations and their interpretation. This work was supported by the U. S. Department of Energy.

8. References

[1] A. J. Sierk, R. J. Hughes, and J. R. Nix, in *Proc. 6th Winter Workshop on Nuclear Dynamics, Jackson Hole, Wyoming, 1990*, Lawrence Berkeley Laboratory Report No. LBL-28709 (1990), p. 119.

[2] B. W. Bush, J. R. Nix, and A. J. Sierk, in *Advances in Nuclear Dynamics, Proc. 7th Winter Workshop on Nuclear Dynamics, Key West, Florida, 1991* (World Scientific, Singapore, 1991), p. 282.

[3] B. W. Bush and J. R. Nix, in *Advances in Nuclear Dynamics, Proc. 8th Winter Workshop on Nuclear Dynamics, Jackson Hole, Wyoming, 1992* (World Scientific, Singapore, 1992), p. 311.

[4] B. W. Bush and J. R. Nix, *Ann. Phys. (N. Y.)* **227**, 97 (1993).

[5] B. W. Bush and J. R. Nix, *Nucl. Phys. A* **560**, 586 (1993).

[6] D. H. Perkins, *Introduction to High Energy Physics*, Third Edition (Addison-Wesley, Menlo Park, 1987), p. 142.

[7] G. G. Simon, F. Borkowski, C. Schmitt, and V. H. Walther, *Z. Naturforsch. A* **35** (1980), 1.

[8] R. K. Bhaduri, *Models of the Nucleon* (Addison-Wesley, New York, 1988).

[9] R. Machleidt, *Adv. Nucl. Phys.* **19**, 189 (1989).

[10] E. J. Moniz and D. H. Sharp, *Phys. Rev. D* **15**, 2850 (1977).

COLLECTIVE FLOW IN AU+AU COLLISIONS

H. G. Ritter[1] for the EOS Collaboration:

S. Albergo[6], F. Bieser[1], F. P. Brady[4], Z. Caccia[6], D. A. Cebra[4],
A. D. Chacon[5], J. L. Chance[4], Y. Choi[3], S. Costa[6], J. Elliott[3], M. Gilkes[3],
J. A. Hauger[3], A. Hirsch[3], E. L. Hjort[3], A. Insolia[6], M. Justice[2], D. Keane[2],
V. Lindenstruth[7], H. S. Matis[1], M. McMahan[1], C. McParland[1],
W. F. J. Mueller[7], D. L. Olson[1], M. Partlan[4], N. Porile[3], R. Potenza[6],
G. Rai[1], J. Rasmussen[1], J. Romanski[6], J. L. Romero[4], G. V. Russo[6],
H. Sann[7], R. Scharenberg[3], A. Scott[2], Y. Shao[2], B. Srivastava[3],
T. J. M. Symons[1], M. Tincknell[3], C. Tuvè[6], S. Wang[2], P. Warren[3],
D. Weerasundara[2], H. H. Wieman[1], and K. L. Wolf[5]

[1] Lawrence Berkeley Laboratory, Berkeley, California 94720
[2] Kent State University, Kent, Ohio 44242
[3] Purdue University, West Lafayette, Indiana 47907
[4] University of California, Davis, California 95616
[5] Texas A&M University, College Station, Texas 77843
[6] Università di Catania and INFN-Sezione di Catania, 95129 Catania, Italy
[7] Gesellschaft für Schwerionenforschung, D-64220 Darmstadt 11, Germany

ABSTRACT

Based on a preliminary sample of Au + Au collisions in the EOS time projection chamber at the Bevalac, we study sideward flow as a function of bombarding energy between $0.25A$ GeV and $1.2A$ GeV. We focus on the increase in in-plane transverse momentum per nucleon with fragment mass. We also find event shapes to be close to spherical in the most central collisions, independent of bombarding energy and fragment mass up to ^{4}He.

1. Introduction

Collective effects have played an important role in the study of nuclear collisions. As far back as the 1950s, hydrodynamic models have been used to predict various kinds of collective behavior.[1] For non-zero impact parameters, the predominant fluid dynamic effect is a sideward deflection of the participant matter in the reaction plane. Such collective correlations, caused by the release of compressional energy, provide a measure of the nuclear pressure generated in the collision. Models indicate that collective correlations are established during the early, high density stage of the collision. They are minimally distorted during the subsequent expansion process. Thus, fluid-like correlations are regarded as being among the most appropriate observables for studying the equation of state of the compressed nuclear matter. Experiments with large solid angle acceptance detectors have confirmed the existence of collective correlations, and have provided measurements of many aspects of the phenomenon. The EOS Time Projection Chamber is a new 4π detector which was designed to continue the progress made during the earlier phases of the Bevalac program. It offers a simple and seamless acceptance, good particle identification, and adequate statistics for a comprehensive characterization of the relevant physics. In these proceedings, we report flow results from preliminary EOS data and preliminary data on event shapes in very central collisions of Au + Au.

2. The EOS Detector

The EOS Time Projection Chamber has a rectangular geometry, and operates in a 1.3 T dipole field provided by a superconducting magnet at the Bevalac's Heavy Ion Spectrometer System (HISS) facility. Unlike previous TPCs, EOS relies solely on pads for readout. The pad plane covers an area of 1.54×0.96 m^2, with 128 pad rows along the longer dimension, and 120 pads per row. Details about the chamber, the electronics and the data acquisition have been reported previously.[2] The standard EOS detector configuration includes the TPC, a multiple sampling ionization chamber (MUSIC II) positioned to intercept projectile spectator fragments, an array of scintillator slats to provide time-of-flight information at small polar angles, and a high efficiency neutron detector (MUFFINS). Only data from the TPC have been used in the current analysis.

3. In-plane Transverse Momentum

Our initial investigation of collective effects in Au + Au collisions includes the same in-plane transverse momentum analysis with essentially the same data selection criteria as used by the Plastic Ball group.[3] In particular, all nuclear fragment species up to ^{4}He are included, and we select an interval of multiplicity centered about the value where the flow has its maximum. The transverse momentum method[4] has been used to calculate the quantity $\langle p^x(y')/A \rangle$, the mean transverse momentum per nucleon in the reaction plane.

Fig. 1 presents $\langle p^x(y')/A \rangle$ as a function of the normalized rapidity y' for each of the six bombarding energies under investigation. We observe the classic "S"-shaped curve which changes sign at $y' = 0$. Although projectile-target symmetry dictates $p^x(y') = -p^x(-y')$, even an ideal 4π detector cannot satisfy this condition because absorption and energy loss in the target introduce distortions for y' approaching -1. In the EOS detector, the target was located about 14 cm upstream from the active volume of the TPC, leading to optimized performance near mid-rapidity and above, at the expense of a progressive loss of acceptance approaching target rapidity. At each beam energy, we fit the $\langle p^x(y')/A \rangle$ curves over the region indicated by the solid lines

222

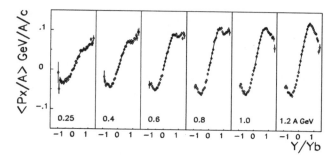

Fig. 1: Transverse momentum per nucleon projected on the reaction plane, as a function of normalized rapidity, for Au + Au collisions at six beam energies.

in Fig. 1 with a function of the form $my' - m_3y'^3$; the fitted values of m characterize the overall magnitude of the sideward flow effect among participant fragments, and these slopes are known simply as "flow" in the literature.

All fragment species up to ^{4}He were included when computing the results. In this work, we focus on the dependence of collective effects on fragment mass, and Fig. 2 presents preliminary flow excitation functions separately for protons, deuterons and alphas. The measurements in Fig. 2 show a consistent pattern of increase in sideward flow per nucleon with increasing fragment mass number A. Such an increase was previously reported by the Plastic Ball group[5] for collisions of Au + Au at a beam energy of $200A$ MeV. Fig. 2 also confirms the previously observed increase in flow with beam energy.

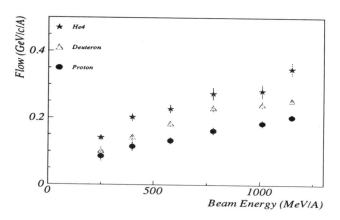

Fig. 2: Flow excitation function for protons, deuterons and alphas from Au + Au collisions.

4. Orthogonal Components of Sideward Flow

The observed transverse momentum $\mathbf{p}^\perp$ of a fragment has an uncorrelated ($\mathbf{p}^\perp_{uncor}$) and a correlated ($\mathbf{p}^\perp_{cor}$) component. $\mathbf{p}^\perp_{cor}$ is an effect of directed flow. We can approximate $\mathbf{p}^\perp_{uncor}$ using mixed events, $i.e.$, events generated by randomly selecting M tracks, each from a different observed event with multiplicity M. The correlated motion can be decomposed into two orthogonal parts: the azimuthal component, associated with rotations of $\mathbf{p}^\perp$ relative to an uncorrelated distribution, and the radial component of sideward flow, associated with changes in the magnitude of $\mathbf{p}^\perp$ relative to an uncorrelated distribution. Measurements of these two components complement each other, and together, they place more complete and stringent constraints on dynamical models. Radial flow, however, can not be distinguished by this method and would be treated as part of the uncorrelated motion.

The azimuthal pair correlation function[6] makes use of the variable ψ, the smaller angle between the transverse momenta of two fragments, and is defined as

$$C(\psi) = \frac{P_{cor}(\psi)}{P_{uncor}(\psi)}, \qquad (1)$$

where $P_{cor}(\psi)$ is the ψ distribution for observed pairs and $P_{uncor}(\psi)$ is the ψ distribution for pairs from mixed events. Sideward flow leads to an enhanced probability for fragments to be emitted with azimuths close to each other, near the reaction plane orientation; thus, if $C(\psi)$ is plotted for a rapidity interval that is not centered on mid-rapidity, we observe $C(\psi) > 1$ at small ψ and $C(\psi) < 1$ at large ψ. If fragments within a given rapidity interval are distributed in azimuthal angle φ according to $P(\varphi) \propto 1 + \lambda \cos\varphi$, then $C(\psi) = 1 + 0.5\lambda^2 \cos\psi$ (see Ref. 6). Fitted λ values provide a dimensionless measure of the azimuthal flow component. The azimuthal pair correlation function offers several advantages over previous flow analyses: it circumvents the need for event-by-event estimates of the reaction plane and the need to correct for dispersion in these estimates, it allows flow measurements in different rapidity intervals to be completely independent of each other, and the denominator in $C(\psi)$ automatically corrects for any azimuthal asymmetry introduced by the detector. Only forward rapidities are studied in the current analysis.

To characterize the radial component of sideward flow, we introduce a new quantity which we call the radial pair variance function:

$$\sigma^2(\psi) = \langle p^2_{sum}(\psi) \rangle - \langle p_{sum}(\psi) \rangle^2, \qquad (2)$$

where $p_{sum} = p^\perp_i / A_i + p^\perp_j / A_j$ is the sum of the $\mathbf{p}^\perp$ magnitudes per nucleon for the pair. In this case there is no reason to compute a ratio like P_{cor} / P_{uncor}, because $\sigma^2(\psi)$ is flat for mixed events. The fact that $\mathbf{p}^\perp$ magnitudes tend to be larger when a fragment's azimuth is parallel to the flow direction, and tend to be smaller when antiparallel, leads to an inequality $\sigma^2(\psi \sim 0°) > \sigma^2(\psi \sim 180°)$. σ^2 decreases linearly with increasing ψ. This can be demonstrated analytically for an idealized example where there are no thermal fluctuation in $\mathbf{p}^{\perp}$, and simulations with realistic momenta also indicate linear $\sigma^2(\psi)$. To characterize the magnitude of the radial component of sideward flow in momentum units, we define $S = \sqrt{(d\sigma^2/d\psi)}$.

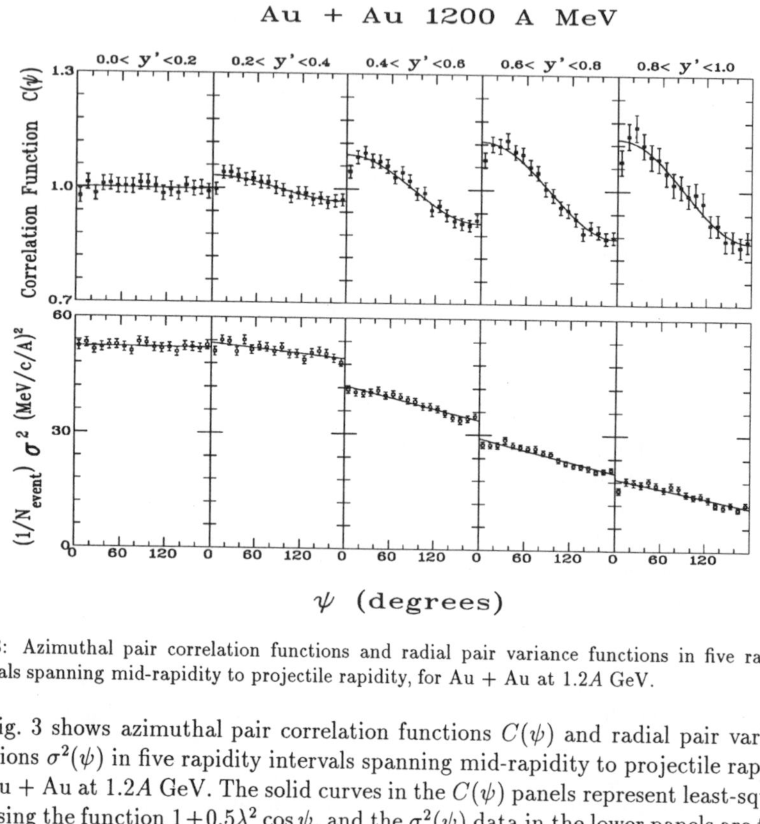

Au + Au 1200 A MeV

Fig. 3: Azimuthal pair correlation functions and radial pair variance functions in five rapidity intervals spanning mid-rapidity to projectile rapidity, for Au + Au at $1.2A$ GeV.

Fig. 3 shows azimuthal pair correlation functions $C(\psi)$ and radial pair variance functions $\sigma^2(\psi)$ in five rapidity intervals spanning mid-rapidity to projectile rapidity, for Au + Au at $1.2A$ GeV. The solid curves in the $C(\psi)$ panels represent least-squares fits using the function $1 + 0.5\lambda^2 \cos\psi$, and the $\sigma^2(\psi)$ data in the lower panels are fitted using a straight line. In this analysis of azimuthal and radial components of sideward flow, only protons and deuterons from events with $M > 0.4M^{max}$ have been included.

5. Dependence of Components of Sideward Flow on Fragment Mass

In order to investigate the fragment mass dependence, we have sorted protons and deuterons into separate samples and computed the same quantities as plotted in Fig. 3. The system Au + Au at $1.2A$ GeV currently provides the best data on fragment flow, but still does not yield useful same-fragment pair correlation statistics for $A \geq 3$. Fig. 4 shows the fitted sideward flow components λ and S for our standard rapidity intervals. At all of the rapidities studied, the observed pattern is consistent with a simple coalescence picture:[7] the radial component of sideward flow S is the same for p and d, while $\lambda_d = 2\lambda_p$ within experimental uncertainties.

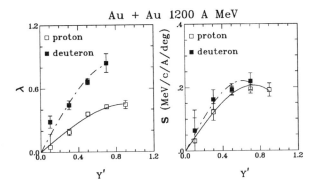

Fig. 4: Azimuthal flow component $\lambda(y')$ and radial component of sideward flow $S(y')$ separately evaluated for protons and deuterons from collisions of $1.2A$ GeV Au + Au.

An important advantage of the EOS TPC is its seamless acceptance, which is simple enough to be simulated with good accuracy. To estimate the effect of detector distortion on the observables under investigation, we use events from a version of the *FREESCO* statistical event generator[8] to which a phenomenological flow correlation[9] has been added. The simulations indicate that any detector-related distortion of the measured sideward flow components is not larger than the statistical uncertainty in samples of several thousand events.

6. Event Shapes in Central Collisions

The energy contained in directed flow is only a few percent of the total available kinetic energy.[3,10] From entropy considerations[11] and from general energy estimates,[12] we would expect to see a much larger fraction of the total energy contained in collective flow. From a systematic study of intermediate mass fragments the FOPI collaboration has obtained evidence for a large amount of "radial" flow.[13,10] The analysis has been done with the most violent events with very stringent cuts on centrality and within a limited range of emission angle in the center of mass system. Therefore, it is not possible to distinguish experimentally between radial (isotropic) flow and azimuthally symmetric flow perpendicular to the beam axis, the limiting case of directed flow for very central events. The EOS data for $y' > 0$ are well-suited for testing event shapes and elucidating these issues.

The thermalization ratio $R = 2\sum |\mathbf{p}^\perp| \, / \, \pi \sum |p_{cm}^z|$ has been widely used since the time when the first 4π measurements became available. An isotropic source implies $R = 1$ and measurements of R can provide information about event shapes. However, directed collective flow effects, kinematic cuts, and detector distortions make the interpretation of R complicated. The presence of spectator matter can bias the thermalization ratio R towards lower values. In order to avoid the possibility of such bias, we include only fragments which satisfy $p_b/A > 0.27$ GeV/c, where p_b denotes fragment momentum transformed into the rest frame of the projectile. This cut biases R in the opposite direction, but it is relatively easy to correct this bias by using simulated events from the *FREESCO* event generator[8] with the same cut.

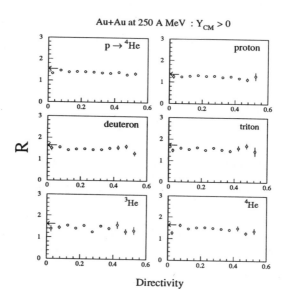

Fig. 5: Average thermalization ratio R versus directivity D for different light fragment species from $0.25A$ GeV Au + Au events with more than 30 detected baryonic fragments at forward rapidities $(y' > 0)$. Note that R values in the vicinity of the horizontal arrow are consistent with a spherical event shape.

As a measure for the centrality of the collision, we use the directivity variable[14] $\mathbf{D} = \sum \mathbf{p}^{\perp} / \sum |\mathbf{p}^{\perp}|$. Fig. 5 presents average thermalization ratios R in bins of directivity magnitude D, for different light fragment species from $0.25A$ GeV Au + Au events with more than 30 detected baryonic fragments at $y' > 0$. Results for the other beam energies show similar behavior. The arrows on the vertical scales mark the R values for an isotropic source with the same spectator cut ($p_b/A > 0.27$ GeV/c) as applied to the EOS data. The main conclusion based on these preliminary data is that the most central collisions (lowest D) are consistent with being close to spherical in shape for all light fragment species at all beam energies studied. Thus, we would expect that the flow observed in the FOPI data is spherically symmetric radial flow, predicted a long time ago.[15] It should be seen at all impact parameters and it should agree well with model calculations[16] that reproduce the directed flow since it is generated by the same mechanism.

This work was supported by the Director, Office of Energy Research, Office of High Energy and Nuclear Physics, Division of Nuclear Physics of the U.S. Department of Energy under Contract DE-AC03-76SF00098.

7. References

[1] S.Z. Belen'kji and L.D. Landau, Nuovo Cimento, 1, 15 (1956); G.F. Chapline, M.H. Johnson, E. Teller, and M.C. Weiss, Phys. Rev. D 8, 4302 (1973); W. Scheid, H. Müller, and W. Greiner, Phys. Rev. Lett. 32, 741 (1974).

[2] G. Rai, A. Arthur, F. Bieser, C.W. Harnden, R. Jones, S. Kleinfelder, K. Lee, H.S. Matis, M. Nakamura, C. McParland, D. Nesbitt, G. Odyniec, D. Olson, H.G. Pugh, H.G. Ritter, T.J.M. Symons, H. Wieman, M. Wright, R. Wright, and A. Rudge, IEEE Trans. Nucl. Sci. 37, 56 (1990).

[3] K.G.R. Doss, H.-Å. Gustafsson, H.H. Gutbrod, K.H. Kampert, B. Kolb, H. Löhner, B. Ludewigt, A.M. Poskanzer, H.G. Ritter, H.R. Schmidt, and H. Wieman, Phys. Rev. Lett. 57, 302 (1986); H.H. Gutbrod, A.M. Poskanzer, and H.G. Ritter, Rep. Prog. Phys. 52, 1267 (1989).

[4] P. Danielewicz and G. Odyniec, Phys. Lett. 157B, 146 (1985).

[5] K.G.R. Doss, H.-Å. Gustafsson, H.H. Gutbrod, J.W. Harris, B.V. Jacak, K.-H. Kampert, B. Kolb, A.M. Poskanzer, H.G. Ritter, H.R. Schmidt, L. Teitelbaum, M. Tincknell, S. Weiss, and H. Wieman, Phys. Rev. Lett. 59, 2720 (1987).

[6] S. Wang, Y.Z. Jiang, Y.M. Liu, D. Keane, D. Beavis, S.Y. Chu, S.Y. Fung, M. Vient, C. Hartnack, and H. Stöcker, Phys. Rev. C 44, 1091 (1991).

[7] Y. Shao, Ph.D. thesis, Kent State U. (1994).

[8] G. Fai and J. Randrup, Nucl. Phys. A404, 551 (1983); Comp. Phys. Comm. 42, 385 (1986).

[9] A.F. Barghouty, G. Fai, and D. Keane, Nucl. Phys. A535, 715 (1991).

[10] T. Wienold et al., these proceedings.

[11] K.G.R. Doss, H.A. Gustafsson, H.H. Gutbrod, B. Kolb, H. Löhner, B. Ludewigt, A.M. Poskanzer, T. Renner, H. Riedesel, H.G. Ritter, A. Warwick, and H. Wieman, Phys. Rev. C 32, 116 (1985).

[12] R. Stock, Phys. Rep. 135, 259 (1986).

[13] S.C. Jeong et al., GSI Preprint 93-38.

[14] J.P. Alard et al., Phys. Rev. Lett. 69, 889 (1992);

[15] P.J. Siemens and J.O. Rasmussen, Phys. Rev. Lett. 42, 880 (1979).

[16] P. Danielewicz and Q. Pan, Phys. Rev. C 46, 2002 (1993).

IS THERE FLOW AT THE AGS?

Johannes P. Wessels and Yingchao Zhang
Department of Physics
SUNY at Stony Brook
Stony Brook, NY 11794-3800, USA

The E877 Collaboration:
J. Barrette[4], R. Bellwied[8], S. Bennett[8], P. Braun-Munzinger[6], W. E. Cleland[5], M. Clemen[5], J. Cole[3], T. M. Cormier[8], G. David[1], J. Dee[6], O. Dietzsch[7], M. Drigert[3], S. Gilbert[4], J. R. Hall[8], T. K. Hemmick[6], N. Herrmann[2], B. Hong[6], Y. Kwon[6], R. Lacasse[4], A. Lukaszew[8], Q. Li[8], T. W. Ludlam[1], S. McCorkle[1], S. K. Mark[4], R. Matheus[8], E. O'Brien[1], S. Panitkin[6], T. Piazza[6], C. Pruneau[8], M. N. Rao[6], M. Rosati[4], N. C. daSilva[8], S. Sedykh[6], U. Sonnadara[5], J. Stachel[6], H. Takai[1], E. M. Takagui[7], S. Voloshin[5], G. Wang[4], J. P. Wessels[6], C. L. Woody[1], N. Xu[7], Y. Zhang[6], Z. Zhang[5], C. Zou[6]

[1] Brookhaven National Laboratory, Upton, NY 11973
[2] Gesellschaft für Schwerionenforschung, Darmstadt, Germany
[3] Idaho National Engineering Laboratory, Idaho Falls, ID 83402
[4] McGill University, Montreal, Canada
[5] University of Pittsburgh, Pittsburgh, PA 15260
[6] SUNY, Stony Brook, NY 11794
[7] University of São Paulo, Brazil
[8] Wayne State University, Detroit, MI 48202

ABSTRACT

We have employed the nearly 4π–calorimetric coverage of the E877 apparatus in order to determine flow in different regions of pseudo-rapidity from the measured transverse energy in Au+Au collisions at 11.4 A·GeV/c. Signatures for the side-splash in the reaction plane at forward and backward rapidities have been established. Indications of a non-zero eccentricity of particles in the reaction plane at mid-rapidity are also found. These observations complement analyses deriving collective longitudinal motion from dN/dy-spectra of protons, kaons, and pions as well as collective transverse motion of the same particles from m_t-spectra at midrapidity.

1. Introduction

The study of flow in nuclear collisions has been carried out at a variety of energies over the past fifteen years. It has been motivated by hydrodynamical calculations of heavy-ion collisions [1] in the quest for the compressibility modulus determining the nuclear equation-of-state. Investigations of the monopole resonance explore the shape of the equation-of-state in a very narrow region around the ground state density of nuclear matter. Experiments in search for flow phenomena in the Fermi energy regime [2,3] probe the balance between the attractive nuclear force and the repulsive Coulomb force comprising the mean field. The first systematic investigations of flow have been carried out at the BEVALAC [4] at several hundred A·MeV, an energy region which is still of great interest. The main focus of these experiments was on the measurement of momentum distributions of charged particles perpendicular to the beam-axis, either in the reaction plane (*side-splash*) or perpendicular to the reaction plane (*squeeze-out*). In order to describe the measured rapidity distributions in ultra-relativistic heavy-ion collisions at the AGS and at CERN [5] collective motion along the longitudinal direction has also been incorporated into the studies of flow.

While most of the above mentioned studies were done using the momenta of identified particles, we will present, for the first time, measurements of flow phenomena at 11.4 A·GeV/c using the calorimeters of the E877 apparatus at the AGS.

2. Experimental Setup

The E877-apparatus, shown below, is an upgrade to the previous E814-apparatus, which is now adapted to the high multiplicity environment of Au+Au collisions which have only recently become available at the AGS.

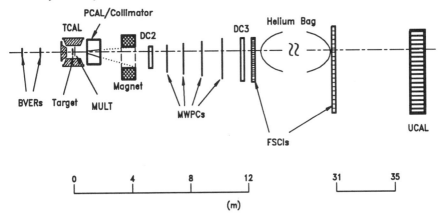

Figure 1. Layout of the E877-apparatus at the AGS

It consists of a beam telescope with the ability to determine the beam trajectory for each incoming particle (BVERs), three calorimeters (target calorimeter(TCAL), participant calorimeter(PCAL), and a forward uranium calorimeter(UCAL)). The TCAL covers $-0.5 \leq \eta \leq 0.8$, PCAL covers $0.83 \leq \eta \leq 4.5$ in pseudo-rapidity, while the UCAL is used as a measure of the forward energy. It is not used in the flow analysis since it is mostly hit by non-interacting nucleons from the projectile. The TCAL is made of 992 NaI-crystals each 5.3 radiation length deep. The PCAL is a lead-iron-scintillator sampling calorimeter subdivided into four depth sections, eight radial sections, and sixteen azimuthal sections. The spectrometer arm of the setup consists of a bending magnet, a set of two drift/pad-chambers(DC2, DC3), four multi-wire proportional-counters(MWPCs), and two time-of-flight walls(FSCI) allowing for particle identification in the forward rapidity region for low and medium values of the transverse momentum p_t.

3. Measuring Flow with Calorimeters

3.1. General Considerations

As mentioned above, most analyses searching for flow effects use the transverse momentum of identified charged particles in order to reconstruct what has become known as the transverse momentum tensor [6]. However, the calorimeters of the E877-apparatus measure the transverse energy (E_t) of all emitted particles. For pions this corresponds to their total transverse energy. For baryons it is their transverse kinetic energy. About 60% of the total measured E_t in the PCAL is due to pions. The effect this has on the reconstruction of something similar to the transverse momentum tensor has been discussed by Gavron [8] for Si+Pb collisions. There it is shown that especially for low multiplicity events it is important to correct for finite particle multiplicities, but it is also pointed out that the sensitivity to study flow phenomena is preserved using calorimetric information.

3.2. Construction of Azimuthal Transverse Energy Distributions

Studying the azimuthal transverse energy distributions will shed some light on the question: How much of the produced transverse energy is directed, i.e. non-isotropic? For this we consider the following:

From the energy (E_i) deposited in each calorimeter cell(i) we define $\varepsilon_t^i = E_i \sin(\vartheta_i)$, the sum of which corresponds to the total transverse energy $E_t = \sum_i \varepsilon_t^i$ for this event. The ε_t^i can be expanded into multipole components perpendicular to the beam direction (x_n, y_n)

$$x_n = \frac{\sum_i \varepsilon_t^i \cos(n\phi_i)}{\sum_i \varepsilon_t^i} \; ; \; y_n = \frac{\sum_i \varepsilon_t^i \sin(n\phi_i)}{\sum_i \varepsilon_t^i} ,$$

with integers n. The magnitude (v_n) and direction (Φ_n) of the moments of the azimuthal E_t-distribution for each event are given by:

$$v_n = \sqrt{x_n^2 + y_n^2}; \; \Phi_n = \frac{1}{n} \tan^{-1}\left(\frac{y_n}{x_n}\right).$$

For $n = 1$ we define v_1, the amount of "directed" transverse energy, as *directivity (D)* (compares to Q in [7]). The distribution of $d^2N/dD_x dD_y$ is symmetric about zero, when averaged over many events. For $n = 2$ in the above equations v_2 probes the *eccentricity* (denoted by α instead of D), i.e. how circular the distributions is. It is similar to $\alpha = (f_1 - f_2)/(f_1 + f_2)$ where the f_i are the eigenvalues of the two-dimensional sphericity tensor[9].

Below are discussed measurements of these quantities determined for the forward hemisphere covered by the PCAL$(1.85 \leq \eta \leq 4.5)$ for different bins of the transverse energy observed in the whole calorimeter (PCAL) referenced as E_t in all following plots. This is about 90% of the total E_t (from PCAL and TCAL). In Au+Au collisions at these energies the total E_t anti-correlates tightly with the impact parameter.

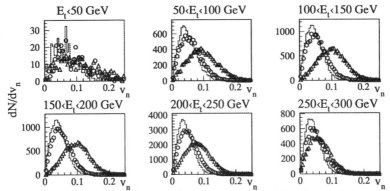

Figure 2. Distribution of multipole moments. Directivity D, n=1($\triangle$), and eccentricity α, n=2($\bigcirc$), measured by the PCAL in forward direction (1.85 $\leq \eta \leq$ 4.5) for different bins of total E_t. The histogram shows the azimuthally symmetrized dN/dD-distribution (cf. text).

Let us first look at the directivity distribution dN/dD. In order to see how much of the produced transverse energy is directed, we plot for each bin also an azimuthally symmetrized distribution (histogram), using the measured ε_t^i and randomizing ϕ_i taking into account the segmentation of the calorimeter. This indicates what would be observed if the production of E_t was isotropically symmetric. One can see a clear evolution from peripheral (low E_t) to central (high E_t) collisions. For the most peripheral as well as for the most central collisions the distributions are closer to isotropic distributions, while at mid-impact parameters the directivity distributions are non-isotropic. The largest difference between the symmetrized and the observed distribution is for $150 < E_t < 200$ GeV. From the comparison of dN/dD- and the dN/dα-distributions it can be noted that the overall shape of the event is close to circular in the forward direction, in fact, the dN/dα- and the symmetrized dN/dD-distribution are strikingly similar.

3.3. Assessing the Shape of the Azimuthal Distributions

A difference of the data from the symmetrized dN/dD-distribution in itself does not necessarily mean that we are observing flow, because finite multiplicity effects have not yet been addressed. Suppose directivity was due to a random process governed by fluctuations in finite particle multiplicity. Then the shape of the distribution, containing N_0 events properly normalized, would be Gaussian:

$$\frac{1}{D}\frac{dN}{dD} = \frac{N_0}{\sigma^2}\exp^{-\frac{D^2}{2\sigma^2}}, \text{ for which } < \frac{dN}{dD} >= \sqrt{\frac{\pi}{2}}\sigma, \text{ and } rms(\frac{dN}{dD}) = \sqrt{\frac{4-\pi}{2}}\sigma.$$

In this case the $\frac{rms}{mean}$ of the directivity distribution would be constant ($\simeq 0.52$). (Similar for the $dN/d\alpha$-distribution.) Below are graphs showing the mean values of the dN/dD- and $dN/d\alpha$-distribution and the ratio of their $rms/mean$ as a function of E_t.

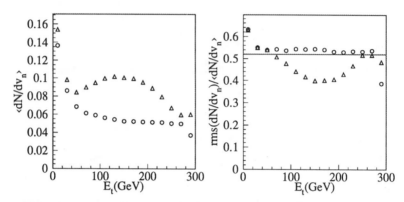

Figure 3. *Mean values(left) and $\frac{rms}{mean}$(right) of the directivity($\triangle$)- and eccentricity($\bigcirc$)-distribution as a function of E_t in the forward hemisphere.*

The fact that $\frac{rms}{mean}$ is close to 0.52 for the $dN/d\alpha$-distribution at all impact parameters tells us that the distribution in this pseudo-rapidity interval is mostly random governed by finite particle numbers and possibly nothing else. The same is true for the dN/dD-distribution in peripheral and the most central collisions. Note that in both cases no flow is expected. However, at mid-impact parameters where the directivity signal shows its maximum this distribution is much narrower than can be explained based on statistical arguments. The narrowing of the distribution can be attributed to the presence of directed sidewards flow in this impact parameter range.

3.4. Generalized Description of Azimuthal Distributions

In a more generalized treatment of azimuthal distributions [10] it can be shown, that x_n, y_n, and v_n can indeed be constructed from any observable, whose azimuthal distribution can be measured. The distribution of all n poles of the $dN/(v_n dv_n)$ over many events is described by:

$$\frac{dN}{v_n dv_n} = \frac{N_0}{\sigma^2} \exp\left(-\frac{\tilde{v}_n^2 + v_n^2}{2\sigma^2}\right) \int_0^{2\pi} d(n\Phi_n) \exp\left(\frac{\tilde{v}_n v_n \cos(n\Phi_n)}{\sigma^2}\right)$$

$$= \frac{N_0}{\sigma^2} \exp\left(-\frac{\tilde{v}_n^2 + v_n^2}{2\sigma^2}\right) I_0\left(\frac{\tilde{v}_n v_n}{\sigma^2}\right); \quad I_0 : \text{modified Bessel} - \text{function}$$

Here $\tilde{v}_n$ is the centroid or the nth-pole anisotropy. $\tilde{v}_1$ corresponds to the value for the directivity (D), $\tilde{v}_2$ corresponds to the value for the eccentricity (α) of the event. The width σ is assumed to be the same for all distributions evaluated in the same E_t-interval.

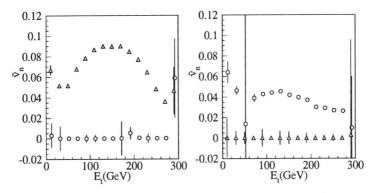

Figure 4. Results for fits to the azimuthal E_t-distributions as a function of total E_t. Shown are results for the fit-parameters $\tilde{v}_n$ to the directivity-distribution (n=1) plotted as triangles, and to the eccentricity-distribution (n=2) plotted as circles for the forward hemisphere(left)(1.85 $\leq \eta \leq$ 4.5) and at mid-rapidity(right)(1.0 $\leq \eta \leq$ 1.85).

Above are the results of fits to the data using the above Bessel-function for two different regions in rapidity. Again we see a pronounced peak in the directivity at mid-impact parameters in forward direction, while the eccentricity shows no dependence on E_t in this range of rapidity. Conversely, the directivity signal remains at zero for the measurement at mid-rapidity, while $\tilde{v}_2$ shows a slight signal.

3.5. Reaction Plane Analysis

In the directivity analysis the vector pointing into the direction of Φ and the beam-axis span the reaction plane. Φ can be measured in forward direction by PCAL and in backward direction by TCAL. An analysis of these angles is shown in the following figure. Plotting the two angles versus each other shows the expected forward-backward anti-correlation. The angle between Φ in the forward and in the backward direction can be fitted by a constant (for the background) plus a Gaussian with its maximum at 180° and its variance corresponding to the precision of the reaction plane determination. The integral under the Gaussian divided by the constant term for different values of E_t shows the forward-backward azimuthal correlation. The variance of the Gaussian corresponds to roughly twice the resolution of the reaction plane measurement achieved in each detector, assuming equal resolution in TCAL and PCAL. From the plot we extract that over most of the impact parameter range we are able to extract the width with a precision of ±35°. This will provide an excellent tool in order to study the emission of certain particles species (π^+, π^-, p) as analyzed in the spectrometer with respect to the reaction plane.

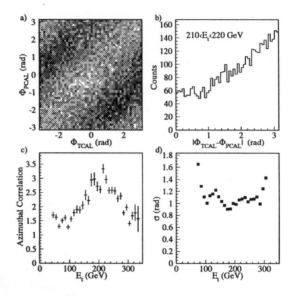

Figure 5. a) Correlation of Φ_{TCAL} vs. Φ_{PCAL}, b) relative angle between Φ_{TCAL} and Φ_{PCAL} at $210 < E_t < 220$ GeV, c) azimuthal correlation of Φ_{TCAL} and Φ_{PCAL} (cf. text), d) width of the combined reaction plane reconstruction.

Since TCAL has fairly large leakage fluctuations for energetic particles, the reaction plane determination from PCAL alone is actually better than the number quoted.

4. Conclusions and Outlook

As a result of the above discussion, we have uniquely identified the *side-splash* signal at mid-impact parameters in symmetric Au+Au collisions. Furthermore, from the analysis at mid-rapidity there seem to be first indications of a non-zero eccentricity signal. In conjunction with the results of an analysis of the rapidity distributions for protons, kaons, and pions in central Au+Au collisions in the framework of a rapidly expanding thermal source, which at the same time is also able to reproduce the measured transverse-mass spectra [5], the above results seem to call for a rephrazing of the title, namely, why shouldn't there be any flow at the AGS? It is certainly in this energy domain that, with heavy projectiles such as Au-nuclei, very high densities can be achieved. In return this would lead to a build-up of pressure gradients, that could act as the driving force for the discussed collective flow phenomena. It will now be important to examine the origin of these phenomena, in order to determine whether indeed the initial pressure is the driving force behind the collective behaviour or, whether in fact the effects are predominantly governed by absorption. In the same light, it will be interesting to address predictions for the flow of, e.g. pions of different charge [11], or, more speculative, anti-protons[12] by combining the spectrometer and the calorimeter data.

5. Acknowledgements

We are greatful for financial support from the U.S. DoE, the U.S. NSF, the Canadian NSERC, and CNPq Brazil. One of us (J.P.W.) would like to acknowledge support by the Alexander-von-Humboldt Foundation.

6. References

[1] W. Scheid, H. Müller, W. Greiner, *Phys. Rev. Lett.* **32**, 741 (1974).

[2] C.A. Ogilvie et al., *Phys. Rev. C* **42**, R10 (1990).

[3] G.D. Westfall et al., *Proc. 8th Winter Workshop on Nucl. Dynamics*, World Scientific, p.196, 1992.

[4] Review by H.H. Gutbrod, A.M. Poskanzer, H.G. Ritter, *Rep. Prog. Phys.* **52**, 1267 (1989).

[5] N. Xu, contribution to these proceedings. J. Stachel, P. Braun-Munzinger, *Phys. Lett. B* **216**, 1 (1989), and *Nucl. Phys. A* **498**, 577c (1989).

[6] G. Hanson et al., *Phys. Rev. Lett.* **35**, 1609 (1975)

[7] P. Danielewicz, G. Odyniec, *Phys. Lett. B* **157**, 146 (1985).

[8] A. Gavron, *Nucl. Instr. and Meth. A* **273**, 371 (1988).

[9] J.Y. Ollitrault, *Phys. Rev. D* **46**, 229 (1992).

[10] S. Voloshin, Y.C. Zhang, to be published.

[11] S.A. Bass, C. Hartnack, H. Stöcker, W. Greiner, *Phys. Rev. Lett.* **71**, 1144 (1993).

[12] A. Jahns,C. Spieles, H. Sorge, H. Stöcker, W. Greiner, UFTP preprint 348/1993.

Study of Dielectron Production in Nucleon-Nucleon and Nucleus-Nucleus Collisions at the Bevalac

Huan Z. Huang
Physics Department
Purdue University
West Lafayette, Indiana 47907-1396, USA

ABSTRACT

Recent measurements of dielectron production in nucleon-nucleon and nucleus-nucleus collisions from the Dilepton Spectrometer at the Bevalac are summarized. The p+d and p+p data at 2.1 and 4.9 GeV show that the dielectron yields from p+n and p+p collisions are similar, indicating that dielectrons from inelastic channels are dominant at these beam energies. The ratio of dielectrons from p+d to that from p+p at the beam energy of 1.0 GeV was measured to be ten, consistent with the theoretical calculation that the exclusive bremsstrahlung of $pn \rightarrow pne^+e^-$ is a dominant process at that energy. The subthreshold production of η in p+d collisions may also contribute to the large ratio. Our nucleus-nucleus measurements show that dielectrons could be a sensitive probe of pion form factor in nuclear medium.

1. Introduction

The Dilepton Spectrometer collaboration (DLS) at the Bevalac at Lawrence Berkeley Laboratory has undertaken two programs to systematically study dieletron production in nuclear collisions since 1990. The nucleon-nucleon program aimed at studying dielectron production mechanisms in nucleon-nucleon collisions. To this end we have measured dielectron production with unprecedented statistics from p+d and p+p collisions at beam kinetic energies ranging from 1 to 4.9 GeV. The nucleus-nucleus program is an extension of our previous study of dielectron production in Ca+Ca collisions at 1 GeV/A beam incident energy. I shall summarize the recent data from the DLS and review progress in understanding production mechanisms of dielectrons.

2. Experimental Apparatus and Analysis Procedure

The Dilepton Spectrometer system consisted of two identical spectrometer arms with respect to beam axes. Each arm was a dipole system with three drift chambers, two scintillation hodoscopes and two Cerenkov counters. A dielectron trigger was constructed with an eight-fold coincidence between four hodoscopes and four Cerenkov counters, two sets per arm. The Cerenkov counters were used to identify electrons after track reconstruction. Details of the DLS apparatus can be found in reference [1].

The combinatorial background in the measured electron-positron (Opposite-Sign, OS) pairs was constructed from measured electron-electron and positron-positron (Same-Sign, SS) pairs. The mass, transverse momentum and rapidity distribution of the OS background was obtained by randomly selecting an electron and a positron each from the SS sample of two spectrometer arms. The number of the background pairs was normalized to the square root of the product of the number of electron-electron and positron-positron pairs. The uncertainties in the background subtraction scheme led to an estimated error in the cross section of up to 20% for masses below 0.3 GeV/c^2, but was negligible for higher masses.

The acceptance of the DLS was calculated using a GEANT simulation package. We assumed an azimuthal symmetry in ϕ of produced pairs and an isotropic decay of the pair in the center-of-mass frame of the virtual photon. The DLS acceptance table was computed as a function of the pair mass (M), the transverse momentum (p_t) and the laboratory rapidity (y). The DLS acceptance covers approximately M of 0.05 to 1.0 GeV/c^2, p_t of 0.0 to 1.0 GeV/c and y of 0.45 to 1.95. This acceptance has to be used to filter theoretical models for meaningful comparisons. Details of the acceptance calculation and analysis procedures have been described in reference [2].

3. Nucleon-Nucleon Program

The objective of the DLS nucleon-nucleon program is to study dielectron production mechanisms in p+p and p+n collisions. Early theoretical calculations have proposed processes of parton-parton annihilation[3] and virtual pion-pion annihilation[4] to explain the observed dielectron yield in nuclear collisions at beam energies above 10 GeV. At the Bevalac beam energy, however, it is not clear that the parton degree of freedom is necessarily excited in nuclear collisions. Theoretical calculations for the Bevalac data have included processes of p+n bremsstrahlung, Dalitz decay of Δ, pion-pion annihilation and pion-nucleon annihilation.[5] The Dalitz decays of η mesons have also been recognized as one of the important sources of dielectrons at this beam energy.[6]

Prior to the nucleon-nucleon program, DLS has measured dielectron production in p+Be collisions at beam incident energies of 1.05, 2.1 and 4.9 GeV.[7] It was hoped that these p+Be data would provide essential constraints for a comprehensive description of dielectron production mechanisms in elementary collisions. Theoretical calculations,[5] [8] in comparison with p+Be data, showed that p+n bremsstrahlung is the most significant source of dielectrons, especially at the beam energy of 4.9 GeV. In spite of the fact that these calculations were in reasonable agreement with the DLS data, the production mechanisms of dielectrons in the elementary collisions were not well understood. Questions regarding the soft-photon approximation for bremsstrahlung, the beam-energy dependence of p+n bremsstrahlung and dielectron production from inelastic channels, have not been addressed satisfactorily. Our systematic measurements of dielectron production in p+p and p+d collisions at beam energies of 1.0 to 4.9 GeV were intended to provide essential experimental data for deepening theoretical understanding of the dielectron production mechanisms.

Figure 1 shows the cross section of dielectrons as a function of the pair mass for p+d and p+p collisions at 4.9 GeV beam energy. The ratio of the integrated cross section for masses above 0.2 GeV/c² is 1.9±0.1, where the error is statistical only. The similarity of the dielectron spectra from p+d and p+p collisions and the ratio of approximately two suggest that the dielectron production from p+p and p+n collisions may be close in magnitude, in contradiction with the model expectation of p+n bremsstrahlung dominance.

Recently Haglin and Gale[9] have shown that p+n bremsstrahlung at a beam energy of 4.9 GeV is strongly suppressed, by a factor of about four, due to the asymmetrical angular distribution of the elastic scattering. As a result, dielectron contributions from simple p+n bremsstrahlung and p+p bremsstrahlung at 4.9 GeV are close in magnitude. The asymmetrical angular distribution of p+n elastic scattering was included also in early calculations by Winckelmann et al.[10] and by Schafer et al.,[11] and a similar suppression factor of p+n bremsstrahlung was observed.

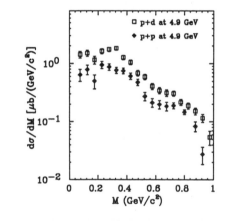

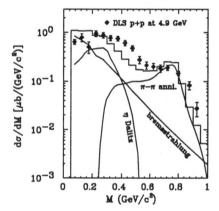

Fig. 1: Dielectron production cross sections from p+d and p+p collisions at the beam energy of 4.9 GeV. Those with masses below 0.15 GeV/c² are mostly from the Dalitz decay of neutral pions. Errors are statistical only.

Fig. 2: Dielectron cross section from p+p collisions at 4.9 GeV in comparison with the calculation by Haglin and Gale.[9] The histogram is the sum over all sources. Sources of Δ Dalitz and pion-pion bremsstrahlung are small by orders of magnitude compared to those shown.

The recent DLS measurement at 4.9 GeV also brought about new understanding of dielectron production mechanisms. Haglin and Gale[9] found that the dielectron contribution from simple two-body p+p and p+n bremsstrahlung processes is too small to account for the measured DLS cross sections. Inelastic scattering channels, p+p→pnπ⁺e⁺e⁻ in p+p for example, are an important source of dielectron production. Figure 2 shows a comparison of their calculation with the DLS data. Both processes of η Dalitz decay and bremsstrahlung are significant sources for dielectron mass below 0.5 GeV/c², where the bremsstrahlung contribution is mostly from inelastic channels. The pion-pion annihilation is considered the dominant source for dielectron masses above 0.5 GeV/c². The theoretical calculation significantly underestimated the dielectron cross section around the mass region of 0.4 GeV/c². Possible remedies would be 1) inclusion of form factors in the bremsstrahlung, 2) extension to higher order in the soft-photon approximation.[12] We also note that a recent calculation of ρ form factor showed that there may be an enhancement of the in-medium form factor over that in vacuum for mass region around 0.3-0.4 GeV/c² at normal nuclear density.[13]

We have, in addition, also measured dielectron production in p+d and p+p collisions at beam kinetic energies of 1.03, 1.26, 1.60, 1.85 and 2.07 GeV. Preliminary analysis indicated that there is a dramatic change in the shape of the dielectron mass spectrum, perhaps due to the onset of a production mechanism in this energy region. Figure 3, taken from reference [14], shows the ratio of dielectron production in p+d to that in p+p as a function of beam energy for pairs with masses greater than 0.15 GeV/c². The ratio is about 10 at the beam energy of 1.0 GeV and monotonically decreases to about 2.0 at the beam energy of 2.0 GeV. The steep decrease of the ratio to 2.0 is consistent with model calculations that inelastic channels are a significant source of dielectrons at high energies.

The large ratio at 1.0 GeV suggests that the p+n bremsstrahlung may be much stronger than the p+p bremsstrahlung, and that the p+n bremsstrahlung is a dominant source. Subthreshold production of η meson in p+d collisions due to FERMI motion may also contribute to the large ratio at 1.0 GeV. Near the η threshold of 1.26 GeV the η production cross section in p+n collisions may be larger than that in p+p collisions according to models[15] based on isospin arguments. Preliminary reports from the PINOT spectrometer[16] at SATURNE indicated a pn/pp ratio of four or more for η production near the threshold.

4. Nucleus-Nucleus Program

The dielectron production mechanisms in nucleus-nucleus collisions at the Bevalac are the same as that in nucleon-nucleon collisions—bremsstrahlung emission, Dalitz decay of η and Δ, direct decay of vector mesons and pion-pion annihilation, but the characteristics of these processes could be modified due to the dense nuclear medium formed in nucleus-nucleus collisions. Among these mechanisms, the pion-pion annihilation process is considered a promising probe of nuclear medium effect because, experimentally, the mass of dielectrons from pion-pion annihilation could populate a high mass region where contributions from other mechanisms are small. Of course, an optimal beam energy is necessary to achieve that condition. Previous DLS measurement of dielectron production in Ca+Ca and Nb+Nb collisions[17] indicated that pion-pion annihilation processes are responsible for dielectrons with masses above 0.5 GeV/c². We will review the progress in understanding of pion-pion annihilation and the status of the DLS recent high statistical measurement.

Two aspects, the pion dispersion relation and the vector meson form factor, have been extensively discussed in the context of pion-pion annihilation. The pion dispersion relation governs the propagation of pions, which in nuclear medium could be significantly softened at high nuclear density. C. Gale and J. Kapusta[18] firstly pointed out that the softness of the dispersion relation could lead to a drastic structure in the dielectron yield near the mass region of two pions. L. H. Xia et al.[19] modelled this effect in a cascade approach and included the width of Δ resonances. They observed a reduced effect, but nevertheless a significant enhancement in the mass spectrum. C. L. Korpa and S. Pratt[20] showed that inclusion of the pion annihilation vertex would strongly reduce the medium effect. A more precise calculation by C. L. Korpa et al.,[21] including the vertex correction and Δ width, reported a substantial enhancement in the dielectron yield near the mass region of two-pion threshold, where the threshold value may also shift depending on medium density. Experimentally, however, any dielectron signal sensitive to nuclear medium effect in the mass region below 0.5 GeV cannot be disentangled from other sources of dielectrons. Both bremsstrahlung emission and η Dalitz decay are dominant sources of dielectrons in that mass region.[22]

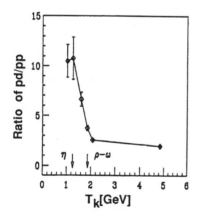

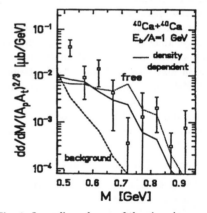

Fig. 3: The pd/pp ratio as a function of the beam energy for pairs with m>0.15 GeV/c². The threshold energies of η and ρ-ω production are indicated by the arrows. The line is only to guide the eye.

Fig. 4: In-medium change of the pion-pion annihilation for Ca+Ca collisions at 1 GeV/A(Gy. Wolf et al.[29]). The dotted and solid lines are obtained with free and with a medium-dependent pion form factor, respectively.

Before we discuss the role of the pion form factor in pion-pion annihilation processes, a few comments on the Vector Dominance Model(VDM) are necessary. The VDM states that virtual photons, which decay into massive lepton pairs, are coupled to hadrons through vector mesons.[23] This is because the vector mesons have the same quantum numbers as photons do. The VDM has been very successful in explaining

the dilepton spectrum from Dalitz decays of pions and etas.[24] The pion-pion annihilation processes proceed through the formation of virtual ρ mesons because of the
form factor. Thus the property of ρ mesons in dense nuclear medium could have a
crucial effect on the spectrum of dielectrons.

The characteristics of ρ mesons in nuclear matter have been calculated by several
groups and their results are not in agreement. G. E. Brown and M. Rho,[25] based on
a scaling effective Lagrangian approach, argued that the mass of the vector mesons
decreases with the nuclear medium density. Hatsuda and Lee[26] used quantum chro-
modynamical sum rules to show that the vector meson masses decrease linearly as a
function of density. Using the vector dominance model M. Herrmann et al.,[27] however, found that the width of the ρ mesons increases drastically with the medium
density with the peak value only slightly shifting towards high mass. Calculations by
M. Asakawa et al.[13] showed that the ρ peak moves to smaller masses with diminishing
strength when the nuclear density increases. Both Herrmann et al.[27] and Asakawa
et al.[13] showed that there may be another structure in the pion form factor in the
mass region around 2-3 pion masses. The exact location and the spectral strength of
the structure depends greatly on the nuclear matter density. Differences among these
calculations have not been resolved yet. Experimental measurement of pion-pion
annihilation processes would shed light on the ρ property in nuclear medium.

Theoretical calculations with detailed description of collision dynamics for nucleus-
nucleus interactions have been performed by several groups.[28] Recent calculations by
Wolf et al.[29] used a density-dependent pion form factor from M. Herrmann et al.[27]
Figure 4 shows their calculation in comparison to previously published DLS data of
Ca+Ca collisions at 1.0 GeV/A beam energy. The dashed line labelled 'background'
is the sum of all contributions other than pion-pion annihilation; the dotted and
the solid lines are from the annihilation process with free space and with a density-
dependent form factor, respectively. The production cross section of dielectrons in the
ρ mass region is reduced by a factor of three when nuclear medium effect is included.
Thus dielectron production in that mass region could be a sensitive probe of dense
nuclear medium.

Reference [29] also showed that the optimal beam energy of studying medium effect
via pion-pion annihilation process is around 1.0 GeV/A incident energy. At lower
beam energy the cross section for the annihilation is too small to be measured accurately; at higher beam energy the bremsstrahlung contribution extends significantly
to the ρ mass region.

The DLS data in Figure 4 are qualitatively consistent with a nuclear medium
effect, but much more statistics are needed to be conclusive. The DLS data of Ca+Ca
collisions at 1.0 GeV/A, included in Figure 4, have about 250 pairs. Recently DLS
has extended the Ca+Ca measurement with a factor of ten in statistical improvement.
Data analysis is currently in progress. To complete our nucleus-nucleus program we
also made measurement of dielectron production in d+Ca, α+Ca and C+C collisions
at the beam energy of 1.0 GeV/A. A systematic study of dielectron production in these

collisions will firmly establish the experimental evidence of pion-pion annihilation and allow for a careful investigation of density-dependent form factors.

5. Conclusions

DLS has systematically measured dielectron production in p+p and p+d collisions at beam incident energies from 1.0 to 4.9 GeV. Preliminary results showed that there is a drastic change in the shape of the dielectron mass spectrum from 1.0 GeV to 1.6 GeV beam energy, indicative of a threshold behavior in the dielectron production. The large pd/pp ratio of ten at 1.0 GeV is consistent with the model calculation that simple p+n bremsstrahlung is much stronger than p+p bremsstrahlung, and that p+n bremsstrahlung is a dominant source at that energy. Subthreshold production of η mesons in p+d collisions due to FERMI motion may also contribute to the large ratio.

The dielectron production in p+d and p+p collisions at 4.9 GeV indicated that dielectron production cross section in p+n is similar in magnitude to that in p+p collisions. The cross section can be explained mostly by Dalitz decays of η mesons, pion-pion annihilation and many-body bremsstrahlung from inelastic channels. This many-body bremsstrahlung is particularly important in accounting for the measured cross section, which has previously been neglected.

Although a reasonable agreement with the DLS data at 4.9 GeV has been achieved by recent calculations of Haglin and Gale,[9] the discrepancy around dielectron mass of 0.4 GeV is still substantial. More work on the extension of the soft-photon approximation to higher orders and on the inclusion of proton form factors with and without nuclear medium effects in bremsstrahlung are needed. The large amount of data with unprecedented statistics and the continuous theoretical effort should lead to an adequate understanding of dielectron production mechanisms in elementary collisions.

The pion-pion annihilation in nuclear collisions is a sensitive probe of nuclear medium effect. This sensitivity arises from the property of ρ form factor in dense nuclear matter, although the exact dependence of the form factor on density has not be unambiguously established. Previous DLS data of Ca+Ca at 1.0 GeV/A qualitatively support a nuclear medium effect, but more statistics are needed. To this end, DLS has collected new set of Ca+Ca data with a factor of ten improvement in statistics. In addition, DLS has also measured dielectron production in d+Ca, α+Ca and C+C collisions at 1.0 GeV/A beam energy. Analyses of these data sets are currently in progress.

6. Acknowledgement

The author wishes to thank other members of the DLS collaboration for helpful discussions, in particular, J. Carroll, K. Wilson, C. Naudet, L. S. Schroeder, R. J. Porter, R. Welsh and L. Madansky. He also thanks Kevin Haglin for explaining his calculations. Comments from Andrew Hirsch are appreciated.

7. References

[1] A. Yegneswaran et al., *Nucl. Inst. and Meth.* **A290**, 61(1990).

[2] H. Z. Huang et al., *Phys. Rev.* **C49**, 314(1994); H. Z. Huang et al., *Phys. Lett.* **B297**, 233(1992).

[3] J. D. Bjorken and H. Weisberg, *Phys. Rev.* **D13**, 1405(1976); V. Cerny et al., *Phys. Rev* **D24**, 652(1981).

[4] T. Goldman et al., *Phys. Rev.* **D20**, 619(1979).

[5] L. Xiong et al., *Phys. Rev* **C41**, R1355(1990); Gy. Wolf et al., *Nucl. Phys.* **A517**, 615(1990).

[6] Gy. Wolf et al., *Phys. Lett.* **B271**, 43(1991); V. D. Toneev et al., *GSI-92-05* Preprint(1992).

[7] C. Naudet et al., *Phys. Rev. Lett.* **62**, 2652(1989); A. Lettessier-Selvon et al., *Phys. Rev.* **C40**, 1513(1989).

[8] C. Gale and J. Kapusta, *Phys. Rev.* **C40**, 2397(1989).

[9] K. Haglin and C. Gale, *Phys. Rev.* **C49**, 401(1994).

[10] L. Winckelmann et al., *Phys. Lett.* **B298**, 22(1993).

[11] M. Schafer et al., *Phys. Lett.* **B221**, 1(1989).

[12] Kevin Haglin, private communication.

[13] M. Asakawa et al., *Phys. Rev.* **C46**, R1159(1992).

[14] W. K. Wilson et al., *Phys. Lett.* **B316**, 245(1993).

[15] J. M. Laget et al., *Phys. Lett.* **B257**, 254(1991); T. Vetter et al., *Phys. Lett.* **B263**, 153(1991).

[16] The PINOT Collaboration, *NEWS from SATURNE* **16**, 37(1994).

[17] G. Roche et al., *Phys. Rev. Lett.* **61**, 1069(1988); G. Roche et al., *Phys. Letts.* **B226**, 228(1989); S. Beedoe et al., *Phys. Rev.* **C47**, 2840(1993).

[18] C. Gale and J. Kapusta, *Phys. Rev.* **C35**, 2107(1987).

[19] L. H. Xia et al., *Nucl. Phys.* **A485**, 721(1988).

[20] C. L. Korpa and S. Pratt, *Phys. Rev. Lett.* **64**, 1502(199).

[21] C. L. Korpa et al., *Phys. Lett.* **B246**, 333(1990).

[22] W. Ehehalt et al., *Phys. Lett.* **B298**, 31(1993).

[23] R. P. Feynmann, Photon-Hadron Interactions, W. A. Benjamin, Inc. (1972) and references therein.

[24] L. G. Landsberg, *Phys. Rep.* **128**, 301(1985).

[25] G. E. Brown and M. Rho, *Phys. Rev. Lett.* **66**, 2720(1991).

[26] T. Hatsuda and S. H. Lee, *Phys. Rev.* **C46**,R34(1992).

[27] M. Herrmann, B. L. Friman and W. Norenberg, *Nucl. Phys.* **A560**, 411(1993).

[28] L. H. Xia et al., *Nucl. Phys.* **A485**, 721(1988); Gy. Wolf et al., *Nucl. Phys.* **A517**, 615(1990).

[29] Gy. Wolf et al., *Prog. Part. Nucl. Phys.* **30**, 273(1993).

NEUTRAL MESON PRODUCTION IN RELATIVISTIC HEAVY ION COLLISIONS

WOLFGANG KÜHN

II. Physikalisches Institut, Universität Gießen, Germany

for the

TAPS COLLABORATION

(GANIL-Gießen-GSI-KVI-Münster-München-Valencia)

and the

HADES COLLABORATION

(Bratislava - Clermont-Fd - GSI Darmstadt - Frankfurt - Gießen - Krakow -
ITEP-LPI-MEPI Moscow - TU München - Nikosia - Rez - Valencia)

ABSTRACT

Recent results for (π^0, η) production in relativistic heavy ion collisions at SIS are discussed. Meson production probabilities follow simple scaling laws with available energy and number of participants. In contrast, the pion transverse momentum spectra show strong variations with system size, pointing out the importance of multi-step processes and final state interaction.

1. Introduction

A major fraction of the physics programme in the Bevalac/SIS energy regime is devoted to the study of dense, hot hadronic matter. At GSI, several new dedicated systems are now operational for the detection of hadrons (FOPI, KAOS, LAND) as well as photons (TAPS). A common goal of experiments in this context is the investigation of the equation of state (EOS), which is of fundamental importance not only to nuclear physics but also required for a deeper understanding of astrophysical processes (supernovae, neutron stars) related to the final phase of stellar evolution. The derivation of the EOS has been a central topic addressed by many experimental [1] and theoretical [2, 3, 4, 5] efforts since the beginning of relativistic heavy ion physics in the pioneering experiments at the Berkeley Bevalac. From the experimental point of view, information about the equation of state involves the combination of a large number of observables (collective phenomena such as flow, squeeze-out as well as meson and anti-proton production). These experimental results are compared to predictions of quantal phase space models (BUU, QMD) , which are sensitive to the EOS.

Of particular interest are particles which are created during the collision, since they carry information on the dynamics of the hot hadronic medium where they were produced. For energies up to 2 AGeV, light pseudoscalar mesons (pions[6, 7], etas[8] and Kaons[9]) as well as antiprotons[10] have been investigated. At lower incident energies, below the π^0 - threshold in the nucleon-nucleon CM system, direct photons from incoherent proton-neutron bremsstrahlung have been observed[11].

Photons, and to some extent K^+ mesons do not undergo significant final state interaction and permit direct experimental access to the production dynamics. In contrast, the information from pions and eta mesons is strongly influenced by the hadronic final state interaction. As a consequence, the analysis is more complicated but might provide additional information about the expansion dynamics. Density-dependent effects are expected to be sizable, since large compression factors (up to 3 times normal nuclear matter density for central 1 AGeV Au + Au collisions) are predicted [3]. Moreover, since meson production is strongly linked to resonance decay, information on the resonance content of the collision zone can be obtained.

This paper presents results of a series of experiments with TAPS (Two-Arm Photon-Spectrometer) studying (π^0, η) production with relativistic heavy ion beams from SIS/GSI Darmstadt. Charged pion spectra have been extensively studied at the Bevalac[6]. With TAPS, we focus on neutral mesons decaying into 2 γ rays. This opens up the possibility to study η-mesons. Moreover, the experiments are sensitive to mesons with very small transverse momentum p_T, since the decay photons have sufficient p_T to be easily detected even for mesons with $p_T \approx 0$.

2. Experimental setup

TAPS (Two-Arm Photon-Spectrometer) is a modular detector system[12] consisting of 256 BaF_2-detectors arranged in 4 blocks with individual charged particle plastic veto detectors. The blocks are mounted in two movable towers positioned symmetrically with respect to the beam direction to detect the decay of neutral mesons emitted near mid-rapidity. Following shower reconstruction, meson identification is done via invariant mass analysis from the measured angles and energies of coincident photons. Photon/hadron discrimination is performed exploiting time-of-flight with respect to a start detector as well as by pulse shape analysis. The combinatorial background from uncorrelated photons is obtained from event mixing and subtracted to derive transverse momentum (p_T) - spectra. The typical mass resolution is about 5% (σ). As an example, fig. 1 shows invariant mass distributions for 4 different reactions. Two structures, corresponding to the 2-photon decay of π^0 and η are visible.

Angle-integrated cross sections are determined from an extrapolation to the full solid angle assuming an isotropic "thermal" source at mid-rapidity with a slope parameter determined from the experiment.

246

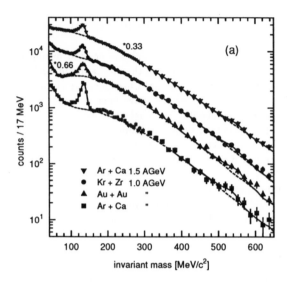

Fig. 1: Spectrum of the invariant mass (2γ) in various reactions at 1 AGeV. Two structures corresponding to the neutral pion and eta decays, respectively, are observed.

3. Experimental Results

The inclusive production cross section can be factorized into the reaction cross section, the average number of participants, as obtained from a geometrical model and the meson production probability per participant nucleon. The meson production probabilities follow a simple energy scaling law[13], as is shown in fig. 2. The relevant scaling variable is the bombarding energy per nucleon divided by the meson production threshold. This quantity can be regarded as a measure of the available phase space for a particular production channel. Considering the fact that elementary production cross sections and final state interaction for the various mesons shown in fig. 2 are quite different, the scaling works very well. It should be noted, however, that most of the data in fig. 2 was obtained within limited detector acceptance and extrapolated to the full solid angle making certain assumptions about the shape of rapidity - and p_T - distributions.

Similar scaling features are observed when comparing pion and eta spectra. A suitable variable for such a comparison is the transverse mass $m_T = \sqrt{p_T^2 + m^2}$. Fig. 3. shows the transverse mass distributions for π^0 and η mesons. m_T - scaling is observed, a well known feature at much higher energies. From the systematics in fig. 2, the total meson production probability turns out to be close to 30 % at 2 AGeV. Most of these mesons originate from resonance decays, pointing to the importance of baryon resonances in understanding the dynamics of hot compressed hadronic matter.

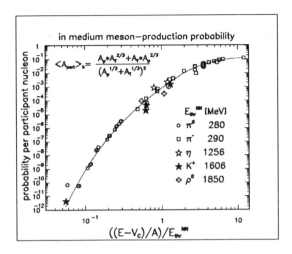

Fig. 2: Meson production probability per participant nucleon as a function of the bombarding energy per nucleon (corrected for the Coulomb barrier) normalized to the respective meson production threshold in the nucleon nucleon system.

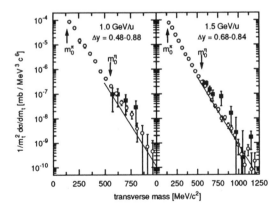

Fig. 3: Transverse mass spectra for π^0 and η mesons in the reaction $Ar + Ca$ at 1 and 1.5 AGeV, showing m_T - scaling.

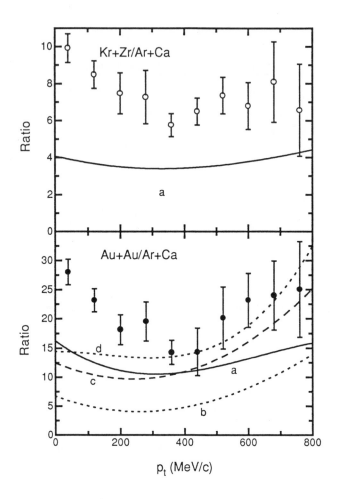

Fig. 4: Ratios of p_T spectra for the systems 1 AGeV Kr+Zr (top) and 1AGeV Au+Au (bottom), normalized the p_T spectrum of 1AGeV Ar+Ca, respectively. For comparison, the results of various mode calculations are shown : (a) BUU[14], (b) IQMD[15], (c) RQMD[16], (d) QGSM[17].

Experimentally, a comparison of transverse momentum spectra for systems of different size should be sensitive to absorption effects. The corresponding data is shown in fig. 4, which plots ratios of p_T - spectra from Au+Au/Ar+Ca and Kr+Zr/Ar+Ca at 1 AGeV. The ratio shows deviations from a constant at small transverse momenta (for both Au and Kr) and at large transverse momenta (for Au). The latter is related to a high momentum tail which is observed in the Au data.

The larger deviation at small transverse momenta cannot be described by current model calculations. For reference, calculations with BUU [14], IQMD [15], RQMD [16] and QGSM [17] are shown. This failure might be due to an insufficient description of the absorption in the final phase of the nucleus-nucleus collision. It should be explored to what extent the model results are sensitive to the parameters of the equation of state.

4. Outlook

A major uncertainty in model calculation arises from in - medium properties such as masses, resonance widths, cross sections of hadrons which are not well understood theoretically. Moreover, there is only few experimental data available to test the various theoretical predictions. It is definitely one of the major challenges in this field, to design experiments which are sensitive to such in-medium effects of hadrons. By observing the unperturbed dielectron branch of short-lived vector mesons which decay inside the hot dense reaction zone, it would be possible to gain direct experimental access to hadron masses and life times in the hadronic medium.

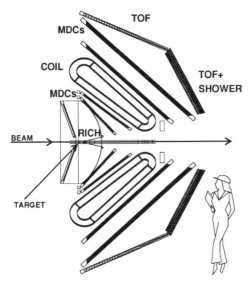

Fig. 5: Schematic view of the planned dilepton spectrometer HADES.

The pioneering experiments on dilepton production by the DLS collaboration[19] have produced important results, but are limited to light systems. To explore the heaviest collision systems, a novel approach going beyond the design limitations of the DLS is needed. A new international collaboration is currently planning a second-generation dilepton spectrometer, HADES (High Acceptance Di-Electron Spectrometer) for SIS/GSI.

This spectrometer is optimized to detect the di-electron decay of light vector mesons (ρ, ω, ϕ) with superior efficiency and resolution. A schematic view of HADES is shown in fig. 5. It consists of a ring imaging Cherenkov detector with solid photocathode, serving for electron identification, a superconducting spectrometer with toroidal geometry and mini - drift chambers for momentum measurement and a general purpose shower/tof - detector for event characterization and second electron identification. The system employs a powerful trigger processor which enables operation at interaction rates up to 10^6 /s for the heaviest systems at the highest SIS energies. The system has sufficient mass resolution (1%) to discriminate ρ and ω - mesons.

The TAPS data presented here are the result of a group effort. The contributions of V.Metag, H.Löhner, R.Novotny, R.S.Simon, Y.Schutz and H.Wilschut as well as many discussions with W.Cassing, U.Mosel and H.Stöcker are gratefully acknowledged. In particular, I would like to thank F.D.Berg, M.Pfeiffer, A.Raschke, J. Ritman, O.Schwalb, and L.Venema, who carried the major load of data analysis. Last - not least - I would like to thank the GSI accelerator crew for providing high energy and high intensity beams and H.Folger (GSI) for producing excellent targets.

This work was supported in part by Gesellschaft für Schwerionenforschung (GSI), Deutsches Bundesministerium für Forschung und Technologie (BMFT) under contract No 06 GI 174 I and 06 MS 145, Fundamenteel Onderzoek der Materie (FOM),the Netherlands and Grand Accélérateur d'Ions Lourds (GANIL), France.

References

[1] R. Stock, Phys.Rep. 202(1991)233

[2] H.Stöcker and W.Greiner, Phys.Rep. 137(1986)277

[3] W.Cassing et al., Phys.Rep. 188(1990)363

[4] U.Mosel, Ann.Rev.Nucl.Part.Sci. 41(1991)29

[5] J.Aichelin, Phys.Rep. 202(1991)233

[6] J.W.Harris et al., Phys.Rev.Lett. 58(1987)377

[7] F.D.Berg et al. Z.Physik A340(1991)297

[8] F.D.Berg et al. Phys.Rev.Lett. 14(1994)977

[9] D.Brill et al., Phys.Rev.Lett. 71(1993)336

[10] A.Schröter et al., Nucl.Phys. A553(1993)775c

[11] R.Hingmann et al., Phys.Rev.Lett. 58(1987)759

[12] R.Novotny, IEE Trans.Nuc.Sci. 38(1991)379

[13] V.Metag, Prog.Part.Nucl.Phys. 30(1993)75

[14] W.Ehehalt, private communication

[15] S.Bass, C.Hartnack, contribution to the Workshop on Meson Production in Nuclear Collisions, GSI Darmstadt 1993

[16] H.Sorge et al., private communication

[17] V.Toneev et al., private communication

[18] D.Ashery and J.P.Schiffer, Ann.Rev.Nucl.Part.Sci 36(19986)207

[19] G.Roche et al., Phys.Lett. B226(1989)228

ON THE MEAN FREE PATH OF PIONS AND KAONS
IN HOT HADRONIC MATTER

K. HAGLIN and S. PRATT

National Superconducting Cyclotron Laboratory
and Department of Physics and Astronomy
Michigan State University, East Lansing MI 48824-1321, USA

ABSTRACT

Pion and kaon mean free paths within a thermal hadronic background are calculated using relativistic kinetic theory. Free cross sections are used which include contributions from ρ, K^*, Δ and heavier resonances. Given pion to baryon ratios appropriate for the breakup stage of a 200·AGeV relativistic-heavy-ion collision, we find for temperatures less than 100 MeV, kaons have a shorter mean free path than pions, while for higher breakup temperatures the reverse is true. Since breakup temperatures should be in the neighborhood of 100 MeV, this suggests that kaon interferometry samples the same emission distribution as pion interferometry.

1. Introduction

By measuring outgoing hadrons in a heavy-ion collision one can reconstruct a reaction's breakup stage. The momenta of outgoing particles can be measured directly and a space-time picture of the last collisions may be constructed using the techniques of two-particle interferometry[1]. Space-time information is especially useful. For instance, if the time a reaction takes to proceed is much longer than 10 fm/c, it would signal a reduction in pressure, inferring a first-order phase transition[2].

Unfortunately, such information regarding the lifetime of the emission can be masked by the presence of long-lived resonances[3], particularly the ω which has a lifetime of 20 fm/c. For this reason Padula and Gyulassy campaigned for two-kaon interferometry. The only long-lived resonance responsible for a significant portion of kaons is the $K^*(892)$ which has a lifetime of 8 fm/c, smaller than characteristic times of even an explosive collision.

In this letter we study mean free paths of pions and kaons at temperatures and densities characteristic of hadronic matter from the breakup stage of a relativistic heavy-ion collision at CERN. We find that in this environment mean free paths are remarkably similar and conclude that information from kaon interferometry describes the dissolution of the entire system, as kaons should escape at the same time as pions which comprise the bulk of the matter.

In baryon-rich matter, such as is characteristic of heavy-ion collisions at the AGS, most charged kaons are negative since s quarks are absorbed into baryons, leaving $\bar{s}$ quarks to form hadrons. Negative kaons interact little with baryons as opposed to pions which interact vigorously due to the Δ resonance. Thus negative kaons escape baryonic matter much more easily than pions and the results of kaon interferometry indeed reflect this fact[4]. For sulfur projectiles at 200 A· GeV incident on heavy targets, resulting pion multiplicities are ten times as large as baryon multiplicities at mid-rapidity. Thus a meson's escape probability depends principally on its cross section with pions.

2. The Model

In the context of a thermodynamic model we estimate mean free paths of a given meson of type a given its momentum. Assuming the meson interacts with an assortment of hadrons of type b whose density is given according to a relativistic Boltzmann distribution and assuming we know the cross sections σ_{ab}, the mean free path is:

$$\lambda_a(p_a) = \frac{p_a}{E_a} \frac{1}{R_a^{net}(p_a)}.$$

Here p_a and E_a represent the meson's momentum and energy while R_a is the collision rate. The net collision rate includes contributions from all different species.

$$R_a^{net}(p_a) = \sum_b R_{ab}(p_a)$$

$$R_{ab}(p_a) = \int ds \frac{d^3 p_b}{(2\pi)^3} f(p_b) \sigma_{ab}(s) v_{rel} \delta \left(s - (p_a + p_b)^2 \right) \qquad (1)$$

where

$$v_{rel} = \frac{\sqrt{(p_a \cdot p_b)^2 - m_a^2 m_b^2}}{E_a E_b}, \qquad f(p_b) = (2s_b + 1)e^{-(E_b - \mu)/T}.$$

Only baryons are given a chemical potential which is chosen to result in a pion to proton ratio of 10. Relative densities of hadron species are dependent on the temperature. They are illustrated in Fig. 1. Species included in the calculation are π, K, ρ, η, ω and K^*.

Cross-sections are dominated by contributions of resonances. For pions the largest contributor to the collision rate is the resonant reaction through the ρ while for kaons the most common collision is with pions through the K^* resonance. Resonant cross-sections are assumed to have the form:

$$\sigma_{ab}(s) \propto \frac{s - (m_a - m_b)^2}{(s - m_R^2)^2 + m_R^2 \Gamma_R^2}.$$

The mass and width of the resonance are m_R and Γ_R respectively. Normalizations to this form are chosen to yield maximum cross sections consistent with unitarity

254

limits. The unitarity limit for cross sections through a resonance are proportional to $1/k^2$ where k is the relative momentum of the scattering. The density of states scales as k^2/v_{rel}, therefore for a given width in energy, lower energy resonances will contribute to scattering rates more than higher energy resonances by a factor of $1/v_{rel}$. The Boltzmann factor also reduces the contribution of high-energy resonances to the rate. Narrow-width resonances contribute to the scattering rates proportional to their widths since resonances with larger widths scatter particles over a greater range of relative energies.

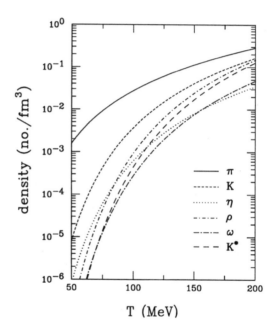

Fig. 1: Thermal ensemble of hadrons and their number densities as a function of temperature.

One expects $\pi\pi$ scattering to be weaker than πK because the K^* has less decay energy than a ρ, hence the relative velocity is smaller while the Boltzmann factor is stronger for the K^* case. However, the width of the K^*, 50 MeV, is one third the width of the ρ. At sufficiently high temperatures pions scatter more readily when travelling through a pion gas than do kaons. At low temperatures the Boltzmann factor is the dominant factor, and kaons scatter more readily.

Fig. 2 shows the mean free paths of particles travelling through a hadron gas at a temperature of 120 MeV as a function of their momentum. The dotted lines show the mean free paths of a pion (Fig. 2a) and a kaon (Fig. 2b) in a pion gas where only interactions through the ρ and K^* resonances are considered. These results are consistent with previous calculations which used similar assumptions[5, 6]. The mean free paths go to zero with zero momentum due to the velocity term in Eq. (1). The rise of the mean free path for pions at low momentum demonstrates how pions without much energy have difficulty colliding through the ρ resonance since it is difficult to find a second pion with sufficient energy to produce a ρ. For kaons this is not so difficult, since K^* s do not require so much center-of-mass energy; and since the kaon is heavier, less of a colliding particle's energy is lost to center-of-mass motion.

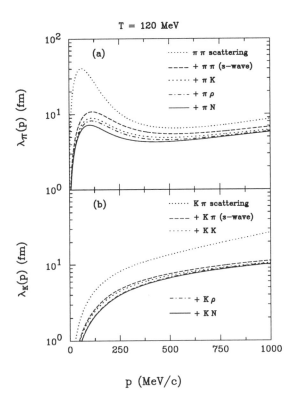

Fig. 2: Mean free path of pions (a) and kaons (b) as a function of their momentum. Resonant ρ and K^* cross sections are included in the dotted curves. The the effects of adding to this the s-wave (long-dashed curves) into the cross section are shown. Finally, K^*, $a_1(1260)$ and Δ into the pion results and ϕ, $K_1(1270)$, $\Lambda(1520)$ and $\Lambda(1800)$ are shown as short-dashed, dot-dashed and solid lines, respectively.

The long-dashed curves result from adding to the resonant ρ and K^* an s-wave contribution. Boltzmann weighting favors near-threshold interactions and therefore enhances the effect of the relatively small s-wave for which a constant value of 8.0 mb is taken. Next we add the heavier resonances. For the pion results (Fig. 2a) we add K^*, $a_1(1260)$ and Δ for pion-nucleon interactions. They are presented as short-dashed, dot-dashed and solid lines, respectively. We also studied the effect of even heavier resonances but conclude they are of no importance for these temperatures. The effect of the resonances beyond ρ is to reduce the mean free path of order 30% at low pion momentum and 15% at high momentum.

To the kaon interactions we add ϕ and $K_1(1270)$ for interactions with other kaons and with ρ s. We also add $\Lambda(1520)$ and $\Lambda(1800)$ for $K^{-,0}$ s interacting with nucleons. The results are again shown as long-dashed, short-dashed, dot-dashed and solid lines, respectively. The effect of the ϕ is quite small as expected. The $K_1(1270)$ is near enough to threshold that the unitarity limit is relatively large resulting in a rather strong interaction and a noticeable change in the kaon mean free path. Finally, the presence of nucleons further reduces the kaon mean free path but only slightly. Overall these heavier resonances play a modest role. On the other hand, quantitative comparisons between experimental data and model calculations that neglect these heavier resonances such as refs. 7,8 and 9, are only reliable, as we have shown, up to tens of percent. Many authors have studied mean collision times for pions[10, 11, 12, 13]. but we choose to compute average mean free paths of pions and compare to kaons. The result is shown in Fig. 3 as a function of the temperature. The average mean free path is defined

$$\bar{\lambda} = \frac{\int d^3 p f(p) \lambda(p)}{\int d^3 p f(p)}.$$

All resonances and s-wave contributions are included in the calculation. The reaction should end when the mean free path is near the size of the system. From interferometry the size of the dissolving system appears similar to the size of a lead nucleus which has a radius of seven fm. A mean free path of seven fermi corresponds to a temperature of approximately 110 MeV. One must be cautious of such conclusions both because of the lack of geometric detail involved in the inference and the questionable assumption of zero chemical potential. Rapid expansion can outrun a system's ability to stay in chemical equilibrium, resulting in large chemical potentials[14, 15, 16], higher densities and therefore shorter mean free paths. This would allow the system to stay together longer, resulting in lower breakup temperatures.

3. Conclusions

The most remarkable aspect of Fig. 3 is that the mean free paths of kaons and pions are so similar. This means that kaons can be used to view the final stage of the collision without the qualification that they have escaped prematurely. This does not mean that correlation functions from kaons and pions should have the same apparent source sizes, and indeed at CERN preliminary measurements point to smaller sizes for kaons than pions[17].

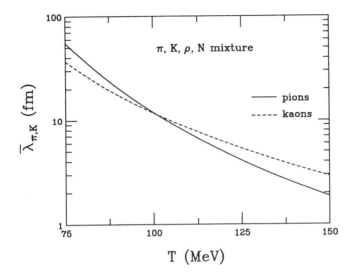

Fig. 3: Average mean free paths of pions (solid curve) and kaons (dashed curve) as they depend on the temperature.

Even if one can account for pions from long-lived resonances, kaon sources can appear smaller due to collective expansion. A heavier particle with a given velocity is more confined to the region with the same collective velocity[18,2]. Given that pions and kaons have such similar escape probabilities, one can then compare and interpret correlation results from kaons and pions. This clarifies the meaning of both measurements. In addition the effect of heavier resonances and s-wave contributions has been cataloged, showing they matter at the level of 10–20 percent. These calculations can provide guidance to those constructing transport models. By using the results of Fig. 2, one can see which channels should be included and which reactions can safely be neglected.

4. Acknowledgements

This work was supported by W. Bauer's Presidential Faculty Fellow Award, NSF grant no. PHY-925355, and from NSF grant no. PHY-9017077.

5. References

1. S. Pratt *et al.*, Nucl. Phys. A566 (1994) 103c.
2. S. Pratt, Phys. Rev. D 33 (1986) 1314.
3. M. Gyulassy and S. S. Padula, Phys. Rev. D 41 (1990) R21.
4. O. Vossnack, HIPAGS Proceedings, ed. G. S. F. Stephans, S. G. Steadman and W. L. Kehue, MITLNS-2158 (1993).
5. J. L. Goity and H. Leutwyler, Phys. Lett. B 228 (1989) 517.
6. M. Prakash, M. Prakash, R. Venugopalan and G. Welke, Phys. Rep. 227 (1993) 321.
7. H. Sorge, R. Mattielo, A. Jahns, H. Stöcher and W. Greiner, Phys. Lett. B271 (1991) 37.
8. Y. Pang, T. J. Schlagel and S. H. Kahana, Phys. Rev. Lett. 68 (1992) 2743.
9. L. V. Bravina, N.S. Amelin, L. P. Csernai, P. Levai and D. Strottman, Nucl. Phys. A566 (1994) 461c.
10. S. Gavin, Nucl. Phys. A435 (1985) 826.
11. E. V. Shuryak, Phys. Lett. B207 (1988) 345.
12. P. Levai and B. Müller, Phys. Rev. Lett. 67 (1991) 1519.
13. M. Prakash, M. Prakash, R. Venugopalan and G. Welke, Phys. Rev. Lett. 70 (1993) 1228.
14. S. Gavin and P. V. Ruuskanen, Phys. Lett. B262 (1991) 326.
15. G. M. Welke and G. F. Bertsch, Phys. Rev. C 45 (1992) 1403.
16. P. Gerber, H. Leutwyler and J. L. Goity, Phys. Lett. B243 (1990) 18.
17. T. J. Humanic (NA44 Collaboration), Nucl. Phys. A566 (1994) 115c.
18. S. Pratt, Phys. Rev. Lett. 53 (1984) 1219.

Heavy Ion Reaction Measurements with the EOS TPC
(Looking for Central Collisions with Missing Energy)

H. H. Wieman[1] for the EOS Collaboration:
S. Albergo[6], F. Bieser[1], F. P. Brady[4], Z. Caccia[6], D. A. Cebra[4],
A. D. Chacon[5], J. L. Chance[4], Y. Choi[3], S. Costa[6], J. Elliott[3], M. Gilkes[3],
J. A. Hauger[3], A. Hirsch[3], E. L. Hjort[3], A. Insolia[6], M. Justice[2], D. Keane[2],
V. Lindenstruth[7], H. S. Matis[1], M. McMahan[1], C. McParland[1],
W. F. J. Mueller[7], D. L. Olson[1], M. Partlan[4], N. Porile[3], R. Potenza[6],
G. Rai[1], J. Rasmussen[1], H. G. Ritter[1], J. Romanski[6], J. L. Romero[4],
G. V. Russo[6], H. Sann[7], R. Scharenberg[3], A. Scott[2], Y. Shao[2],
B. Srivastava[3], T. J. M. Symons[1], M. Tincknell[3], C. Tuvè[6], S. Wang[2]ph,
P. Warren[3], D. Weerasundara[2], H. H. Wieman[1], and K. L. Wolf[5]

[1] *Lawrence Berkeley Laboratory, Berkeley, California 94720*
[2] *Kent State University, Kent, Ohio 44242*
[3] *Purdue University, West Lafayette, Indiana 47907*
[4] *University of California, Davis, California 95616*
[5] *Texas A&M University, College Station, Texas 77843*
[6] *Università di Catania and INFN-Sezione di Catania, 95129 Catania, Italy*
[7] *Gesellschaft für Schwerionenforschung, D-64220 Darmstadt 11, Germany*

ABSTRACT

The EOS TPC was constructed for complete event measurements of heavy ion
collisions at the Bevalac. We report here on the TPC design and some preliminary
measurements of conserved event quantities such as total invariant mass, total momentum, total A and Z.

1. Introduction

The EOS TPC does a very complete measurement of relativistic heavy ion reactions. Nearly all of the charged particles in a central collision are identified and their
momenta are accurately measured. This provides a new, unique way to study heavy
reactions. By measuring the total invariant mass of all the charged particles in an
event, we can search for events with missing energy to look for exotic phenomena.
Clearly, the chances of discovering interesting physics are remote, but, in any case,
the measurement of conserved quantities, such as invariant mass and total event momentum, provide an excellent check of detector performance. If exotic events should

appear, the highly redundant nature of TPC measurements allows easy rejection of false signals.

2. Experimental Setup

The arrangement of the detectors in the EOS experiment is shown in Figure 1. The central detector is the EOS TPC, followed by MUSIC, a multisampling ionization detector for detecting fragments with Z>6, and a time of flight wall (TOF). There is also a neutron detector called "muffins". This report will, however, be limited to the TPC portion of the experiment. The target is located just inside of the HISS dipole magnet, 10-20 cm upstream from the TPC. The TPC provides true 3D tracking in the magnetic field for most of the particles exiting the downstream side of the target. The active tracking region in the TPC is a rectangular box 1.5 m long in the beam direction and ˜1 m in the directions perpendicular to the beam. The beam passes directly through the center of the TPC volume. This arrangement provides excellent coverage for scattered particles in the forward half of the lab frame. The tracking in the magnetic field of 13 kG measures particle rigidity and the multisampling of dE/dx along the track provides particle ID.

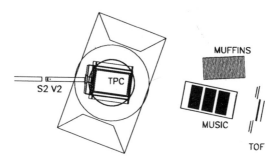

Fig. 1: Detector setup for EOS experiment. The beam enters from the left and the target is located at the entrance to the TPC.

The TPC is constructed with a rectangular field cage sitting over a multiwire proportional chamber (MWPC)-pad plane structure. Electron tracks left by the charged particles passing through the TPC gas of Ar 90% + CH$_4$ 10% drift down to the pad plane where they are amplified by the MWPC and recorded as a function of time. The MWPC pad plane provides a full 2D readout and the drift time provides the third dimension. The MWPC pad plane system is composed of 20 μm anode wires on a 4 mm pitch located 4 mm above a continuous rectangular array of 8 mm by 12 mm pads. The electrons create avalanches as they drift to the anode wires. The positive ions created in the avalanche induce signals in the pads as they drift from the anode region. The induced signal is spread over 2 to 3 pads allowing accurate (600 μm) position determination through centroid reconstruction. The TPC volume is divided up into a 2.5 million pixel volume (15360 pads X 160 time buckets). The high pixel density is necessary to isolate and record the 200 tracks that can result from heavy ion collisions at Bevalac energies.

There are additional wires in the MWPC for field shaping and there is a gating grid that only passes tracks for amplification from triggered events. This gate also prevents avalanche-generated positive ions from entering the drift volume where their space charge field can distort the drift path of the electrons left by the particles of interest. This gate is essential in our configuration where heavy ion beams pass through the drift volume and leave ionization densities up to 6000 times minimum ionizing tracks. If left ungated, the beam load would quickly lead to wire aging and possible sparking and glow discharge. Using the gate we routinely operate the TPC with Au beams in excess of 1000 beam particles per spill with no observable degradation in performance. A summary of the TPC characteristics are listed in Table 1.

HISS TPC Characteristics	
Pad Plane Area	1.5m × 1.0m
Number of Pads	15360 (120 × 128)
Pad Size	12mm × 8mm
Drift Distance	75 cm
Time Sampling Freq.	10 MHz
Signal Shaping Time	250 ns
Electronic Noise	700 e
Gas Gain	3000
Gas Composition	$90\% Ar + 10\% CH_4$
Pressure	1 Atmosphere
B Field	13 kG
E Field	120 V/cm
Drift Velocity	5cm/μ s
Event Rate	10-80 events/ 1 sec spill
dE/dx range	$Z = 1\text{-}8, \Lambda, \pi, p, d, t, He, Li - O$
Two Track Resolution	2.5cm
Multiplicity Limit	≈ 200

Table 1: TPC detector specifications.

Special electronics were developed for the TPC to read out the 15360 pads. The electronics design was driven by the need for low noise (700 e rms), high dynamic range, high density and low cost. The first amplifier stage had to be located on the chamber to reduce the noise (low noise is required for position resolution). Since space is at a premium in the magnetic field volume, it was decided to design a highly multiplexed system with local, on-chamber, digitization to avoid potentially impossible cabling problems. The first element of the electronics chain is a 4 channel, integrated CMOS preamp circuit developed at LBL for this project by Michael Wright. This is followed by a shaper amplifier built with surface mount technology. The shaper signal is time sampled with a custom integrated analog storage circuit, also developed at LBL for this project. This circuit, a switched capacitor analog transient waveform recorder (SCA), was developed by Stuart Kleinfelder. The SCA has 16 channels with 256 sampling capacitors per channel. The signals are written into the SCA at 10 MHz and they are readout and digitized at a lower rate with a commercial 12 bit ADC chip. The number of time buckets is selectable, and most of our data was recorded with 160 time buckets. One ADC reads out the up to 256 time samples for 60 channels and the digital output from two ADC's is transmitted off the chamber over an optical fiber serial data link. There are a total of 128 boards with one fiber each on the chamber. A time of 10 ms is required to digitize and transfer the data off the chamber. The data is passed to a receiver card where pedestal subtraction, gain factor multiplication and zero suppression take place. Pedestal subtraction is done with a separate pedestal value for each time bucket to remove transient background effects that arise from opening the gating grid and from clock signal pickup. During the time that data is recorded into the SCA, the electronics operation is fully synchronous to maintain event independent pedestals. The zero suppressed and gain corrected data from the receiver card is collected by the event builder and written to Exabyte tape.

The TPC is also equipped with a Nd-Yag laser system that generates 17 straight ionized tracks in the tracking volume for checking the distortion corrections. This distortion correction is applied to correct for the non-uniform B field. The Nd-Yag laser is frequency quadrupled to produce 5 ns wide pulses of 266 nm UV light. The UV pulses produce tracks through 2 photon-ionization of contaminants in the drift gas.

The MWPC is constructed with tight mechanical tolerances and operated at low gas gain to provide good energy resolution for dE/dx. Particle identification is obtained from the truncated mean of multiple energy loss samples along the ionization track. Under ideal conditions with low multiplicity, the full track length through the TPC (160 cm) is available with 128 samples for the truncated mean determination. In practice, the available track length is somewhat less due to track overlap in the high track density region near the target. A measured dE/dx spectrum is shown as a function of rigidity for high multiplicity events of 0.8 GeV Au on Au in Figure 2.

Good isotopic identification is achieved for charge 1 and 2 baryons and the pions are clearly identified. At higher momentum the isotopes merge and require the time of flight wall for proper identification. Charge is identified in the TPC for the heavier elements from Li through O, but without isotopic identification.

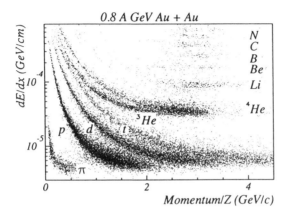

Fig. 2: Measured dE/dx truncated mean with the EOS TPC as a function of particle rigidity.

3. Measurements With the TPC

A summary of the large number of beam target systems measured at the Bevalac by the EOS collaboration is shown in Figure 3. This lego plot shows the number of events recorded for each system. The preliminary measurements to be shown here are for only two systems, 800 MeV*A La + La and 1000 MeV*A Au + C. We have constructed several conserved quantities for the events from the TPC measurements. The total invariant mass minus the rest mass, the total momentum, the total charge and the total mass were determined for the baryons detected in the TPC. For the most central collisions, there are few heavy ion remnants with Z>8. Almost all of the charged baryon matter appears as either hydrogen or helium isotopes so we expect the TPC to lose very little when constructing these totals for the event. The free neutrons should account for the only significant loss in the total momentum, total invariant mass and total mass. The pions can and will be included in future analyses.

In the first example we measured the total invariant mass minus the rest mass $(M_{inv} - M_0)$ for 800 MeV*A La + La. This quantity is simply the total kinetic energy in the center of mass reference frame, but it can be calculated without transforming to the center of mass frame. We can compare this with the total available invariant mass less rest mass calculated with just the beam nucleus and the target nucleus:

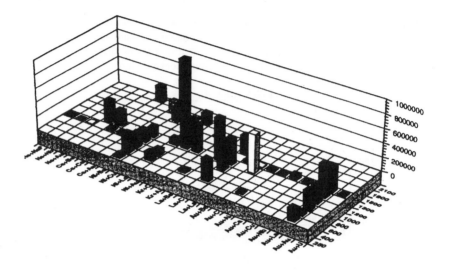

Fig. 3: Summary of beam-target systems measured with EOS at the Bevalac.

$\sqrt{(E_{beam} + E_{target})^2 - P_{beam}^2} - M_{0beam} - M_{0target}$. At the highest multiplicity the measured $(M_{inv} - M_0)$ accounts for 37% of the total available for the 800 MeV*A La + La system. The missing 63% is carried by the undetected particles and the energy required to break up the projectile and target. We have also looked at $(M_{inv} - M_0)/A_{measured}$. We can compare this directly with the total available kinetic energy in the CM frame divided by the mass of the target and the projectile. This will normalize out the effect of the undetected neutrons if the neutrons carry the same energy per nucleon as the rest of the baryons. One could expect this if the expansion is totally hydrodynamic. If it is strictly thermal expansion, then the lighter neutrons will carry more energy per nucleon than the heavier isotopes and the measured ratio will be lower than the total ratio calculated from the beam energy and mass and target mass. At the highest multiplicity, $(M_{inv} - M_0)/A_{measured}$ is 68% of the total expected value (0.18 GeV/A). This measurement is low for the system chosen because the geometric acceptance of the TPC misses some of the target rapidity nucleons. This is clear when viewing pperp vs rapidity plots.

To correct for the missing target rapidity particles, we have done the same analysis using only the data in the forward half of the CM frame by reflecting all detected particles in the forward half to the back half before computing $(M_{inv} - M_0)/A_{measured}$ for the event. This result is shown in Figure 4. The measured ratio in this case is now 89% of the total available for a symmetric target - beam system with a beam energy of 800 MeV*Abeam. This now leaves only 20 MeV/nucleon missing. Part of this missing energy goes into breaking up the nucleus. Total disintegration of the La nuclei into free protons and neutrons is 8.5 MeV/nucleon. This is an upper limit since there are many deuterons, tritons and helium isotopes. In a future, more careful analysis, we can calculate this quantity directly because we have complete particle ID for Z=1 and 2. However, using the number of 8.5 MeV/nucleon, this leaves 11.5 MeV/nucleon unaccounted for, or a total of 3.2 GeV, enough for 23 pions. In order to pursue this analysis more thoroughly, we must now include the pions and small affects such as energy loss in the target to get a more precise accounting of the total available kinetic energy in the CM. It is clear at this point that the detector is coming close to a complete measurement in the forward half of the CM frame and that, in a complete analysis, we will have a sensitive indicator for special events with missing energy. It should also be possible to compare $(M_{inv} - M_0)/A_{measured}$ for different mass fragments to distinguish between thermal expansion and collective expansion just has been done previously [1] with single particle kinetic energy spectra.

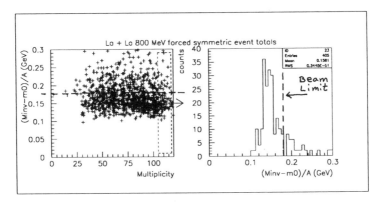

Fig. 4: The quantity $(M_{inv} - M_0)/A_{measured}$ as a function of the event multiplicity. Only the particles in the forward half of the CM frame have been used and reflected into the back half to create a full event. The right hand frame shows a histogram of the highest multiplicity portion of the scatter plot projected onto the $(M_{inv} - M_0)/A_{measured}$ axis.

We have also looked at several other conserved event quantities. As an example, we show (see Figure 5) $p_{lab}/A_{measured}$ as a function of multiplicity for the 800 MeV*A La + La system. At the highest multiplicity $p_{lab}/A_{measured}$ is 107% of the beam $p_{labbeam}/(A_{beam} + A_{target})$. This a little high due to missing target rapidity particles. But this agreement to within 7% is close and, as a consistency check, shows that the TPC is correctly measuring Z , A and momentum.

We also studied another system, 1000 MeV*A Au + C. In this case the CM velocity

266

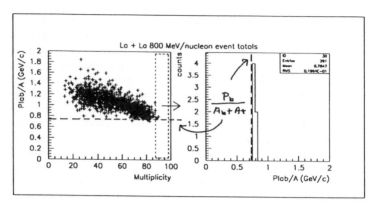

Fig. 5: The measured quantity, $p_{lab}/A_{measured}$ is shown as a function of multiplicity. The limiting value, $p_{labbeam}/(A_{beam} + A_{target})$, that would be seen if all the particles were detected is shown for comparison.

is larger so that the TPC has better geometric acceptance, but isotope resolution is not as good for the highest momentum particles. Figure 6 shows $(M_{inv} - M_0)/A_{measured}$ for the Au + C measurement. This measurement of $(M_{inv} - M_0)/A_{measured}$ is 138% of the initial value defined by the beam and target. The high value of 138% may be due to isotope mis-identification. Including the TOF wall in the analysis will improve the isotope identification at high momentum and possibly resolve this problem. A summary of the various measurements, selected on high multiplicity, are given in Table 2 as a fraction of the totals available in the reaction. The quantity, n, gives the fraction of the total neutrons bound in the measured fragments.

	La + La 800 MeV/A	La + La 800 MeV/A (forward half)	Au + C 1000 MeV/A
	% detected	% detected	% detected
$M_{inv} - m_0$	37	62	82
p_{lab}	59	73	60
A	55	73	65
Z	85	109	87
n	36	48	50
$(M_{inv} - m_0)/A$	68	89	128
p_{lab}/A	107	98	91

Table 2: Summary of total event quantity measurements for the most central, highest multiplicity events. The "forward half" column shows results for events constructed from forward half particles in the CM plus their reflection to the back half. The quantity "n" is the number of bound neutrons in the detected isotopes.

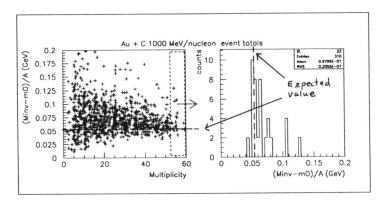

Fig. 6: The quantity, $(M_{inv} - M_0)/A_{measured}$, is shown as a function of multiplicity for the 1000 MeV*A Au + C system. The expected value, if all the energy is accounted for, is shown for comparison.

4. Conclusion

Measurements of total event quantities such as $(M_{inv} - M_0)/A_{measured}$ and $p_{lab}/A_{measured}$ show that the TPC provides very good coverage and could be useful for searching for exotic missing energy events. The $(M_{inv} - M_0)/A_{measured}$ analysis for different isotopes may be an interesting tool for studying the collective versus thermal expansion issue.

5. References

[1] D.Pelte, W. Reisdorf, T. Wienold , FOPI collaboration, *GSI-Nachrichten* 09-93 (1993).

FRAGMENTATION OF COLLIDING DROPS: SYMMETRIC SYSTEMS

A. MENCHACA-ROCHA, F. HUIDOBRO, K. MICHAELIAN
and
V. RODRIGUEZ
Instituto de Física, Universidad Nacional Autónoma de México
A.P. 20-364, 01000 México D.F., México

ABSTRACT

Collisions between pairs of $1g$ mercury drops as a function of impact parameter and relative velocity, in the range $20\ cm/s \le v_r \le 90\ cm/s$, have been studied experimentally. The aim was to illustrate some collective (rotational and vibrational) aspects of the atomic nucleus as well as to provide some new grounds to test the *bare* hydrodynamics of nuclear reaction models. Our measurements include velocity, mass, multiplicity and angular distributions of the residues.

1. Introduction

Hydrodynamical models are known to be of great help in understanding some of the collective aspects of nuclear phenomena. Under the assumption that the fluid dynamics involved is well understood, most of the effort in modeling nuclear interactions, hydrodynamically, has been theoretical. Since the particular characteristics of nuclear systems (quantum and finite size effects, diffuse potentials, etc.) prevent a direct comparison between nuclear-reaction and drop-collision observables, we are interested in comparing the underlying hydrodynamics of nuclear models with data obtained in macroscopic drop-collision experiments. An illustration of this approach may be found in Ref. 1, where we applied the Rotating-Liquid-Drop model of Blocki and Swiatecki [2] to drop-collision data.

Here we report on experiments aimed at understanding the general breakup processes of the temporarily coalesced system formed in the collision of two equal-size drops, as a function the relative velocity (v_r), and the impact parameter (b). These measurements include velocity, mass, multiplicity, and angular distributions of the drop-collision residues. Since the main interest in presenting these results at a nuclear physics workshop was to gain the interest nuclear hydrodynamicists, this contribution is mainly devoted to presenting data with only a few simple model calculations where the temptation was too much to bear.

2. Experiment

The measurements were carried out with the aid of a liquid-drop collider, the "gotatron", in which we observe the interactions of equal mass mercury drops sliding along a flat, horizontal, glass surface, specially treated [3] to minimize the drag induced by wetting.

Two drops, of mass $m_{1,2}$ (where $m_1 \approx m_2$), are "accelerate" to approximately equal and opposite, velocities $v_{1,2}$, with the aid of plastic ramps fixed on two extremes of the glass surface. A groove in each ramp surface guides the drops smoothly down the slopes into parallel trajectories separated by an impact parameter b. In this manner, the outcome of the drop collisions can be studied as a function of $m_{1,2}$, $|v_{1,2}|$ and b. Measurements are obtained by recording the action with a fast shutter (1/4000 s) video system having a 30 frame/s recording frequency. The final multiplicity is measured just by counting the number of residual droplets; however, since secondary scattering (leading to coalescence) among the primary fragments is not infrequent, a "primary" multiplicity can also be extracted by replaying the video images. The position vs time information, needed to determine the relative velocity v_r and b, as well as the speed $|v_i|$ and direction of motion θ_i of each residue, is also extracted from a frame-by-frame analysis of the video-recorded data. The drop initial $(m_{1,2})$ and final (m_i) masses are measured with a $0.1mg$ precision analytic scale. More information about our experimental techniques may be found in refs. 1 and 3.

3. Results

The data result from the analysis of 400 collisions, taken within the 20 $cm/s \leq v_r \leq$ 90 cm/s range. No systematic $v_r < 20$ cm/s measurements were done since that region is dominated by permanent coalescence (fusion). The maximum of $v_r =$ 90 cm/s corresponds to the operational upper limit of the gotatron [3]. For the sake of generality, the results will be presented in terms of dimensionless (hopefully scaling-) variables, such as the collisional Weber number [4] $W_e = \rho D v_r^2/\sigma$, where D is the drops' diameter, ρ is the density, and σ the surface tension of the liquid. This number is the ratio of the available kinetic energy to the energy gained by the change in surface area in coalescence (i.e., the fusion "Q-value").

The distribution of normalized impact parameters (b/D) and W_e values covered in the experiments is shown in Figs. 1a-1f, where they are classified by the primary multiplicity of the collision. The solid curve in Fig. 1a represents the limiting impact parameter for symmetric fission, $b_c/D = 2.51$ $W_e^{-1/2}$, which results from equating the rotational energy to the above mentioned "Q-value" [5]. As can be observed, one characteristic of our data is a direct correlation between impact parameter and final multiplicity. The increase in the number of residues for the more peripheral interactions is associated to a corresponding increase in the length of the neck formed. Hence, within the W_e regime studied here, the most probable outcome corresponds to interactions having two intermediate mass residues plus a number of smaller drops.

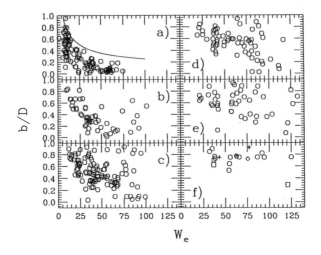

Fig. 1: Impact parameter vs Weber number, for primary multiplicities 1 to 5, in (a) to (e), respectively. In (f) the circles, squares, stars and diamonds represent mult=6,7,8 and 9, respectively. The solid curve in (a) represents the b_c/D rotational limit for symmetric fission (see text).

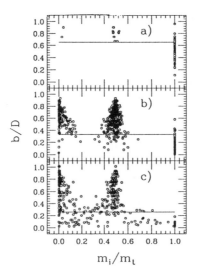

Fig. 2: Normalized mass of the residues as a function of the normalized impact parameter for three W_e number regions: $W_e = 0 - 15$ ($v_r = 0 - 30$ cm/s) in (a); $W_e = 15 - 60$ ($v_r = 30 - 60$ cm/s) in (b); and $W_e = 60 - 130$ ($v_r = 60 - 90$ cm/s) in (c). The horizontal lines represent the lowest b_c/D symmetric fission limit of the corresponding W_e region.

Beyond multiplicity 3, the largest number of the drops produced have small masses (≤ 0.2 of the total, which we shall call "tiny" drops).

Figures 2a-2c show how the mass m_i of the residues (normalized to the total mass of the system $m_t = m_1 + m_2$) vary with impact parameter for three W_e regions (see caption). As can be appreciated in Fig. 2a, at the lowest W_e coalescence dominates at all but the highest b/D values, where rotational limitations lead to final states characterized by two intermediate-mass-fragments ($m/m_t \approx 0.5$) and an occasional tiny drop originating in the neck region. At the intermediate W_e values (Fig. 2b), coalescence is reduced to the $b/D \leq 0.4$ region, above which the two-body final states predominate. Note how the width of the intermediate-mass-fragment group increases for smaller b/D values. At the highest W_e, an interesting change occurs for impact parameters below $b/D \approx 0.4$, where coelescence seems to give way to a breakup mechanism characterized by a broad mass distribution having no memory of the incident channel's mass. The video images reveal that these asymmetric final states (residues of a variety of sizes) often result from fairly symmetric three-lobed string configurations, in which the symmetry is lost when only one of the two intermediate necks breaks, leaving a large drop, a smaller drop and, occasionally, one (or a few) tiny residues from the broken neck. Incidentally, this mechanism is similar to one proposed by Nix to explain asymmetric fission [6].

In Figs. 2a-2c the b_c/D predictions of Fig. 1a, have been indicated by horizontal lines. It is also interesting to note that "inclomplete fusion", a mechanism of low energy heavy ion reactions, also produces asymmetric final states. However, since in the region under the b_c/D limit the systems are expected to be rotationally stable, unlike the heavy ion reaction case, the asymmetric breakup we observe seems to result from vibrational rather than rotational instabilities.

The fractional kinetic energy loss E_l (initial minus final, normalized to initial, kinetic energies) final extracted from our data shows (Figs. 3a-3c) a definitive correlation with the impact parameter throughout the whole range of W_e values explored. Since a fraction of the energy lost corresponds to a surface energy change, in Figs. 3d-3f we have subtracted that contribution from the corresponding data on the right-hand side. This "Q-value" correction has the effect of increasing the E_l vs b/D correlation, particularly in the lowest W_e windows, where the surface energy change is important compared to the available kinetic energy. It is interesting to note that the shape of the E_l vs b/D correlation is similar to that of the geometrical overlap between the cross sectional areas of the incident drops, represented by the solid curve in Fig. 3f.

We would like to conclude this review of our results with a picture which should be more familiar to nuclear physicists, given in figs. 4a-4c. By selecting on final multiplicities ≥ 2 these figures plot the speed of the residues, normalized to the average speed of the incident drops (roughly $v_r/2$), as a function of their deflection angle, for the same three W_e windows used previously. This modified Wilczynski plot shows a transition from an isotropic to a foward peaked distribution which is reminiscent of the evolution of heavy ion reactions with incident energy.

272

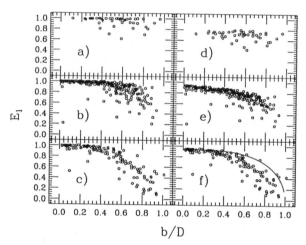

Fig. 3: Fractional energy loss vs normalized impact parameter for the same three W_e regions (same order) of Fig. 2. In (a)-(c), the data is presented without (and in (d)-(f) with) surface energy "Q-value" correction. The solid curve in (f) shows the dependence of the (fractional) geometric cross-sectional overlap of the incident drops with b/D.

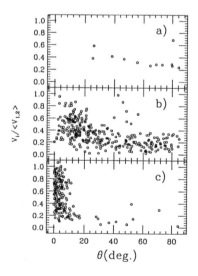

Fig. 4: Residue speed (normalized to the average speed of the incident residues) as a function of its deflection angle, for the same W_e regions defined (and shown in the same order as) in Fig. 2.

4. Conclusion

Collisions of $1g$ mercury-drop pairs have been studied as a function of impact parameter, in a relative-velocity range going from a coalescence-dominated region to interactions yielding up to 9 residues, the highest multiplicities corresponding to large impact parameter interactions. At the highest relative velocities, as the impact parameter is reduced, a marked transition appears between a narrow and a broad mass distribution. Simple calculations suggest that this transition corresponds to the onset of vibrational instabilities. The energy loss is found to be a well defined function of the impact parameter, a correlation which is further improved when surface-energy "Q-value" corrections are taken into account. Finally we show how some of the gross features of our data, such as the angular dependence of the residues' velocity, are reminiscent of low energy heavy ion reactions. We hope that these results motivate nuclear hydrodynamicists to strip their models of the nuclear details and try them on our data.

5. References

[1] A. Menchaca-Rocha, A. Cuevas, M. Chapa and M. Silva, *Phys. Rev.* **E47**, 1433 (1993); A. Cuevas, M. Chapa, M. Silva and A. Menchaca-Rocha, *Rev. Mex. Fis.* **39**, 428 (1993).

[2] J. Blocki and W.J. Swiatecki, L.B.L. preprint 12811-UC-34d, May 1982, unpublished.

[3] A. Menchaca-Rocha, *J. Coll. and Int. Sc.* **114**, 472 (1992).

[4] H.R. Pruppacher and J.D. Klett, *Microphysics of Clouds and Precipitation*, D. Reidel Publ. Co. Dordrect, Holland, 1978.; and references therein.

[5] A. Menchaca-Rocha, Disruption of Colliding Drops, *Proc. Workshop on Fragmentation Phenomena*, Les Houches, France, 1993, Nova Publ. Co., in press.

[6] J.R. Nix, *Nucl. Phys.* **A130**, 241 (1969).

THE DISAPPEARANCE OF FLOW

G. D. WESTFALL for the 4π Collaboration

National Superconducting Cyclotron Laboratory
and Department of Physics and Astronomy
Michigan State University, East Lansing, MI 48824-1321, USA

ABSTRACT

The disappearance of flow has been measured for five projectile/target combinations, C+C, Ne+Al, Ar+Sc, Ar+V, and Kr+Nb. The measured balance energies scale approximately as $A^{-1/3}$. BUU calculations under predict the observed balance energies and are not sensitive to the incompressibility of nuclear matter, K. A density dependent reduction of the nucleon-nucleon cross sections of about 20% provides qualitative agreement with the observed balance energies.

1. Collective Flow

When two nuclei collide at sufficiently high energy, an interaction zone is formed between the two nuclei in which hot, compressed nuclear matter is created. By studying the formation and decay of this hot matter, information about the nuclear equation of state, EOS, can be extracted. One direct measure of the formation of the interaction region is the study of collective flow. Collective flow is defined as the non-thermal transport of mass, energy, and momentum. The extent to which nuclear matter can absorb energy through compression will influence greatly the emission patterns of the measured particles.

1.1. Extraction of Flow

Flow is usually defined in terms of a reaction plane which includes the beam vector and the impact parameter vector. The impact parameter vector (chosen to be in the x direction) and the beam velocity vector (taken to be the z direction) define a reaction plane for the collision. We expect collective phenomena to occur in this plane. The method of Danielewicz and Odyniec[1] is the standard method of determining the reaction plane. The method involves the transverse momentum, $\vec{p}_\perp$. If collective flow exists, then the vector $\vec{Q}$ will lie in the reaction plane where

$$\vec{Q} = \sum_i \omega_i \vec{p}_\perp^i \tag{1}$$

and ω_i is a weighting factor usually defined as +1 for particle going forward in the center of mass frame and -1 for particles going in the backward direction. The vector points in the direction of the projectile. Each transverse momentum vector is then projected on the reaction plane taking care to exclude the particle of interest when $\vec{Q}$ is determined to avoid undesirable auto correlation effects.

The average transverse momentum per nucleon projected on the reaction plane, $< p_x >$, is then determined as a function of a variable in the direction parallel to the beam such as rapidity. The distribution of $< p_x >$ produces a characteristic S shaped curve centered around the center of mass rapidity. Collective flow is defined as the slope of this distribution at midrapidity. More recently another method has been developed to determine the reaction plane which is more suitable for lower energy reactions. This method is called the azimuthal correlation method.[2]

An alternative way presenting flow results is to divide p_x by the total transverse momentum, $p_\perp$. This method has the advantage of minimizing distortions due to detector acceptance and directly provides information concerning the fraction of the total transverse momentum that can be identified with collective flow. An example of this type of analysis is shown in Figure 1 for reactions of 55 MeV/nucleon C+C and 35 MeV/nucleon Kr+Nb[3] The slope of this curve at midrapidity is defined to be the reduced flow.

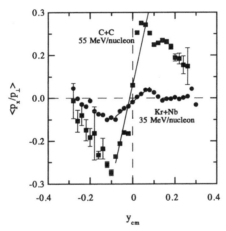

Fig. 1: Average transverse momentum in the reaction plane divided by the total transverse momentum for C+C and Kr+Nb. Solid lines represent the reduced flow.[3]

Much of the current understanding of the EOS comes from collective flow measurements at incident energies ranging from 150 to 1000 MeV/nucleon[4,5]. By comparing these results for collective flow with transport theories such as the Boltzmann-Uehling-Uhlenbeck (BUU) model, information such as the incompressibility of nuclear matter, K, was inferred. The BUU model is a microscopic theory that incorporates the EOS through the nuclear mean field and explicitly treats nucleon-nucleon collisions using a Monte Carlo technique.[6] Comparisons between BUU and flow results indicate that the most likely value of K is around 215 MeV.[7]

2. The Disappearance of Flow

At lower incident energies, additional information about the EOS can be extracted by studying the disappearance of collective flow.[3,8,9,10] In the reactions discussed above, the dominant interaction between the two nuclei is repulsive. At incident energies around 10 MeV/nucleon, the interaction between the two nuclei is dominantly attractive. The incident energy at which the attractive interaction balances the repulsive interaction is called the balance energy, E_{bal}.[9]

2.1. Extraction of Balance Energy

In Figure 2, the reduced flow is plotted for protons from Ar+Sc at incident energies ranging from 35 to 115 MeV/nucleon. The reduced flow decreases with increasing incident energy up to about 85 MeV/nucleon after which the reduced flow begins to increase again. The incident energy at which the reduced flow disappears, E_{bal}, is readily apparent in this figure.

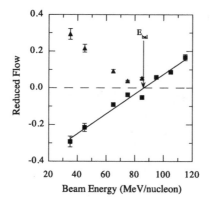

Fig. 2: Reduced flow values for protons from Ar+Sc. The squares represent reduced flow values that were inverted below the balance energy as described in the text[3]

We associate the reduced flow values at incident energies below E_{bal} with attractive scattering and flow values above E_{bal} with repulsive scattering according to expectations from BUU calculations. To extract a quantitative value for E_{bal}, we employ the method of inverting the flow values at low energies. A straight line is then fit to the resulting points as shown in Figure 2. This method depends on the choice of which points to flip. The systematic error caused by the ambiguity in flipping the reduced flow values is included in the systematic error for the extracted E_{bal}.

2.2. Independence of Fragment Type

In Figure 3, the independence of the balance energy on particle type is demonstrated for reduced flow values for p,d,t, and He fragments from Ar+Sc. The method described above was employed and one can see that the resulting values of E_{bal} are nearly independent of the particle type.

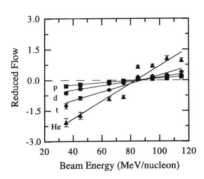

Fig. 3: Reduced flow values for p, d, t, and He fragments from Ar+Sc. The lower energy points are inverted as described in the text.[3]

This observable has the advantage that a comparison can be made directly with BUU calculations without the need of filtering the calculations through the acceptance of the experimental device. In BUU, the attractive component is incorporated through the mean field which contains information about the EOS. The repulsive component is represented by nucleon-nucleon scattering. The mass dependence of E_{bal} may provide information about nuclear matter because of the delicate interplay between the mean field and nucleon-nucleon scattering.

2.3. Comparison with BUU

In Figure 4, the balance energies for five systems are shown along with BUU calculations for soft and stiff EOS.

It is clear from Figure 4 that BUU underpredicts the observed balance energies and that K does not make a dramatic difference in the predictions. However, the BUU calculations do reproduce the overall trend of the measured balance energies with the mass of the system. This agreement affirms the interpretation of the balance energy as resulting from competition between the attractive mean field interaction and the repulsive nucleon-nucleon scattering.

One way to raise the predicted balance energies is to reduce the nucleon-nucleon scattering cross sections in BUU. These cross sections may be reduced from the free

278

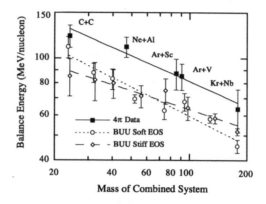

Fig. 4: Measured balance energies as a function of the mass of the combined projectile-target system compared to BUU calculations with a soft and a stiff equation of state.[3]

space values in a dense nuclear medium. To demonstrate this effect, the free space cross sections were reduced in a density dependent manner and the resulting BUU calculations are shown in Figure 5. In this figure are shown the experimental result for the balance energy as a function of mass along with the BUU predictions for a soft EOS, a soft EOS with a 10% density dependent reduction of the nucleon-nucleon cross sections, and a soft EOS with a 20% density dependent reduction.

The BUU calculations with the 20% density dependent reduction of the cross sections qualitatively reproduce the observed balance energies. However, the mass dependence is not completely understood in terms of BUU. To understand the behavior of nuclear matter under these conditions, one needs to complete the systematics of the balance energy measurements. In addition, the impact parameter dependence of the balance energy has not been addressed here.

2.4. Azimuthal Correlations

A promising new set of observables is multiparticle azimuthal correlations that can be used to study emission lifetimes and collective flow.[11] One difficulty with the disappearance of collective flow measurements is that the resulting errors are relatively large. The determination of E_{bal} using multiparticle azimuthal correlations does not require reaction plane determination and promises to allow a much better determination of variables of the EOS for heavier systems.

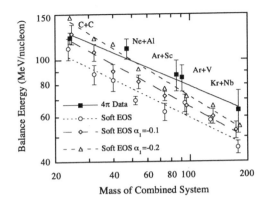

Fig. 5: Measured balance energies as a function of the mass of the combined projectile-target system compared to BUU calculations with a soft equation of state using the free nucleon-nucleon cross sections and calculations with a 10% and 20% density dependent reduction of the cross sections.[3]

3. References

[1] P. Danielewicz and G. Odyniec, Phys. Lett. B157, 146 (1985).

[2] W.K. Wilson, R. Lacey, C.A. Ogilvie, and G.D. Westfall, Phys. Rev. C45, 738 (1992).

[3] G.D. Westfall, W. Bauer, D. Craig, M. Cronqvist, E. Gualtieri, S. Hannuschke, D. Klakow, T. Li, T. Reposeur, A.M. Vander Molen, W.K. Wilson, J.S. Winfield, Y. Yee, S.J. Yennello, R. Lacey, A. Elmaani, J. Lauret, A. Nadasen, and E. Norbeck, Phys. Rev. Lett. 71, 1986 (1993).

[4] H.H. Gutbrod, A.M. Poskanzer, and H.G. Ritter, Rep. Prog. Phys. 52, 1267 (1989).

[5] P. Danielewicz, H. Ströbele, G. Odyniec, D. Bangert, R. Bock, R. Brockmann, J.W. Harris, H.G. Pugh, W. Rauch, R.E. Renfordt, A. Sandoval, D. Schall, L.S. Schroeder and R. Stock, Phys. Rev. C38, 120 (1988).

[6] W. Bauer, Phys. Rev. Lett. 61, 2534 (1988) and W. Bauer, C.K. Gelbke, and S. Pratt, Ann. Rev. Nucl. Part. Sci. 42, 77 (1992).

[7] C. Gale, G. M. Welke, M. Prakash, S. J. Lee and S. Das Gupta, Phys. Rev. C41, 1545 (1990).

[8] J.P. Sullivan, J. Péter, D. Cussol, G. Bizard, R. Brou, M. Louvel, J.P. Patry, R. Regimbart, J.C. Steckmeyer, B. Tamain, E. Crema, H. Doubre, K. Hagel, G.M. Jin, A. Péghaire, F. Saint-Laurent, Y. Cassagnou, R. Lebrun, E. Rosato, R. Macgrath, S.C. Jeong, S.M. Lee, Y. Nagashima, T. Nakagawa, M. Ogihara, J. Kasagi, and T. Motobayashi, Phys. Lett. B249, 8 (1990).

[9] C.A. Ogilvie, W. Bauer, D.A. Cebra, J. Clayton, S. Howden, J. Karn, A. Nadasen, A. Vander Molen, G.D. Westfall, W.K. Wilson, and J.S. Winfield, Phys. Rev. **C42**, R10 (1990).

[10] D. Krofcheck, D.A. Cebra, M. Cronqvist, R. Lacey, T. Li, C.A. Ogilvie, A. Vander Molen, K. Tyson, G.D. Westfall, W.K. Wilson, J.S. Winfield, A. Nadasen, and E. Norbeck, Phys. Rev. **C43**, 350 (1991).

[11] E. Bauge, A. Elmaani, Roy A. Lacey, J. Lauret, N.N. Ajitanand, D. Craig, M. Cronqvist, E. Gualtieri, S. Hannuschke, T. Li, B. Llope, T. Reposeur, A. Vander Molen, G.D. Westfall, J.S. Winfield, J. Yee, S. Yennello, A. Nadasen, R.S. Tickle, and E. Norbeck, Phys. Rev. Lett. **70**, 3705 (1993).

MULTIFRAGMENTATION IN PERIPHERAL COLLISIONS OF HEAVY NUCLEI

U.LYNEN

Gesellschaft für Schwerionenforschung, D-64220 Darmstadt, Germany
for the ALADIN and ALADIN/MINIBALL collaborations

Univ. Catania, GSI Darmstadt, Univ. Frankfurt, Univ. Milano, FZ Rossendorf,
SINS Warszawa,
and
MSU East Lansing, WU St.Louis

ABSTRACT

Using an Au-beam of 600 MeV/u of the SIS accelerator at GSI multifragmentation of the outgoing projectile spectator was investigated with the ALADIN-spectrometer. A remarkable absence of preequilibrium effects as compared to similar reactions at lower energies was found. The results can be reproduced by statistical model calculations.

For an investigation of the symmetric system Au + Au at energies between 100 and 400 MeV/u close to 4π coverage was achieved by adding the MSU-MINIBALL to the experimental setup. With increasing energy the maximum of the fragment multiplicity was found to shift from central to peripheral collisions and at the same time from the participant to the spectator region. The observed differences in the fragment distributions indicate the importance of dynamical contributions to the mechanism of fragment production.

1. Introduction

The disassembly of a nucleus into several fragments, the so called multifragmentation, is a sensitive probe for the dynamics of heavy-ion collisions at intermediate and high energies. Beyond this general interest it is expected that the investigation of multifragmentation might reveal signatures for the predicted liquid-gas phase transition in nuclear matter [1].

Initially it was expected that an initial compression of a nucleus is necessary in order to drive the system into the spinodal region so that multifragmentation sets in. The experimental investigation, therefore, concentrated on central collisions between symmetric projectile-target combinations. The first experiments with the ALADIN spectrometer at GSI, however, have shown that at incident energies of several hundred MeV per nucleon also in peripheral collisions, resp. with asymmetric projectile-target combinations an equilibrated, highly excited nucleus can be formed which decays via multifragmentation without initial compression [2,3]. Whereas in central collisions the interaction region is the source decaying via multifragmentation, in the latter case it is the spectator part of the collision. Since collective momentum components, e.g. radial flow, are expected to be very different in the two cases a comparison of the two reactions will allow to investigate the influence of the expansion dynamics on fragment formation.

In the first part of this report results obtained for multifragmentation of the projectile spectator at an incident energy of 600 MeV/u will be summarized. In the second part first results obtained for the reaction Au + Au at energies between 100 and 400 MeV/u will be presented. In the latter experiment the ALADIN spectrometer was used in combination with the MSU-MINIBALL.

2. Equilibration in Peripheral Collisions

Events which showed the typical pattern of a peripheral collision and where the projectile remnant was breaking up into several fragments of intermediate masses have first been seen in emulsion experiments [4]. It was, however, not clear whether the production of these fragments could be attributed to a well defined, equilibrated source or whether it was dominated by preequilibrium effects, as was found for asymmetric systems at incident energies around 100 MeV/u [5]. The first experiments performed with the ALADIN spectrometer at GSI have shown that the spectator remnants emerging from a collision at 600 MeV/u form highly equilibrated sources. This equilibration is apparent in several observables, e.g.:

- The velocity distributions of all fragments heavier than Li are symmetric around the beam velocity, indicating that their F/B-ratio in the spectator system is equal to one. Only for the lightest fragments, Li and He, some enhancement towards midrapidity is observed [6].
- For all fragments with $Z{\geq}8$ and independent of the impact parameter the widths of the velocity distributions in transverse direction are equal to those in longitudinal direction as shown in the lower part of fig.1. At present no statement can be made for lighter fragments since for them tracking in the MUSIC-detector has been achieved only recently.
- The observed multiplicities of intermediate mass fragments (IMF) plotted as a function of Z_{bound} are independent of the target as shown in fig.2, although the relation between Z_{bound} and impact parameter is strongly target dependent so that central collisions with a light target are to be compared with peripheral collisions with a heavy target [2].

Here and for most of the discussion the results will be shown as a function of Z_{bound}, where Z_{bound} is defined as the sum of all fragments with charges $Z{\geq}2$ and with velocities close to that of the incident beam. Compared to the multiplicity of light charged particles which is generally used to determine the impact parameter, Z_{bound}

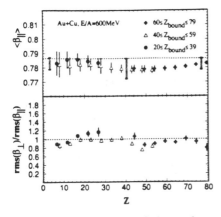

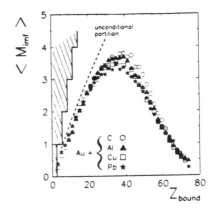

Fig. 1: Top: Mean fragment velocity as a function of fragment charge. The different symbols correspond to different gates in Z_{bound} resp. different impact parameters. The horizontal line marks the beam velocity and the hatched bars indicate systematic uncertainties in the walk correction.

Bottom: Ratio of the rms velocities in transverse and longitudinal direction in the source system.

Fig. 2: Correlation between the mean IMF multiplicity and Z_{bound} for reactions of Au on different targets at an incident energy of 600 MeV/u. The hatched area marks the region which is excluded due to the lower limit of the IMF charge of $Z=3$. The dashed line is the result of an unconditional Z partition of finite nuclear systems with charges $Z<40$.

has the advantage that it is deduced from the projectile residue alone and thus can be used if results from different targets are to be compared.

A possible explanation for the high degree of equilibration observed for the decay of the projectile spectator can be found in the enhanced forward peaking of N-N scattering at high energies [7], due to which predominantly the rather low-energetic recoil nucleons will penetrate into the spectator regions and deposit their energy there. Whereas at lower energies the deposited energy is correlated with the *linear* momentum transfer, at these energies the *transverse* momentum transfer is of importance. Therefore the mean velocity of the fragments is independent of the impact parameter, resp. the violence of the collision as shown in the upper part of fig.1. The reduced importance of the linear momentum transfer would also explain why at even higher energies the signatures and cross sections for spectator break-up remain nearly unchanged [8].

3. Charge Correlations

In addition to the IMF-multiplicity discussed in the previous section several other quantities that can be derived from the charges of the produced fragments have been investigated as a function of Z_{bound} [9]. Among them were:

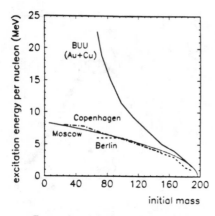

Fig. 3: Excitation energy per nucleon of the decaying spectator nucleus as a function of its mass. The upper solid line represents the deposited energies predicted by BUU simulations. The solid, dashed and dotted lines are the theoretical inputs required by the Moscow [10], Berlin [11] and Copenhagen [12] multifragmentation codes in order to reproduce best the observed charge distributions.

- Z_{max}, the charge of the largest fragment.
- The shape of the fragment distribution characterized by the exponent τ of a power law $\sigma(Z)=Z^{-\tau}$.
- Correlations between the largest and second-largest, as well as between the second- and third-largest fragments.
- Fluctuations of the fragment charges normalized to the mean charge as given by $\gamma_2 = \frac{\sigma_Z^2}{\langle Z \rangle^2} + 1$.

All above quantities showed the same target-independence as was observed for $< M_{IMF} >$. All observed correlations can be reproduced as well by percolation [9] as by statistical model calculations. In the statistical model calculations the experimental results could not be reproduced if excitation energies determined from BUU calculations were used as input [9]. Choosing, however, considerably lower values, as shown in fig.3, a remarkable agreement with the data could be obtained for different model calculations [10,11,12]. The differences as compared to the BUU results can qualitatively be explained by the emission of light particles prior to the formation of the fragments, as e.g. predicted by Friedman [13] and Papp et al. [14].

Whereas the multiplicity of fragments and their charges were found to be very sensitive probes for the excitation energy of the decaying system, the measurement of higher order correlations between several fragments did not provide additional information.

4. Au + Au at 100 and 400 MeV/u

In order to compare multifragmentation in central and multifragmentation in peripheral collisisons we have investigated the system Au + Au at incident energies of 100 and 400 MeV/u. An energy around 100 MeV/u has been chosen since here the maximum fragment multiplicity in central collisions is expected. The energy of 400 MeV/u, chosen for the investigation of the peripheral collision, is rather uncritical,

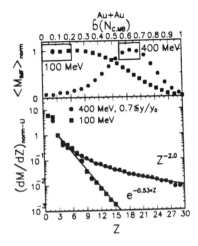

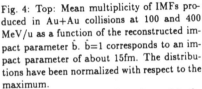

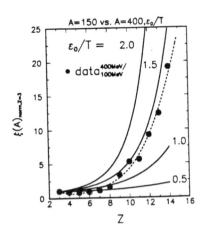

Fig. 4: Top: Mean multiplicity of IMFs produced in Au+Au collisions at 100 and 400 MeV/u as a function of the reconstructed impact parameter b̂. b̂=1 corresponds to an impact parameter of about 15fm. The distributions have been normalized with respect to the maximum.
Bottom: Charge distributions observed in the regions of maximum IMF multiplicity (indicated by the rectangles in the upper part of the figure).

Fig. 5: Suppression of heavy IMFs due to radial flow within a simple coalescence picture. The dots represent the ratio between the measured charge distributions at 400 and 100 MeV/u respectively. The full lines are the result of a simple (analytic) coalescence model. The dashed line is meant to guide the eye.

since in this case the energy dependence over a wide range of incident energies is only weak.

The mean multiplicity of IMFs normalized to its maximum value is shown in the upper part of fig.4 as a function of the reconstructed impact parameter b̂. At 100 MeV/u the maximum mean fragment multiplicity $\langle M_{IMF}\rangle \approx 10$ is observed in central collisions and the majority of fragments originate from the interaction region. At 400 MeV/u, on the other hand, the maximum of the fragment multiplicity is found in peripheral collisions and the main sources are the two spectator nuclei. Since a large fraction of fragments from the target spectator is below the detection threshold of the detector system the *observed* mean fragment multiplicity $\langle M_{IMF}\rangle$ amounts to only 5. If the multiplicities are, however, corrected for the detection efficiency and normalized to the size of the decaying system (2×197 nucleons in central collisions and approximately 2×150 spectator-nucleons in peripheral events) then in both cases nearly the same probability of about *one* IMF per 33 nucleons is obtained.

Despite this similarity, the element distributions are significantly different in central and peripheral collisions as shown in the lower part of fig.4. The yields have been normalized at Z=3 and the used ranges of impact parameters are indicated by the rectangles in the upper part of the figure. Whereas in peripheral collisions the typical power-law behavior with an exponent $\tau=2$ is observed, in central collisions the yield of heavier fragments is strongly reduced and an exponential charge distribution is found. The ratio between these two distributions is shown in fig.5 as a function of the fragment charge.

A spectral change of the fragment distribution from a power law to a more exponential decay has already been suggested [1] for the case that the excitation energy of a decaying nuclear system is once close to the critical point and once further away from it. Compared to the excitation energies shown in fig.3 for peripheral collisions the cm-energy available in central collisions at 100 MeV/u is indeed much higher. A large fraction of this energy is, however, not thermalized during the collision and survives as radial flow. An indication that the excitation energies might not be too different in both reactions comes from the observation that the normalized probabilities of fragment formation are nearly equal.

A quantity in which the two experiments differ considerably is the radial flow. Whereas for peripheral collisions values below 1 MeV per nucleon have been determined [15], in central collisions values above 10 MeV are found [16,17]. We, therefore, try to relate the suppression of fragment formation in central collisions to the reduced nucleon density in momentum space caused by the additional radial flow. Within a coalescence picture [17] which at the same time describes the stability of a nuclear ensemble with respect to a subsequent decay and under the assumption of a fixed freeze-out temperature, the probability that a fragment of mass A is formed in central collisions is reduced by the factor

$$f_A = \left[\frac{p_{rms}^{thermal}}{\sqrt{(p_{rms}^{thermal})^2 + (p_{rms}^{flow})^2}} \right]^{3 \cdot A}$$

The lines in fig.5 show the results of these calculations for various ratios ϵ_0/T, where T is the temperature and ϵ_0 denotes the maximum flow energy per nucleon. A satisfactory description of the measured intensity ratios is obtained for a value $\epsilon_0/T=1.4$, which is in good agreement with the measured flow energy $\langle \epsilon \rangle = \frac{3}{5}\epsilon_0 = 10$ MeV and the above assumption that the temperatures in peripheral and central collisions are similar.

5. Conclusions

At incident energies of several hundred MeV per nucleon heavy spectator nuclei emerging from peripheral collisions form highly equilibrated sources. Depending on the impact parameter the excitation energy and thus the decay features of the spectator nuclei can be varied over a wide range. Although the maximum excitation energies are limited to values around 8 MeV per nucleon, this covers the region where multifragmentation is dominant and where the largest number of IMFs is formed. The multiplicity of fragments, the charge distributions and the correlations between the charges of different fragments are sensitive probes for the excitation energies at the time of fragment formation. The absence of radial flow in peripheral collisions results in comparatively long expansion times, so that these reactions provide favorable conditions for the search for the liquid-gas phase transition.

Increasing the incident energy from 100 MeV/u to 400 MeV/u for the symmetric system Au + Au the peak of the fragment multiplicity was found to shift from central to peripheral collisions and at the same time the source from which the majority of fragments originated changed from the interaction region to the two spectator nuclei. Despite this difference the number of IMF normalized to the size of the decaying system is nearly the same in both reactions and amounts to about *one* IMF per 33 nucleons. The element distributions, on the other hand, are significantly different and change from an exponential shape at 100 MeV/u to a power-law behavior at 400 MeV/u. A consistent description of these differences could be obtained by relating the probability of fragment formation to the nucleon density in momentum space, which is reduced for central collisions at 100 MeV/u because of the high value of radial flow. In order to describe the charge distributions resulting from multifragmentation within statistical decay models a detailed treatment of the expansion dynamics will therefore be necessary.

6. References

[1] P.J. Siemens, Nature (London) **305**, 410 (1983).
[2] C.A. Ogilvie et al., Phys. Rev. Lett. **67**, 1214 (1991).
[3] J. Hubele et al., Phys. Rev. C **46**, R1577 (1992).
[4] C.J. Waddington, P.S. Freier, Phys. Rev. C **31**, 888 (1985).
[5] R. Trockel et al., Phys. Rev. Lett. **59**, 2844 (1987).
[6] J. Hubele et al., Z. Phys. A **340**, 263 (1991).
[7] J. Cugnon, T. Mizutani und J. Vandermeulen, Nucl. Phys. A **352**, 505 (1981).
[8] W.Trautmann, Proceedings International Workshop XXII, Hirschegg 1994, to be published.
[9] P.Kreutz et al., Nucl. Phys. A **556**,672 (1993).
[10] A.S. Botvina and I.N. Mishustin, Phys. Lett. B **294**, 23 (1992).
[11] Bao-An Li, A.R. DeAngelis, D.H.E. Gross, Phys. Lett. B **303**, 225 (1993).
[12] H.W. Barz et al., Nucl. Phys. **A561**, 466 (1993).
[13] W.A. Friedman, Phys.Rev.Lett. **60**, 2125 (1988).
[14] Y.Papp and W.Nörenberg, GSI-Nachrichten GSI-02-94.
[15] V. Lindenstruth, Ph.D. thesis Univ. Frankfurt, GSI-Report 93-18 (1993).
[16] S.C. Jeong et al., GSI-93-38 Preprint (1993).
[17] G.J.Kunde, Ph.D. thesis Univ. Frankfurt, GSI-Report (1994).

FRAGMENT PRODUCTION AND COLLECTIVE EFFECTS IN CENTRAL AU ON AU COLLISIONS

T. WIENOLD for the FOPI - Collaboration [†]

Gesellschaft fuer Schwerionenforschung
64220 Darmstadt, Germany

ABSTRACT

We present data on central Au on Au collsions from 100 to 800 MeV/u measured with the phase 1 of the FOPI - detector at SIS/Darmstadt. The multiplicities of intermediate mass fragments (IMF: $Z > 2$) in the participant region and their flow is discussed. Besides the directed sideward flow a strong flow component connected to the expansion of the hot and dense participant region was observed.

1. Introduction

The formation of IMFs in the participant region is of special interest since the available energy per nucleon in this region is very high compared to the binding energy. A head on collsion of Au on Au at 400MeV/u for example provides an available energy of roughly $100MeV/u$ in the center of mass system. The major part of this energy might be shared between the thermal and collective degrees of freedom. As was shown by the Bevalac - experiments, clusters are a sensitive probe for collective effects like the directed sideward flow [1,2] in semi - central collisions. On the other hand the yield of clusters may serve as a probe for the (local) temperature at the time of freeze out. Therefore the combined study of cluster yields and their momentum distributions in central collisions should give some information about the degree of equilibrium and the collective expansion of the fireball.

2. Clusterproduction

The setup used in the experiment on Au on Au from 100 to 800MeV/u allowed the measurement of nuclear charge and velocity of the reaction products from 1 to 30 degree in the laboratory. Charge identification up to about $Z = 15$ was achieved by a ΔE versus time-of-flight (TOF) method. TOF and ΔE were measured with a highly granular and azimuthally symmetric Plastic Scintillator Wall. The ΔE of slow and heavy particles which were stopped in the Scintillator Wall was detected in an ionization chamber and a thin plastic scintillator array. For technical details see reference [3].

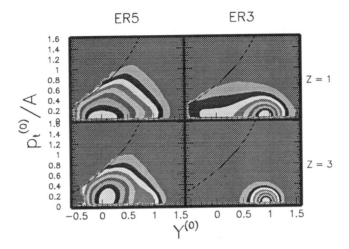

Fig. 1: Experimental transverse momentum versus rapidity distributions (invariant cross sections) at 600 MeV/u for central (left) and semi - central (right) collisions. The index (0) indicates a normalization to the projectile momentum and projectile rapidity.

In the following we sort the events according to a quantity ERAT [4] which is defined as the ratio of the transversal- to longitudinal kinetic energy in the forward hemisphere:

$$ERAT = \frac{\sum p_t^2/2m}{\sum p_z^2/2m}$$

Events with a large ERAT can be asigned to more central collision than those with smaller ERAT. In figure 1 we show the experimental rapidity distribution for Hydrogen and Lithium clusters at 600MeV/u. For the central cut ER5 which corresponds to the upper 200mb of the ERAT - distribution we observe a *midrapidity source* for both elements whereas for the semi - central cut ER3 ($0 < ERAT < 0.37$) a source near the projectile rapidity is visible. In both cases the distribution of the Lithium clusters is narrower than the one of Hydrogen. The production of IMFs in the participant region even at incident energies of 600MeV/u is surprising because of the high available energy in the cm - system. The IMF multiplicities extrapolated to 4π increase from about 4 at 600MeV/u to about 13 at 100MeV/u in this central eventclass [5,6]. An analysis of the cluster multiplicities within the QSM - model [7] indicates rather *low temperatures* for reasonable freeze out densities (between 7 and 12 MeV at $\rho/\rho_0 = 0.3$) [5].

The evolution of the cluster production with the centrality and the incident energy can be illustrated by the change in the slope of the charge distribution. As can be seen from figure 2 this slope is rising fast with the beam energy when central collisions are considered (cuts $ER5, D < 0.15$, $PM5, D < 0.15$, for explanation of cuts see also [12]). When intermediate multiplicities (PM3, PM4) are selected, the beam energy dependence is much weaker (open diamonds and squares).

290

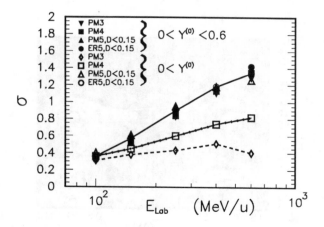

Fig. 2: Slope of the charge distribution as a function of the incident energy. The fit was performed with an exponential $N(Z) \propto e^{-\sigma z}$ for $Z = 3$ to $Z = 7$.

An additional cut on the mid rapidity zone (black symbols) which excludes the 'spectator' contribution gives for the intermediate centralities (PM3, PM4) the same result as for the high centrality cuts. This indicates a *stable chemical composition* of the participant region even when its size is varied considerably. From filtered calculations with the Quantum Molecular Dynamics model [9] we estimate the participant zone (clean cut geometry) to contain ≈ 380 nucleons in the eventclass $ER5, D < 0.15$ and ≈ 160 nucleons in the eventclass $PM3$.

3. Flow

In the Bevalac - experiments nuclear flow was observed in terms of the $< P_x/A >$ versus rapidity representation [10] and was found to increase with the mass of the fragments [1,2]. Recently this finding was qualitatively reproduced by our data [11,12]. Since the slope of the $< P_x/A >$ at midrapidity reflects only a 'flow signal' of the participants, we calculate the directed sideward flow in a complementary way for all nucleons (in one hemisphere). This should give a more quantitative measure of the *global amount of directed sideward flow* (i.e., one number per event). We calculate a quantity F_s (similar as in [10]):

$$ F_s = \frac{N_c}{N_c - 1} * \frac{\sum_i \sum_{j \neq i} | \vec{p}_{t(i)} | \, | \, \vec{p}_{t(j)} | \, cos\varphi_{ij}}{(\frac{P_{proj}}{A_{proj}} * \sum_i A_i)^2} $$

The indices i,j are running over the number of charged particles N_c measured in the forward hemishere. For the momenta we use the assumption $A = 2 * Z$. The normalization was chosen in a way that F_s is dimensionless. The mean of F_s corresponds to the fraction of the available energy which is contained in the sideward motion.

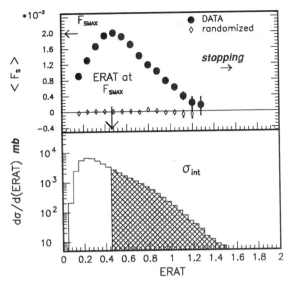

Fig. 3: Correlation of the directed sideward flow (via F_S) with the degree of 'stopping' and the corresponding 'stopping' - distribution at 150 MeV/u.

The evolution of this quantity with the 'degree of stopping' is shown in figure 3. It has a well pronounced maximum at intermediate values of ERAT (semi - central collisions) and vanishes almost completely for events with a high degree of stopping. From this correlation one can extract easily the maximum of the directed sideward flow (F_{SMAX}), the degree of stopping at the point of this maximum ($ERAT$ at F_{SMAX}) and the integrated cross section (σ_{int}), which possibly gives some information about the collision geometry at this distinct point. The excitation function of F_{SMAX} is compared to filtered QMD - calculations [9] in figure 4. We have rescaled F_S to get the absolute magnitude of the global directed sideward flow per nucleon in terms of momenta. This flow is increasing with the incident energy. The best overall agreement is reached using a hard equation of state with momentum dependence (version HM, $K = 380 MeV$) but the other two parametizations (H: hard EOS without momentum dependence, SM: soft EOS with momentum dependence) differ only by about 20 percent. Furthermore version H produces the same F_{SMAX} as SM! This ambiguity can be solved by comparing to the observables $ERAT$ at F_{SMAX} and σ_{int} (see reference [13]).

From this result one can estimate that the kinetic energy connected to the sideward flow represents only a few percent of the available energy in the cm - system. In figure 5 we make an attempt to extract the collective energy which is due to the collective expansion of the fireball. It shows the kinetic energy per nucleon as a function of the fragment mass for a central cut (here $ERAT > 0.67$) [14]. To estimate the amount of collective energy we compare the experimental result to a calculation of the statistical eventgenerator FREESCO [15].

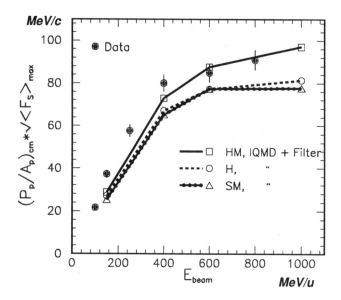

Fig. 4: Excitation function of the rescaled quantity F_{SMAX} in comparison to filtered QMD calculations.

The triangles in figure 5 correspond to the extreme case where all the available energy is converted into heat whereas the dots represent the situation where 16 MeV/u is contained in an isotropic flow with a flow profile increasing linearly from the center to the surface of the source. Obviously the data can't be explained with a simple thermal scenario and the calculation including the *blast* goes in the right direction. On the other hand a more detailed look to the experimental data has shown recently that the exprimental result depends still on the relative orientation to the reaction plane. In the direction of the reaction plane larger flow values are observed than opposite to it. Nevertheless it turned out from a comparison to QMD - calculations that the 'spectator' contribution cannot account for the major part of the flow energy. Also different approaches analyzing transverse momentum spectra [16,17,18] or using likelihood methods [19] which take care of the full experimental information including the detector limitations point towards a strong collective component in the expansion.

4. Conclusion

We have demonstrated that a midrapidity source of clusters exists even at higher bombarding energies. The large cluster multiplicities in the participant region are compatible with low temperatures when they are interpreted with statistical models. The analysis of the momentum distribution of clusters gives evidence for a strong collective expansion.

Au + Au at 150 AMeV

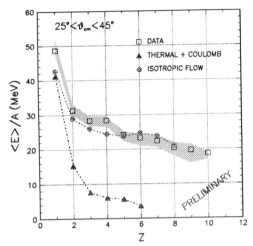

Fig. 5: The mean kinetic energy per nucleon as a function of the fragmentcharge. For comparison the result of a caluclation with the FREESCO - eventgenerator is given: i) for pure thermal spectrum and ii) a scenario with an isotropic expansion of 16MeV/u.

† FOPI - collaboration: IPNE Bucharest, CRIP Budapest, LPC Clermont-Ferrand, GSI Darmstadt, Univ. /INFN Florence, Univ.Heidelberg, Univ. Mainz, ITEP Moscow, IAE Moscow, ZfK Rossendorf, CRN Strasbourg, Univ. Warsaw, RBI Zagreb

5. References

[1] K.G.R.Doss et al., Phys. Rev. Lett. **57**, 302 (1986)

[2] K.G.R.Doss et al., Phys. Rev. Lett. **59**, 2720 (1987)

[3] A.Gobbi et al., Nucl. Inst. Meth. **A324**, 156 (1993)

[4] W.Reisdorf, Proceedings, Hirschegg 1993

[5] C.Kuhn et al., Phys. Rev. C. **48**, 302 (1993)

[6] T.Wienold, Phd thesis, University Heidelberg, GSI - Report **93-28** (1993)

[7] D.Hahn and H.Stöcker, Phys. Rev. C. **A476**, 718 (1988)

[8] J.P.Alard et al., Phys. Rev. Lett. **69**, 889 (1992)

[9] C.Hartnack, Phd thesis, University Frankfurt, GSI - Report **93-05**, (1993)

[10] P.Danielewicz et al., Phys. Lett. B. **157**, 146 (1985)

[11] F.Rami et al., GSI - Report 93 - 1, 40 (1992)

[12] J.P.Alard et al., GSI - Report 93 - 1, 39 (1992)

[13] T.Wienold, proceedings to the Nato Advanced Study Institute on HOT AND DENSE NU-CLEAR MATTER, Bodrum/Turkey (1993)

[14] S.C.Jeong et al., GSI - Preprint 93 - 38, (1993)

[15] J.Randrup, Comp. Phys. Comm. **77**, 153 (1993)

[16] T.Wienold, proceedings to the Mazurian Lake Summer School, Piaski/Poland (1991)

[17] T.Wienold et al., GSI - Report 93 - 1, 34 (1992)

[18] J.P.Coffin, proceedings to the Nato Advanced Study Institute on HOT AND DENSE NU-CLEAR MATTER, Bodrum/Turkey (1993)

[19] W.Reisdorf, proceedings, Hirschegg (1994)

MULTIFRAGMENTATION IN ^{3}He-INDUCED REACTIONS*

K. KWIATKOWSKI,[a] D. S. BRACKEN,[a] H. BREUER,[b] J. BRZYCHCZYK,[a]
E. RENSHAW FOXFORD,[a] W. A. FRIEDMAN[c] R. G. KORTELING,[d]
R. LEGRAIN,[e] K. B. MORLEY,[a] E. C. POLLACCO,[e] V. E. VIOLA,[a]
C. VOLANT[e] and N. R. YODER[a]

*Research supported by the U.S. Department of Energy and the National Science Foundation.

ABSTRACT

Multifragment emission data from the ^{3}He + ^{nat}Ag reaction are examined in the context of an intranuclear cascade code followed by an expanding, emitting source calculation. The Indiana Silicon Sphere 4π detector array is described and first results from ^{3}He-induced reactions at LNS are presented. The IMF multiplicity spectrum decreases monotonically from $M_{IMF} = 1$ for all systems.

Much of the early interest in the emission of intermediate mass fragments (IMFs) was stimulated by studies of proton-induced reactions (See Ref. 1 for a review of these data). More recently, inclusive studies of the ^{3}He + ^{nat}Ag system at energies up to 3.6 GeV have suggested a change in mechanism at bombarding energies in the vicinity of 1.8 GeV total energy.[2] Subsequent coincidence studies[3] with ^{3}He beams have demonstrated the existence of multifragmentation events for this system. In the present work, we first examine these latter results in the context of a model in which the energy dissipation stage is described by the intranuclear cascade code (INC)[4] and

[a]Indiana University, Bloomington, IN 47405
[b]University of Maryland, College Park, MD 20742
[c]University of Wisconsin, Madison, WI 53706
[d]Simon Fraser University, Burnaby, BC, Canada
[e]DAPNIA/SPhN, CEN Saclay, France

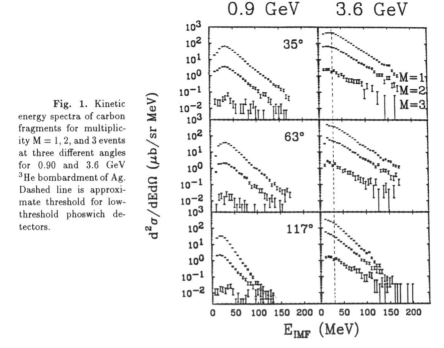

Fig. 1. Kinetic energy spectra of carbon fragments for multiplicity $M = 1, 2$, and 3 events at three different angles for 0.90 and 3.6 GeV ^{3}He bombardment of Ag. Dashed line is approximate threshold for low-threshold phoswich detectors.

the decay by an expanding, emitting source (EES) undergoing statistical decay.[5] Next, the Indiana Silicon Sphere 4π detector array is described, along with preliminary first results from bombardment of ^{nat}Ag and ^{197}Au targets with 1.8 - 4.8 GeV ^{3}He ions from the Saturne II accelerator at Saclay.

I. INC-EES Calculations

Both inclusive and exclusive (8% of 4π) studies of the ^{3}He + ^{nat}Ag and ^{197}Au reactions between 0.20 and 3.6 GeV at IUCF and Saturne II have previously been reported. The inclusive results[2] indicated a distinct change in reaction mechanism above and below a bombarding energy of about 1.8 GeV. The exclusive studies[3] revealed a strong increase in the probability of multifragment events at 3.6 GeV relative to energies below one GeV. One of the most striking aspects of these data is the evolution of the fragment kinetic energy distributions as a function of both bombarding energy and IMF multiplicity, shown in Fig. 1. The spectra become distinctly harder and the Coulomb peaks become broadened toward lower fragment energies as a function of both bombarding energy and IMF multiplicity. A rapidity-

3.6 GeV ^{3}He + ^{107}Ag

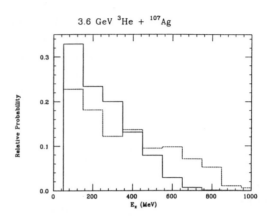

Fig. 2. Excitation energy distribution for collisions of 3.6 GeV ^{3}He ions with ^{nat}Ag. Dashed line includes delta formation cross sections. Solid line is the result when delta formation is suppressed.

like analysis for $Z \leq 6$ fragments indicates that the high-IMF-multiplicity events can be described by an isotropically-emitting source with an average linear velocity, $< v_s > \approx 0.4$ cm/ns, indicative of equilibrated emission. In contrast, for multiplicity $M_{IMF} = 1$ the source velocity increases linearly with IMF kinetic energy up to nearly twice this value.

Two features seem to be essential to explain these data. First, a rapid, efficient energy dissipation mechanism is required to convert projectile kinetic energy into internal excitation energy while at the same time producing on average a low source velocity; i.e., transverse momentum transfer must be large. For light-ion-induced reactions at energies up to a few GeV per nucleon, intranuclear cascade (INC) calculations[4] provide a reasonable approximation for calculating the reaction dynamics. However, since significant excitation due to density compression is unlikely with light-ion probes, an alternative rapid energy dissipation mechanism is required in order to form highly excited final states. Excitation of Δ resonances in central collisions, followed by subsequent rescattering and reabsorption of the decay pions, provides one such possibility—one that is inherent in the physics and contained in the INC codes.

In Fig. 2 we show the ISABEL prediction for the excitation-energy distribution of residual heavy nuclei formed in the 3.6 GeV ^{3}He + ^{nat}Ag reaction. Two calculations have been performed—one which includes cross sections for Δ formation and one in which this cross section is suppressed by a factor of ten. The relative importance of the Δ resonance is apparent in enhancing high excitation-energy events. For those events which deposit more than 50 MeV of excitation energy, inclusion of the Δ produces an excitation energy distribution in which about 33% of the events have excitation energies greater than 500 MeV, the approximate threshold for multifragmentation. Without Δ formation, only 12% reach 500 MeV or more, suggesting that Δ formation and decay strongly contribute to the fast equilibration process.[6]

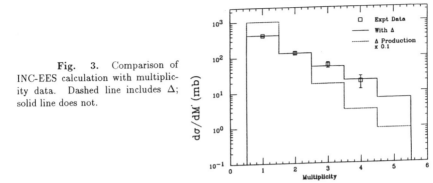

Fig. 3. Comparison of INC-EES calculation with multiplicity data. Dashed line includes Δ; solid line does not.

The second essential feature required to explain the IMF multiplicity data and fragment spectra appears to be expansion of the emitting source. This can be qualitatively understood in the context of the expanding-emitting source (EES) model.[5] In this model, multifragmentation is treated in terms of statistical emission from a hot, expanding nucleus. The expansion is viewed as a giant monopole oscillation driven by thermal pressure. In order to overcome the restoring nuclear force in this model, excitation energies in excess of 500 MeV (T $\gtrsim 12$ MeV) are required for hot residues formed in the ^{3}He + ^{nat}Ag system. Beyond this temperature the nucleus undergoes rapid disintegration, leading to multifragmentation.

In Fig. 3 we compare the combined INC-EES calculation[7] with intermediate-mass fragment (Z ≥ 3) multiplicities inferred from our data. The solid line represents a calculation with Δ formation included in the INC and the EES model with an equation of state having an effective compressibility parameter K = 144. The data have been normalized to multiplicity M = 2, which gives a reaction cross section, σ_R = 1900 mb, consistent with the expected value. The solid line in Fig. 3 shows the result for the INC-EES calculation without Δ formation cross sections, which severely underpredicts the data for large multiplicities and overpredicts the M = 1 yields. Calculations which include the Δ, but use larger values of K (which reduce the expansion rate) give similarly poor fits to the data. In order to describe the data with larger values of K, the probability for Δ excitation and/or pion absorption in the INC code would need to be increased significantly in order to enhance the probability for high excitation energy residues. The importance of employing a realistic excitation energy distribution in multifragmentation calculations is also stressed by this work. If the EES calculations are performed with the average INC excitation energy instead of the full distribution, the data are significantly underpredicted.

Comparison of the INC/EES calculations with experimental kinetic energy spectra provides qualitative agreement for the major features of the data. The calculation

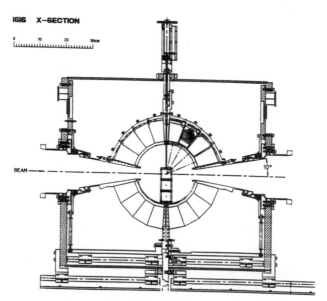

Fig. 4. Schematic design drawing of Indiana Silicon Sphere.

successfully predicts changes in IMF spectral shapes for bombarding energies below and above about 1 GeV, i.e., broadening of the Coulomb peaks toward lower energies and flattening of the spectral tails at high projectile energies, as observed in the data.[2] In the context of the EES model, this behavior is explained in terms of energetic IMFs being emitted early in the expansion, where the expansion velocity is large, and low energy IMFs being emitted from nuclear matter at low density. While the quantitative agreement with the spectra is not perfect, the qualitative features of the data are successfully described. It is also of interest to compare the charge distributions predicted by the INC/EES model with the experimental data. The calculation is quite sensitive to the value of K; a value K = 144 gives the best fit to the data. Thus, the results for the charge distributions are self- consistent with other observables. The INC predicts wide fluctuations in Z and A of the residual nucleus for a given excitation energy bin. Here we have used averages for each bin, rather than the full distribution. Taking this distribution into account on an event-by-event basis would further enhance the effective damping predicted by the code.

II. The Indiana Silicon Sphere 4π Detector Array

In order to provide a more quantitative understanding of energy dissipation and multifragmentation in these light-ion-induced reactions, we have recently completely and successfully implemented the Indiana Silicon Sphere 4π detector array (Fig. 4

shows a design sketch). The detector consists of 162 triple telescopes–90 in the forward hemisphere and 72 in the backward hemisphere–covering the angular ranges from 14°to 86.5°and 93.5°to 166°. The design consists of eight rings, each composed of 18 truncated-pyramid telescope housings. To increase granularity for the most forward angles, the ring nearest 0°is segmented into two components. Each telescope is composed of (1) a gas-ionization chamber (GIC) operated at 12-20 torr of C_3F_8; (2) a 500 μm ion-implanted passivated silicon detector, Si(IP), and (3) a 28 mm thick CsI(Tl) crystal with light guide and photodiode readout. Detectors are operated in a common gas volume; vacuum isolation is provided by a 250 μg/cm^2 polypropylene window supported by a cage-like structure. The telescope dynamic range permits measurement of Z $\approx$ 1 - 16 fragments with discrete charge resolution over the dynamic range $0.8 \leq E/A \leq 96$ MeV. The Si(IP)/CsI(Tl) telescopes also provide particle identification (Z and A) for energetic H, He, Li and Be isotopes (E/A $\geq$ 8 MeV). The Si(IP) detectors constitute a critical component of the array in that they provide both excellent energy resolution and reliable energy calibration for the GIC and CsI(Tl) elements.

The total solid angle is 80% of 4π, as determined by simulations with the GEANT code. The major acceptance advantage of the ISiS array is its very low detector thresholds (E/A $\approx$ 0.8 MeV compared to E/A $\approx$ 2.0 - 3.5 MeV for phoswich-based arrays telescopes). This is illustrated in Fig. 1, where we show carbon spectra obtained at various angles for multiplicity M = 1, 2, and 3 events for the ^{3}He + Ag reaction with similar telescopes.[3] Thresholds (including target thickness and gas-ion-chamber window considerations) are about 0.8 MeV/nucleon, as shown in the data here. For comparison, typical thresholds for existing thin plastic phoswich devices are indicated by the dashed line in Fig. 1. For these light-ion data, the detectors based on phoswich technology would miss about 90% of the cross section in the backward hemisphere. Dynamic range is also important, extending up to 96 MeV/nucleon. With these thresholds and dynamic range, we obtain excellent coverage of LCP and IMF multiplicities and energy spectra for a large fraction of the total yield.

III. 4π Measurements of the ^{3}He + Ag, Au Systems

In its inaugural experimental run with 1.8 - 4.8 GeV ^{3}He ions during October/November 1993 at LNS Saclay, the Silicon Sphere detector modules met design specifications in their fully packed configuration, especially the ability to work at very small angles to the beam. The full electronics were successfully operated and the CAMAC/VME/ethernet data acquisition functioned with 9 CAMAC crates on line and a high event processing capacity. Beam currents were typically $\sim 5 \times 10^7$ particles/spill.

The quality of the data is indicated in Fig. 5, where calibrated 2D bit spectra summed over all 18 IC/Si telescopes in a detector ring (22°- 31°) are shown for a single experimental run. Charge resolution is evident up to Z $\approx$ 15. Of particular importance, separation between Li and He fragments is quite sharp.

300

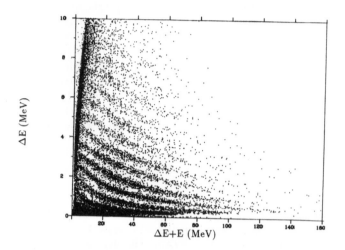

Fig. 5. Calibrated Δ E vs. (E + ΔE) bit spectra for 18 IC/Si telescopes at 21°- 32°, overlaid upon each other.

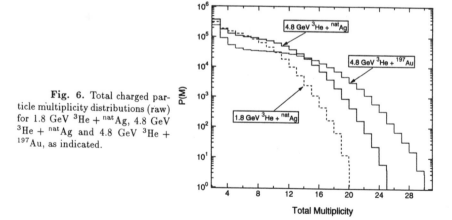

Fig. 6. Total charged particle multiplicity distributions (raw) for 1.8 GeV ^{3}He + ^{nat}Ag, 4.8 GeV ^{3}He + ^{nat}Ag and 4.8 GeV ^{3}He + ^{197}Au, as indicated.

Although results are still preliminary, two conclusions are apparent:

(1) The total charged particle multiplicity distributions are sensitive to both bombarding energy and target type, as shown in Fig. 6. The bombarding energy dependence is reflected by the 1.8 GeV and 4.8 GeV ^{3}He $+$ ^{nat}Ag distributions; the shapes are quite similar, but extended to distinctly higher multiplicities at the higher energy. For the ^{197}Au target at 4.8 GeV the distribution is both extended to higher multiplicities and much flatter in shape. These comparisons presumably reflect differences in the energy dissipation that characterize each of these systems, and must correspondingly constrain the INC codes.

(2) The most probable raw multiplicity for IMFs occurs at $M_{IMF} = 1$ for all systems studied in this work, decreasing monotonically up to $M_{IMF} \sim 8$ - 9. Thus, our results appear to differ significantly from an earlier report that indicated a most probable multiplicity of $M_{IMF} \approx 3.5$ for the 4.0 GeV ^{4}He $+$ ^{197}Au system.[8] However, in Ref. 8, discrete charge identification was not obtained for IMFs.

References

1. W. G. Lynch, Ann. Rev. Nucl. Sci. **37**, 493 (1987).

2. S. J. Yennello et al., Phys. Lett. **246**, 26 (1990).

3. S. J. Yennello et al., Phys. Rev. Lett. **67**, 671 (1991); Phys. Rev. C **48**, 1092 (1993).

4. Y. Yariv and Z. Frankel, Phys. Rev. C **26**, 2138 (1982).

5. W. A. Friedman, Phys. Rev. C **42**, 667 (1990).

6. M. Cubero et al., Nucl. Phys. A **519**, 345 (1990).

7. K. Kwiatkowski et al., Phys. Rev. C **49** (April 1994).

8. V. Lips et al., Workshop on Fragmentation Phenomena, Les Houches, France, April 12 - 17, 1993.

CHAOTICITY AND MULTIFRAGMENTATION

M. Baldo, G. F. Burgio and A. Rapisarda

Istituto Nazionale di Fisica Nucleare
and Dipartimento di Fisica Università di Catania
Corso Italia 57, I-95129 Catania, Italy

ABSTRACT

We review recent results on the chaotic dynamics of nuclear mean field and we discuss the meaning and the possible implications for nuclear multifragmentation mechanism.

1. Introduction

Non-linear dynamics is at the basis of a rich variety of natural phenomena like for example turbulence, weather and climate fluctuations, population evolution, epidemic diseases, economic development and even cardiac arrhythmias or the solar system dynamics[1]. This means that most of the world around us can follow regular and predictable evolution, but also that under certain circumstances non-linearity can prevail and an erratic behavior can follow. The study of non-linear dynamics has seen a strong development in the last two decades and has become a very important issue to understand what was previously rejected as spurious noise. This new way of thinking is not limited to classical mechanics and has already pervaded also quantum physics[2]. It is the hope of many the fact that non-linearity can help in understanding the mysterious and subtle links of the microscopic quantum world to the macroscopic classical one. Several problems in nuclear physics can be interpreted taking into account a chaotic behavior and actually this field has given a lot to chaos science[3]. In this paper we review some recent results obtained about one of the many examples in nuclear physics where chaoticity can be important to understand the main mechanism: nuclear multifragmentation.

During the last decade one of the most debated topics in heavy–ion collisions has been the nuclear disassembly which occurs for energies above 50 MeV/A. For these incident energies many experiments have shown a simultaneous fragmentation of the excited composite system formed during the reaction into several big fragments (> 3). Different models have been advanced in order to explain this phenomenon which is commonly believed to be important to learn fundamental properties of nuclear matter. These approaches range from a stochastic scenario to a dynamical one[4]. A

clear description has not emerged up to now, but some fundamental features have been clearly outlined:
- Multifragmentation cannot be explained with models valid at lower energies;
- It is a fast process with a characteristic time $\tau \sim 100\ fm/c$;
- It shows characteristic similar to other fragmenting objects;
- There are strong indications for a nuclear phase transitions.

In the following we discuss the mean-field chaoticity we found for two-dimensional simulations of nuclear matter dynamics in the spinodal zone, and we argue that this behavior can be one of the most natural to explain nuclear multifragmentation. Part of the material here presented is going to be published elsewhere[5,6].

2. Chaotic dynamics of the mean field

2.1. Theoretical framework

We have taken into account the Vlasov-Nordheim equation

$$\frac{\partial f}{\partial t} - \{H[f], f\} = I[f] \tag{1}$$

which is known to give a good description of the average mean field dynamics in the nuclear case. The latter was solved numerically on a two-dimensional lattice, using the same code of ref.7. In eq. (1) $f(x, y, p_x, p_y, t)$ is the Wigner distribution function, $H[f]$ is the self-consistent effective one-body Hamiltonian and $I[f]$ is the two-body Pauli blocked collision integral. The single particle phase space is divided into cells and nuclear matter is confined in a large torus with periodic boundary conditions. The local momentum distribution was assumed to be the one of a Fermi gas at a temperature $T = 3\ MeV$. For further detail of the calculations see refs. 5, 6

As a first step we analyzed the mean field dynamics neglecting two-body collisions, i.e. solving the left-hand side of eq.(1). We initialized nuclear matter with a sinusoidal profile along the x-direction, characterized by a wave number $k = 2\pi(n_k/L_x)$ and a small amplitude equal to 1% of the local density. In fig.1(a), for a value of the average density equal to $0.5\rho_0$, we plot the evolution of two density profiles with different initial wave numbers, more precisely $n_k=5$ as full curve and $n_k=6$ as dotted curve. The initial distance (see eq.(4)) is in this example $d_0 \sim 10^{-3}\ fm^{-2}$.

In this case the two profiles differ completely between each other after a short time interval until several "fragments" are formed. In our case due to the fact that we are considering a two-dimensional torus of nuclear matter no real fragmentation occurs and we call fragments the macroscopic structures of the density profile. The behavior shown in fig.1 is typical of a dynamical system in a chaotic region, where very small initial perturbations are rapidly amplified and distorted. The relevant feature is that *two initially close density profiles strongly differ not only in their amplitude, but also in their shape*, a fact which can occur only in a non-linear regime. We note

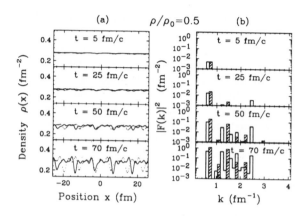

Fig. 1: (a) Non-linear evolution of two density profiles at an average density $\rho/\rho_0 = 0.5$. The two trajectories, full and dotted curve, differ for an initial distance $d_0 \sim 10^{-3} \, fm^{-2}$. (b) The corresponding Fourier transforms are reported as empty and cross-shaded histograms. See text for further details.

that the initial symmetry is spontaneously broken during the evolution, although the equations do not contain any explicit noise term.

This behavior is confirmed by fig.1(b) where we display the corresponding modulus square of the Fourier transforms of the density profiles. The empty histogram is the one relative to the full curve and the cross-shaded one refers to the dotted curve.

One can see an increase of the populated wave numbers together with a strong mixing as the evolution proceeds, despite the fact that only one wave number k is initially occupied. The final frequencies spectrum has little resemblance with the initial one. The evolution of the Fourier transforms for the two intial wave numbers considered is also very different. Note however that in both cases the initial mode is not completely damped and some memory of it is still retained.

A very different behavior occurs outside the unstable *spinodal region*, i.e. for densities greater than $\frac{2}{3}\rho_0$. The initial oscillation in this case is damped and the profiles of two close trajectories remain close to each other[5,6]. This is typical of the *regular* region of a dynamical system, and indicates the stability of the dynamics with respect to small perturbations.

We checked that these trends are very general and do not depend either on the initial shape of the profile or on the accuracy of the calculations[6].

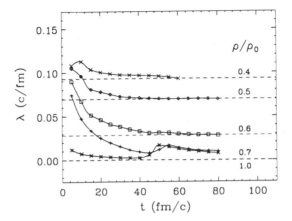

Fig. 2: The quantity $\lambda(t)$ as a function of time for different average densities. The dashed line represent the largest Lyapunov exponent $\overline{\lambda}$ to which $\lambda(t)$ converges.

2.2. Largest Lyapunov exponent

A way to characterize quantitatively the dynamics in a chaotic regime is by means of the rate of divergence of two nearby trajectories. The mean rate is usually called largest Lyapunov exponent[1] and it is calculated by means of the expression

$$\overline{\lambda} = \lim_{t \to \infty} \lim_{d_0 \to 0} \lambda(t) \tag{2}$$

where

$$\lambda(t) = \frac{\log(d(t)/d_0)}{t} . \tag{3}$$

In the latter $d(t)$ is the distance between two phase space trajectories, along an unstable direction. In our notation $d_0 = d(0)$. We have chosen as metric the following one

$$d(t) = \sum_i |\rho_i^{(1)}(t) - \rho_i^{(2)}(t)|/N_c , \tag{4}$$

where the index i runs over the N_c cells in ordinary space, and $\rho_i^{(1)}$, $\rho_i^{(2)}$ are the densities in the cell i for the trajectories 1 and 2 respectively. The definition of eq. (4) represent the difference between the two density profiles and is sufficient for our purpose, although possible differences in momentum space are averaged out. It should include the contribution of all the unstable modes which dynamically grow up during the evolution.

In fig. 2 we show the behavior of $\lambda(t)$ for various average densities and its convergence to a limiting $\overline{\lambda}$ value, represented by a dashed line, which is just the largest Lyapunov exponent. Inside the spinodal region $\overline{\lambda}$ is greater than zero demonstrating that the dynamics is chaotic. We have done calculations including the collision term. The latter have shown only a small change in the Lyapunov exponents, around 10-30% (see ref. 6). We checked that also the temperature dependence is of the same order. It is important to note that if the system is only linearly unstable[8] the Lyapunov exponent would coincide with the growth rate of the mode excited and would depend linearly on this mode. This does not occur in our case.

3. Chaoticity implications for multifragmentation

In this section we discuss what are the consequences of a chaotic behavior in the spinodal zone for the final fragments formation. In our simplified description we miss the initial dynamics which leads the system to a compression and a following expansion. However one can think of initializing our density profile in a random way inside the spinodal region. At a certain time then the freeze-out should finally determine the final fragmentation. In our case we do not have real fragments but a natural criterium would be to consider those close cells whose density exceeds a threshold value as forming a cluster. Adopting a freeze-out time of 100 fm/c, a threshold density of 0.22 fm^{-2} and using an ensemble of 100 events with a random initial profile, whose strength is 0.01 ρ_0, we get the multiplicity distribution and the fragment size distribution reported in fig.3. The figure shows large distributions of a kind very similar to those observed experimentally[4]. In particular a power law with an exponent $\tau=1.92$ comes out. Note that a percolation calculation in 2d using a lattice of the same size (12x12) gives a critical $\tau \sim 2.03$.

Though we do not want to stress too much these similarities since our simulation is at the moment too schematic, we think it is very important the fact that a deterministic chaotic behavior allows for final large distributions. On the contrary we note that a linear unstable evolution[8], with the most unstable mode having the largest growth rate, would lead always to the same final fragmentation even with a random initial condition. Therefore a deterministic chaotic mechanism seems the most natural explanation for the experimental data. In a chaotic regime even selecting events with the same impact parameter, energy, temperature, etc. the small uncertainties which will always be present would lead inevitably to very large fluctuations. Of course at the moment our calculations are too schematic and need a firmer confirmation by more realistic calculations. In particular the Lyapunov exponents or other details could change a little - memory loss is a question which should be further investigated - but deterministic chaos seems the simplest mechanism for nuclear multifragmentation.

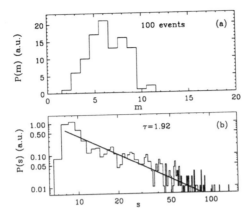

Fig. 3: (a) Multiplicity distribution P(m) for 100 events with random initialization in arbitrary units, see text. (b) Log-log plot of the fragments size distribution P(s) (histogram) corresponding to part (a) and the relative fit with a power law (full line): $P(s) = cs^{-\tau}$. The extracted value τ is also shown. See text for further details.

4. Conclusions

A chaotic behavior seems to be the most natural explanation for the nuclear disassembly. It can provide very large distributions and in particular a power law behavior for the fragment size distribution. These features go in the same direction of the experimental data and seem to strongly indicate a kind of phase transition. A lot of work still need to be done in order to prove unambiguously this scenario in more realistic simulations. It should be also clearly stated that we have *not presented a new model*, but rather *a new strategy* for dynamical models (experimental data) which seems more appropriate. The message to bear in mind is the following: *in a chaotic regime one event (one simulation) can be very much different from another one, even if both of them belong to the same class and their evolution follows deterministic laws.*

5. References

[1] A.J. Lichtenberg and M.A. Lieberman, *Regular and Stochastic motion*, Springer-Verlag (1983);
Hao Bai–Lin, *Chaos*, World Scientific (1984);
M. Tabor, *Chaos and Integrability in Nonlinear Dynamics*, John Wiley (1989);
P. Cvitanovic, *Universality in Chaos*, Adam Hilger (1989).

[2] M.C. Gutzwiller, *Chaos in Classical and Quantum Mechanics*, Springer-Verlag, 1990 and references therein;
R.V. Jensen, *Nature* **355** 311 (1992).

[3] O. Bohigas and H.A. Weidenmüller, *Ann. Rev. Nucl. Part. Sci.* **38** 421 (1988) and references therein;
D.C. Meredith, S.E. Koonin and M.R. Zirnbauer, *Phys. Rev. A* **37**, 3499 (1988);
Y. Alhassid and D. Vretenar, *Phys. Rev. C* **46** 1334 (1992);
J. Blocki, J.-J. Shi and W.J. Swiatecki, *Nucl. Phys. A* **554** 387 (1993);
V.G. Zelevinsky, *Nucl. Phys. A* **553** 125c (1993);
A. Rapisarda and M. Baldo, *Phys. Rev. Lett.* **66** 2581 (1991);
M. Baldo, E.G. Lanza and A. Rapisarda, *Chaos* **3** 691 (1993);
C.H. Dasso, M. Gallardo and M. Saraceno, *Nucl. Phys. A* **549** 265 (1992).

[4] For a general discussion see:
D.H.E. Gross, *Nucl. Phys. A* **553**, 175c (1993) and references therein ;
The proceedings of this conference.

[5] G.F. Burgio, M. Baldo and A. Rapisarda, Phys. Lett. B in press.

[6] G.F. Burgio, M. Baldo and A. Rapisarda, Proceedings of the international conference *Dynamical features of nuclei and finite Fermi systems*, september 13-17,1993, Sitges (Barcelona), Spain, World Scientific in press.

[7] G.F. Burgio, Ph. Chomaz and J. Randrup, *Phys. Rev. Lett.* **69** 885 (1992).

[8] H. Heiselberg C.J. Pethick and D.G. Ravenhall, *Phys. Rev. Lett.* **61** 818 (1988) and *Ann. Phys.* **223**, 37 (1993).

HOT NUCLEI IN 2 GEV P, 2 GEV ^{3}HE AND AND 150 MEV/A ^{84}KR INDUCED REACTIONS

J. GALIN, D. GUERREAU, X. LEDOUX, A. LEPINE-SZILY,
B. LOTT, M. MORJEAN, A. PÉGHAIRE, B.M. QUEDNAU
GANIL, BP 5027, F-14021 Caen Cédex, France
H.G. BOHLEN, H. FUCHS, B. GEBAUER, D. HILSCHER,
U. JAHNKE, G. RÖSCHERT, H. ROSSNER
Hahn-Meitner-Institut, Glienicker Str. 100, D-14109 Berlin, Germany
D. JACQUET, B. GATTY, S. PROSCHITZKI, C. STÉPHAN
IPN Orsay, BP 1, F-91406 Orsay Cédex, France
S. LERAY, L PIENKOWSKI
LN Saturne, F-91191 Gif-sur-Yvette Cédex, France
R.H. SIEMSSEN
KVI, NL-9747 AA Groningen, Netherlands
E. CREMA
IFUSP, DFN CP20516, 01498 São Paolo, S.P., Brazil
J. CUGNON
Univ. de Liège, B-4000 Liège, Belgium

ABSTRACT

Multiplicity distributions of neutrons were measured inclusively with the large-volume neutron calorimeter ORION for proton and ^{3}He induced reactions at an incident energy of 2 GeV. The experimental distributions are found to be in qualitative agreement with model calculations assuming an intranuclear cascade as first reaction stage and, following, statistical emission from a thermalized nucleus. Angular and atomic-number distributions of projectile-like "spectator" fragments from the reaction ^{84}Kr+^{197}Au were measured at E/A=150 MeV using an annular multi-strip ΔE-ΔE Si-detector. The observed correlations between multiplicity of neutrons and projectile-like fragment atomic numbers indicate large deposits of excitation energy in these fragments.

1. Introduction

For many years now, production of hot nuclei[1] and study of their decay have been in the focus of interest of intermediate-energy heavy-ion studies. The fact that large amounts of kinetic energy can be dissipated in heavy-ion collisions and the prospect of being able to study thermal limits of stability of nuclei strongly motivated such studies. For this high excitation-energy regime new decay processes are predicted[2-7], e.g. multi-fragmentation, vaporization, and explosion-like reactions, which, if confirmed, are of great interest to the study of properties of nuclear matter. The use of heavy projectiles, however, does complicate the interpretation of experimental data because of large dynamical effects and the excitation of not only intrinsic degrees of freedom but collective ones as well, e.g. compression, rotation. In this work, first results of an experimental study will be presented using two different approaches to examine separately, purely thermal decay and, by comparison, collective ones.

2. Experimental Setup

First series of experiments have been performed at SATURNE accelerator facility at Saclay, France, using either 475 MeV ^{1}H, 2 GeV ^{1}H, 2GeV ^{3}He, or ^{84}Kr at $E/A=150$ MeV and 400 MeV impinging on a variety of targets (C,Ag,Ho,Au,Bi,U). For all projectile-target combinations inclusive and exclusive neutron multiplicities were measured using the 4π liquid scintillator detector ORION. This detector is well suited to measure essentially thermal energy deposition in the target nucleus close to rest, because of its high efficiency for low-energy neutrons ($\epsilon =0.7$-0.8) and small sensitivity to high-energy neutrons. In addition, a variety of detectors were employed to measure charged reaction products. Light-charged particles (LCP) and intermediate-mass fragments (IMF) were detected with 10 silicon detector telescopes (9 telescopes in the Kr experiments). Fission fragments of the target-like nucleus were measured in coincidence using two parallel-plate avalanche counters. For reactions involving Krypton projectiles, angular and atomic number distributions of projectile-like ("spectator") fragments were measured with an annular multi-strip ΔE-ΔE Si-detector.

3. Reactions With Light Projectiles

Fig. 1 displays inclusively measured neutron multiplicity distributions for 2 GeV p+(U, Bi, Au, Ag) reactions. The distributions shown in this figure are corrected for background contributions but not for the finite detection efficiency of the detector. Cross sections for low neutron multiplicities are not shown because of their large experimental errors. The neutron multiplicity distributions exhibit for all target nuclei a broad maximum whose position shifts with increasing target mass from ca. $m_n=8$ for Ag to $m_n=23$ for Uranium. It is interesting to note that these distributions are strikingly similar to those typically observed for reactions between heavy ions at lower bombarding energies, e.g. ^{40}Ar+^{197}Au at $E/A=44$ MeV[8]. In order to estimate the multiplicity of directly emitted particles and the amount of energy thermalized in the system, theoretical calculations have been performed using an intranuclear cascade (INC) model [9] together with a statistical decay model[10]. The first stage of the reaction was simulated within the INC-model allowing to calculate the multiplicity and velocity distribution of neutrons emitted during the cascade. A thermalization time of 30 fm/c, after which an evaporation stage was assumed to set in, was estimated. The assumption of thermalization is justified by the fact, that at times larger than

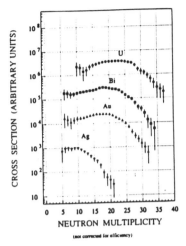

CROSS SECTION (ARBITRARY UNITS)

NEUTRON MULTIPLICITY

(not corrected for efficiency)

Fig. 1: Cross-sections for multiplicities of neutrons measured inclusively for 2 GeV p+(Ag,Au,Bi,U) reactions.

30 fm/c the number of particles emitted in the cascade, their total kinetic energy and the decrease of excitation energy in the residual nucleus exhibit an exponential dependence on time similar to a statistical emission process observed for a thermalized system. To allow a further comparison between simulation and experimental data, the response of the experimental setup to the theoretically calculated neutron distributions was simulated with the Monte-Carlo code DENIS [11].

Results of the simulation calculation folded with detector efficiency for 2 GeV p+Au reactions are plotted in form of a histogram in the upper panel of Fig. 2. The experimentally measured neutron multiplicity distribution is represented by circles. A good qualitative description is achieved except for the highest multiplicities. The lower panel of Fig. 2 shows separately the calculated multiplicity distributions folded with detector efficiency for evaporated and directly emitted neutrons. The simulations predict that most of the detected neutrons are emitted from a thermalized residual nucleus, indicating that measuring the neutron multiplicity is a good approach to measure the amount of energy thermalized even at such high bombarding energies. From these simulations a broad distribution of thermal energy is predicted with an average value of $\langle E^* \rangle = 280$ MeV and sizable cross sections even for excitation energies as large as 600 MeV ($T > 5$MeV). These values are found to be in agreement with LCP-multiplicities and spectral temperatures deduced from coincident light charged particle spectra associated with large neutron multiplicities.

4. Reactions With Heavy Projectiles

The second series of experiments using Kr-projectiles was intended to determine

312

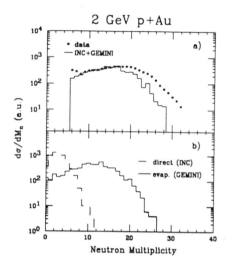

2 GeV p+Au

Fig. 2: Upper panel: Theoretical and experimental neutron multiplicity distributions (arbitrary normalization in ordinate). Results of theoretical calculations combining an intranuclear cascade (INC) model with a statistical decay model (GEMINI) for 2 GeV p+Au collisions are plotted in form of a histogram. The experimental data are represented by circles. Lower panel: Theoretical multiplicity distributions for directly emitted and evaporated neutrons.

the temperature of "spectator" nuclei for peripheral collisions from the associated multiplicity of evaporated neutrons. The influence of angular momentum was to be measured by the correlation between fission-fragment plane and reaction plane, defined by the deflection direction of the coincident projectile-like fragment. Fig. 3 shows double differential cross sections of fragments, produced in Kr+Au collisions at E/A = 150 MeV, detected in the forward-angle multi-strip ΔE-ΔE Si-detector for 6 scattering angle bins. The distribution of fragments detected between $\Theta = 0.5 - 1°$, close to the grazing angle for this reaction ($\Theta_{graz} \approx 0.9°$), is displayed in the upper-left panel. Most prominent feature of this distribution is a group of elastically and quasielastically scattered projectiles, centered at atomic number Z=36 and neutron multiplicity $m_n = 0$, from which a continuous band of cross section evolves with increasing average neutron multiplicity for decreasing Z. This pattern is quite similar to the one already observed for the same system at a much smaller bombarding energy of E/A=32 MeV[12]. No qualitative change of the average correlation between atomic number of the projectile-like fragment and average neutron multiplicity as function of scattering angle is observed. However, a strong change in the production cross-sections is seen. Large Z are strongly suppressed for larger scattering angles whereas low Zs are distributed more evenly in the angular range between $\Theta = 0.5°$ and 5°. The correlation between Z of the projectile-like fragment and neutron multiplicity reflects the strong dependence of the size of the projectile-like fragment and dissipated energy

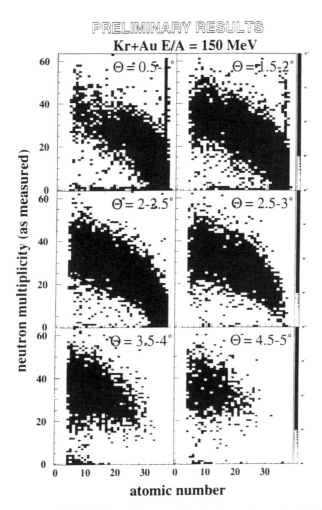

PRELIMINARY RESULTS

Kr+Au E/A = 150 MeV

Fig. 3: Distributions of double-differential cross-sections for fragments produced in Kr+Au collisions at E/A = 150 MeV for six bins of scattering-angle. The logarithmic intensity scale is the same for all distributions and defined by gray levels. The intense line structure at Z=36 is a result of pile-up of reaction events with elastic scattering. The solid line in the middle-left panel represents results of theoretical calculations combining an intranuclear cascade model with a statistical decay model.

on impact parameter. Thus, neutron multiplicity appears to be a good observable to select an impact parameter or a degree of dissipation as it was shown at much smaller bombarding energies[13]. Because of electronic detection thresholds, fragments with atomic numbers of less than ca. 5 were not measured. For these lightest fragments average multiplicities of $m_n \approx 37$ are observed. Results of simulation calculations using the intranuclear cascade code ISABEL[14] are represented by a solid line in the middle left panel of Fig. 3. As seen from this figure the functional dependence of the average neutron multiplicity on Z is reproduced but the absolute values of $\langle m_n \rangle$ are underpredicted by almost a factor of 1.5. Because of the early stage of the analysis no definite answer can be given to why the model calculation cannot reproduce the experimental data even on the average. Possible reasons may be that the model simply fails to describe correctly the collision process, the detection efficiency of energetic neutrons is underestimated, or additional neutrons might be generated in secondary interactions of reaction products with the detector itself and its surrounding. However, despite the lack of a quantitative understanding, the thermal excitation energies produced in Kr+Au collisions at E/A=150 Mev are likely to be much larger than the amount that can be explained by a change in surface energy of the nuclei, as already pointed out by [15-17]. Further work is in progress to obtain a more quantitative understanding of the experimentally observed correlations.

5. Summary

For the first time, excitation energy distributions were determined from neutron multiplicity measurements for light-ion induced reactions at relativistic bombarding energies. The results for 2 GeV p+Au reactions are described qualitatively by calculations assuming an intranuclear cascade and evaporation from thermalized residual nuclei. The experimental distributions show that large amounts of thermal energies can be deposited in such collisions. Average neutron multiplicities measured in coincidence with projectile-like "spectator" fragments produced in Kr+Au reaction at E/A= 150 MeV show a smooth evolution with atomic number and scattering angle. The observed correlations between multiplicity of neutrons and projectile-like fragments indicate large deposits of excitation energy in these fragments.

6. References

[1] D. Guerreau, in: *Nuclear Matter and Heavy-Ion Collisions*, eds. M. Soyeur, H. Flocard, B. Tamain and M. Porneuf (Plenum, New York and London, 1989), and references therein

[2] J.P. Bondorf, R. Donangelo I.N. Mishutsin, C.J. Pethick, H. Schulz, K. Sneppen, *Nucl. Phys.* **A443**, 321 (1985). J.P. Bondorf, R. Donangelo I.N. Mishutsin, H. Schulz, *Nucl. Phys.* **A444**, 460 (1985).

[3] W. Bauer, G.F. Bertsch and S. Das Gupta, *Phys. Rev. Lett.* **58**, 863 (1987).

[4] G. Fai und J. Randrup, *Nucl. Phys.* **A381**, 557 (1982), G. Fai und J. Randrup, *Nucl. Phys.* **A404**, 551 (1983).

[5] D.H.E. Gross, Zhang Xiao-ze und Xu Shu-yan, *Phys. Rev. Lett.* **56**, 1544 (1986), X. Zhang, D.H.E. Gross, S. Xu, Y. Zheng, *Nucl. Phys.* **A461**, 641 (1987), and *Nucl. Phys.* **A461**, 668 (1987).

[6] C. Ngô, *Nucl. Phys.* **A471**, 381c (1987).

[7] E. Suraud,*Nucl. Phys.* **A462**, 109 (1987).

[8] B. Lott *et al.*, *Z. Phys.* **A346**, 201 (1993).

[9] J. Cugnon,*Nucl. Phys.* **A462**, 751 (1987).

[10] R.J. Charity *et al.*, *Nucl. Phys.* **A483**, 371 (1988).

[11] J. Poitou and C. Signarbieux, *Nucl. Instr. Meth.* **114**, 113 (1974).

[12] E. Crema, Compliation GANIL

[13] D.X Jiang, *Nucl. Phys.* **A503**, 566 (1989).

[14] Y. Yariv and Z. Fraenkel, *Phys. Rev.* **C20**, 2227 (1979).

[15] C. Stéphan *et al.*,*Phys. Lett.* **B262**, 6 (1991).

[16] U. Lynen *et al.*, these proceedings.

[17] N. Porile *et al.*, these proceedings.

COEXISTENCE OF CHAOS AND REGULAR UNDAMPED MOTION IN GIANT NUCLEAR OSCILLATIONS

WOLFGANG BAUER, DAVID MCGREW, VLADIMIR ZELEVINSKY
National Superconducting Cyclotron Laboratory
and Department of Physics and Astronomy
Michigan State University, East Lansing, MI 48824-1321, USA

PETER SCHUCK
Institut de Physique Nucléaire
Université de Grenoble
53 avenue des Martyrs, 38026 Grenoble Cedex, France

ABSTRACT

We study the conditions under which the nucleons inside a deformed nucleus can undergo chaotic motion. To do this we perform self-consistent calculations in semiclassical approximation utilizing a multipole-multipole interaction of the Bohr-Mottelson type for quadrupole and octupole deformations. For the case of harmonic and non-harmonic static potentials, we find that both multipole deformations lead to regular motion of the collective coordinate, the multipole moment of deformation. However, despite this regular collective motion, we observe chaotic single particle dynamics.

1. Introduction

The question about the origin of dissipation in collective motion of finite Fermi systems[1] such as atomic nuclei or small metallic clusters is an intriguing and up to now not completely satisfactorily solved problem. For example, the mutual balance of one-body and two-body processes is still a question of debate. For the case of one-body dissipation and friction in nuclear dynamics, Swiatecki and coworkers[2,3,4,5] have developed this picture: particles which move in a shape-deformed container are reflected from the (moving) walls, and due to parts of it having positive curvature (for higher multipole moments) the particles very quickly loose their coherence, thus inducing pseudo-random motion, i.e. heat, into the system. At the same time the shape oscillation is very much slowed down.

Blocki *et al.*[4,5] consider a purely classical gas of particles contained in a deformed billiard. The only similarity with a Fermi gas comes from the fact that initially the particles' momenta are distributed within a Fermi sphere. The walls of the container undergo periodic shape oscillations with a frequency much smaller than a typical single particle frequency. In the interior of the container the particles move on linear trajectories. They study the particle kinetic energy increase as a function of time and find that for ellipsoidal shape deformations ($\ell = 2$) the particles act as a classical Knudsen gas,[6] i.e. the total kinetic energy increase over an entire shape oscillation period is 0. However, for $\ell \geq 3$ the kinetic energy in the single particle motion is not completely 'given back', but rather steadily increases in time. This is explained by the fact that in an $\ell = 2$ potential the motion of the particles remains non-chaotic and therefore coherent, whereas in the $\ell \geq 3$ the scattering of the segments of the wall with positive curvature leads to chaotic motion similar to the one observed in a Sinai[7,8] billiard and thus a destruction of coherence. The fact that deformed nuclear potentials may exhibit chaotic motion was recognized early by Arvieu and co-workers.[9]

This scenario is very similar to the so-called Fermi acceleration, proposed to explain the occurrence of very high energy cosmic radiation.[10,11]

2. A Self-Consistent Model

In this paper we present an attempt to include selfconsistency into the problem of motion in multipole-deformed nuclear potentials. We have chosen a selfconsistent, but schematic, model of separable forces. We chose an interaction of the Bohr-Mottelson type[12] with a static r^2 potential and multipole-multipole interactions as studied, for example, by Stringari et al.[13,14,15,16] In the small amplitude limit, this model has recently been investigated in the semiclassical limit;[16] as is known, the low-lying quadrupole and octupole frequencies come out to be in reasonable agreement with experimental data. Our single-particle Hamiltonian is then

$$
\begin{aligned}
\mathcal{H} &= \mathcal{H}_0 + V^{(\ell)}(\mathbf{r}, t) \\
&= \frac{p^2}{2\,m} + V_0 + V^{(\ell)}(\mathbf{r}, t) ,
\end{aligned}
\tag{1}
$$

where $V^{(\ell)}(\mathbf{r}, t)$ is the potential energy associated with the (separable) multipole-multipole force[13,14,16]

$$
V^{(\ell)}(\mathbf{r}, t) = \lambda_\ell \, q_\ell(\mathbf{r}) \, Q_\ell(t) ,
\tag{2}
$$

and V_0 is the static external potential energy. We take here $V_0 = \frac{1}{2} m \omega_0^2 r^2$, resulting in the Bohr-Mottelson Hamiltonian.[12] However, we have also investigated non-harmonic static potentials (r^6) and obtained similar results.

The coupling constants λ_ℓ can be calculated using a self-consistent normalization condition,[12,16]

$$
\lambda_{2,s} = -\frac{3\,m\,\omega_0^2}{A\langle r^2 \rangle} ,
\tag{3}
$$

$$
\lambda_{3,s} = -\frac{15\,m\,\omega_0^2}{A\langle r^4 \rangle} ,
\tag{4}
$$

where A is the mass number of the nucleus under consideration. $q_\ell(\mathbf{r})$ is given by

$$q_2(\mathbf{r}) = r_y\,r_z \tag{5}$$

$$q_3(\mathbf{r}) = r_x\,r_y\,r_z \tag{6}$$

and the multipole moments $Q_\ell(t)$ are

$$Q_\ell(t) = \int \frac{d^3r\,d^3p}{(2\pi)^3} q_\ell(\mathbf{r}) f(\mathbf{r},\mathbf{p},t) . \tag{7}$$

$f(\mathbf{r},\mathbf{p},t)$ is the one-body phase space distribution function of nucleons, the Wigner transform of the one-body density.

We treat this problem in semi-classical approximation by a Wigner transformation of the von Neumann equation of motion for the density matrix, $i\,\partial_t \rho = [\mathcal{H}, \rho]$, to obtain a Vlasov equation, $\partial_t f = \{\mathcal{H}, f\}$. We then solve the Vlasov equation in the test particle method[17,18] using a fourth-order Runge-Kutta algorithm with typical time step sizes of 1 fm/c. Our numerical simulation is fully selfconsistent and conserves total energy to better than 0.1 %.

In order to generate selfconsistent initial deformations in coordinate and momentum space, we start with a spherically symmetric configuration generated in local Thomas-Fermi approximation without the deformation potential $V^{(\ell)}$.

We now apply a time-dependent external potential of the form

$$V_0^{(\kappa)}(\ell, \mathbf{r}, t) = \kappa_0 \sin(\omega_D t)\, s(t)\, q_\ell(\mathbf{r}) \tag{8}$$

where ω_D is the driving frequency, and where $s(t)$ is a differentiable spline interpolation function on the time interval $[0,\tau]$ with vanishing first derivatives at both ends, which is monotonically increasing from 0 to 1. This procedure results in a giant oscillation of the nucleus at $t = \tau$, provided that τ is chosen $\tau \gg \omega_D^{-1}$. The deformation is dependent on the value of the coupling κ_0 chosen.

3. Results

We now use the initial conditions generated in the above way (with $\tau = 1500$ fm/c) to study the time evolution under the action of our Hamiltonian as defined in Eq. (1). The upper panel (a) of Figure 1 contains the results of our calculations. One can clearly observe a regular undamped oscillation of the quadrupole moment in coordinate space as a function of time. One can observe that the period of oscillation has been stretched from the 0-coupling value of 88.4 fm/c to 128 fm/c. This is consistent with the analytic calculations for infinitesimal deformations,[19,12,16] which yield

$$\begin{aligned}
Q_2(t) &= Q_2(t_0) \exp(i\,\omega_{2+}\,t) \\
\omega_{2+} &= \sqrt{4\omega_0^2 + \lambda_2 \frac{2A\langle r^2\rangle}{3m}} \\
&= \sqrt{2}\,\omega_0
\end{aligned} \tag{9}$$

for the giant quadrupole frequency, and consequently $T = 125$ fm/c for $\omega_0 = 0.0355$ c/fm used in this example.

In the lower panel (b) of Figure 1 we show our calculations for the octupole case. Again we see harmonic oscillations (The small variations in amplitude are in both cases due to beating between the initial driving frequency and the oscillation frequency of the self-consistent calculation.). The observed oscillation period is $T_3 \approx 66$ fm/c, in good agreement with the analytical result of[16]

$$\omega_{3-} = \sqrt{7}\,\omega_0 \tag{10}$$

for the giant octupole oscillation frequency, which results in $T_3 \approx 66.8$ fm/c for our value of ω_0.

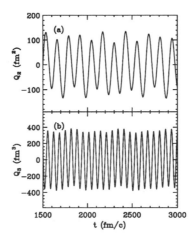

Fig. 1: Time evolution of quadrupole (a) and octupole (b) deformations of an $A=200$ nucleus in coordinate space under propagation with the Hamiltonian of Eq. (1) with a static harmonic oscillator potential and selfconsistent coupling strength $\lambda_2 = \lambda_{2,s} = 8.9 \cdot 10^{-4}$ MeV fm^{-4} (a), and $\lambda_3 = \lambda_{3,s} = 1.3 \cdot 10^{-4}$ MeV fm^{-6} (b).

The most important observation is here, however, that there is no damping of the collective motion apparent in our calculation, thus indicating that no chaoticity is present in the collective multipole coordinates. We have also used slightly different initial conditions (by using a different number of test particles in the simulation) and obtained only slightly different results. This indicates that there is no sensitive dependence on the initial conditions present here as would be the case for chaotic motion. As an additional test, we performed a Fourier transform of the time signal and found one peak at the dominant frequency and no ω^{-1} noise.

To study the single particle dynamics we employ a projection method similar to the method of Poincaré surface of sections. In Poincaré sections, we stroboscopically record the phase space coordinates of particles in time. This 'stroboscope' is triggered whenever the particles cross a certain plane in phase space. A variation of this technique is also applicable to our problem, where we solve the Vlasov equation utilizing the test particle method. Here we choose the plane $r_x = 0$ as the trigger condition.

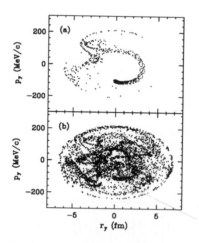

Fig. 2: Poincaré section of a single test particle (a) and an ensemble of 10 test particles (b) from the calculation leading to the time evolution of the octupole moment shown in Fig. 1 (b).

In Figure 2 (a) we show the Poincaré section for one test particle used to generate the octupole motion of Fig. 1 (b). One can clearly see that this Poincaré section will become area-filling in certain regions of phase space, and in the limit of very long time scales. (For the present calculations we ran 10^5 time steps of 1 fm/c each, leading to approximately 550 crossings of the $r_x = 0$ plane from below). In order to enable the reader to better compare the structures, we superimpose in Figure 2 (b) the Poincaré sections of ten test particles with different initial conditions (but parts of the same simulation). We obtain similar results for the collective quadrupole motion.

The above sections are not an exact proof of chaoticity in the single particle motion, because we do not have a 4- but a 6-dimensional phase space for the single particle trajectories. In addition, the total energy of the individual test particles is not conserved, because they constantly exchange energy with the collective degree of freedom. Thus the total single particle energy is not a constant of motion. However, the above projection technique still yields valuable qualitative insight into the character of single particle motion. For comparison, we also conducted the same kind of projections for a system, in which the coupling to the multipole coordinate was set to 0, and for a system in which the collective deformation was not allowed to oscillate, but was kept at maximum deformation. In both cases, the sections looked qualitatively different and did not approach the area-filling limit. The final proof for the single particle chaoticity is obtained from the analysis of Lyapunov exponents. Preliminary analysis on this seems to indicate positive values for the maximum Lyapunov exponent, but work on this is currently in progress.

This chaoticity is not a result of the weak destruction of the integrability of the corresponding static multipole potential, which can be shown to exist for the static octupole (but not for the static quadrupole) potential. Instead we attribute the chaoticity in the single particle motion to the exchange of energy between the motion of the individual test particles and the collective motion of the multipole coordinate. This exchange of energy is possible, because the individual test particles oscillate with frequencies, which do not have a rational ratio with the frequency of the collective coordinate. This results in the particle reaching meta-stable or unstable points in phase space during the course of its time evolution. At these points small changes in the initial conditions will have a large effect on the subsequent dynamics. An example for this would be the decision if the particle will temporarily oscillate in or out of phase with the collective coordinate. Consequently, these points provide large positive contributions to the Kolmogorov entropy, and chaotic single particle dynamics results.

In turn one also expects each single test particle to have a randomly fluctuating effect on the energy contained in the motion of the collective coordinate. However, since there are quite many test particles, these chaotic random fluctuations are averaged out leaving only a smooth sinusoidal oscillation of the collective coordinate. This is qualitatively new in our investigation: the generation of regular dynamics for the collective variable, the multipole moment of the collective oscillation, from the ensemble of single particles with chaotic trajectories. This is an example of how ordered macroscopic motion can result form underlying chaotic microscopic dynamics. (To obtain this result, it was crucial to employ a self-consistent treatment of the dynamics entailing conservation of total energy.)

Our finding are not restricted to the static harmonic potential, which we used as an example here, but they hold for a general class of central potentials, V_0. We have also performed calculations for a $V_0 \propto r^6$ potential with very similar results to the ones presented here.[20]

And, finally, one may speculate that this interplay between chaoticity in individual single particle degrees of freedom and regularity in certain collective coordinates may also play a role in the time evolution of other physical systems. Examples that come to mind as likely candidates are plamas in a tokamak, the human brain wave activity, the weather. Chaos on a microscopic level need not necessarily lead to a catastrophic breakdown of the system on the macroscopic scale.

This work was supported in parts by the US National Science Foundation under grant number PHY 90-17077. W.B. acknowledges support form a US NSF Presidential Faculty Fellow award. We are grateful for useful discussions with J. Blocki, E.

Heller, and A. Rapisarda.

4. References

[1] D.L. Hill and J.A. Wheeler, Phys. Rev. **89**, 1102 (1953).
[2] J. Blocki, Y. Boneh, J.R. Nix, J. Randrup, M. Robel, A.J. Sierk, and W.J. Swiatecki, Ann. Phys. (N.Y.) **113**, 330 (1978).
[3] J. Randrup and W. Swiatecki, Ann. Phys. (N.Y.) **125**, 193 (1980).
[4] J. Blocki, F. Brut, T. Srokowski, and W.J. Swiatecki, Nucl. Phys. **A545**, 511c (1992).
[5] J. Blocki, J.-J. Shi, and W.J. Swiatecki, Nucl. Phys. **A554**, 387 (1993).
[6] M.H.C. Knudsen, *The Kinetic Theory of Gases* (Wiley, New York, 1950).
[7] Ya.G. Sinai, Russ. Math. Surveys **25(2)**, 137 (1970).
[8] W. Bauer and G.F. Bertsch, Phys. Rev. Lett. **65**, 2213 (1990) and *ibid.* **66**, 2172 (1991).
[9] J. Carbonell and R. Arvieu, Proceedings of the topical meeting on nuclear fluid dynamics, Trieste, Oct. 1982, Ed.: M. Di Toro, M. Rosina, and S. Stringari; R. Arvieu, F. Brut, J. Carbonell, and J. Touchard, Phys. Rev. A **35**, 2389 (1987).
[10] E. Fermi, Phys. Rev. **75**, 1169 (1949).
[11] A.J. Lichtenberg and M.A. Lieberman, *Regular and Stochastic Motion (Applied Mathematical Sciences, Vol. 38)* (Springer, New York, 1983).
[12] A. Bohr and B.A. Mottelson, *Nuclear Structure*, Volume II, p. 350 ff (W.A. Benjamin, Reading, Mass., 1975).
[13] S. Stringari, Nucl. Phys. **A325**, 199 (1979).
[14] S. Stringari, Phys. Lett. **103B**, 5 (1981).
[15] H. Reinhardt and H. Schulz, Nucl. Phys. **A391**, 36 (1982).
[16] H. Kohl, P. Schuck, and S. Stringari, Nucl. Phys. **A459**, 265 (1986).
[17] G.P. Maddison and D.M. Brink, Nucl. Phys. **A378**, 566 (1982).
[18] C.Y. Wong, Phys. Rev. C **25**, 1460 (1982).
[19] T. Suzuki, Nucl. Phys. A **217**, 182 (1973).
[20] A more complete study including comparison to the 'wall formula' results as well as a presentation of the Fourier spectra and the discussion of the destruction of integrability in the static octupole potential will be published elsewhere.